Günter Dietmar Roth (Ed.)

Compendium of Practical Astronomy

Volume 2:
Earth and Solar System

Translated and Revised by
Harry J. Augensen and Wulff D. Heintz

With 181 Figs., some in color, and 25 Tables

Springer-Verlag
Berlin Heidelberg New York
London Paris Tokyo
Hong Kong Barcelona Budapest

Dipl.-Kfm. Günter Dietmar Roth
Ulrichstrasse 43, Irschenhausen, D–82057 Icking/Isartal, Germany

Dr. Harry J. Augensen
Department of Physics and Astronomy, Widener University, Chester,
PA 19013, USA

Professor Dr. Wulff-D. Heintz
Department of Physics and Astronomy, Swarthmore College, Swarthmore,
PA 19081, USA

Completely Revised and Enlarged Translation of the 4th German Edition of the title "Roth (Ed.), Handbuch für Sternfreunde, Vols. 1 and 2".

ISBN 3-540-54885-8 Springer-Verlag Berlin Heidelberg New York
ISBN 0-387-54885-8 Springer-Verlag New York Berlin Heidelberg

ISBN 3-540-56273-7 Volumes 1, 2, and 3
ISBN 0-387-56273-7 Volumes 1, 2, and 3

Library of Congress Cataloging-in-Publication Data
Handbuch für Sternfreunde. English
Compendium of practical astronomy / Günter Roth, ed. :
translated by Harry J. Augensen and Wulff D. Heintz.
  p. cm.
Rev. translation of: Handbuch für Sternfreunde (4th ed.).
Includes index.
Contents: v. 1. Instrumentation and reduction techniques -- v.
2. Earth and solar system -- v. 3. Stars and stellar systems.
ISBN 0-387-56273-7 (New York). -- ISBN 3-540-56273-7 (Berlin)
1. Astronomy--Handbooks, manuals, etc. I. Roth, Günter Dietmar.
QB64.H3313 1993 520--dc20 93-27023

This work is subject to copyright. All rights are reserved, whether the whole or part of the material is concerned, specifically the rights of translation, reprinting, reuse of illustrations, recitation, broadcasting, reproduction on microfilm or in any other way, and storage in data banks. Duplication of this publication or parts thereof is permitted only under the provisions of the German Copyright Law of September 9, 1965, in its current version, and permission for use must always be obtained from Springer-Verlag. Violations are liable for prosecution under the German Copyright Law.

© Springer-Verlag Berlin Heidelberg 1994
Printed in Germany

The use of general descriptive names, registered names, trademarks, etc. in this publication does not imply, even in the absence of a specific statement, that such names are exempt from the relevant protective laws and regulations and therefore free for general use.

Cover Design: Erich Kirchner, Heidelberg
Typesetting: Data conversion by Lewis & Leins Buchproduktion, Berlin
Production: PRODUserv Springer Produktions-Gesellschaft, Berlin
SPIN 10018885   55/3020 - 5 4 3 2 1 0 – Printed on acid-free paper

Contributing Authors to Volume 2

Beck, Rainer, Dr.
Max-Planck-Institut für Radioastronomie, Auf dem Hügel 69, D–53121 Bonn, Germany

Gericke, Volker, Dr.
Mühlweg 27, D–71334 Waiblingen, Germany

Häfner, Reinhold, Dr.
Universitäts-Sternwarte, Scheinerstrasse 1, D–81679 München, Germany

Haupt, Hermann, Prof. Dr.
Institut für Astronomie, Karl-Franzens-Universität Graz, Universitätsplatz 5, A–8010 Graz, Austria

Heintz, Wulff-Dieter, Prof. Dr.
Department of Physics and Astronomy, Swarthmore College, Swarthmore, PA 19081, USA

Hilbrecht, Heinz, Dr.
Schubertstrasse 9, D–79761 Waldshut, Germany

Jahn, Cord-Hinrich, Dr.
Rotermundstrasse 24, D–30165 Hannover, Germany

Junker, Elmar, Dr.
Johanneshof 12, D–35578 Wetzlar, Germany

Kresken, Rainer, Dipl.-Ing.
Am Birngarten 20a, D–64372 Ober-Ramstadt, Germany

Leinert, Christoph, Dr.
Max-Planck-Institut für Astronomie, Königstuhl 17, D–69117 Heidelberg, Germany

Petri, Winfried, Prof. Dr.
Postfach 106, D–83722 Schliersee, Germany

Reinsch, Klaus, Dr.
Am Wochenmarkt 22, D–37073 Göttingen, Germany

Roth, Günter Dietmar, Dipl.-Kfm.
Ulrichstrasse 43, Irschenhausen, D–82057 Icking/Isartal, Germany

Schmeidler, Felix, Prof. Dr.
Mauerkircherstrasse 17, D–81679 München, Germany

Völker, Peter
Wilhelm-Foerster-Sternwarte, Munsterdamm 90, D–12169 Berlin, Germany

Preface to the Second English Edition

It has been a particular pleasure to produce this revised English edition of the German *Handbuch für Sternfreunde*, thus making it available to a wider readership. I should like to express my gratitude to the authors and translators, who have contributed invaluably to the process of revision and translation.

I am deeply indebted to Prof. W.D. Heintz and Prof. H.J. Augensen, who not only made the translation but also assisted in the critical reading and improvement of various chapters. Moreover, Prof. Augensen has written for Vol. 1 a chapter which says a great deal about "Astronomy Education and Instructional Aids". This contains full information with regard to the situation in the United Kingdom, Canada, and the United States. It is an extremely helpful survey for both teachers and students in colleges and high schools, and for the staff of planetaria and public observatories.

Welcomed as a new author is R. Kresken, who wrote the chapter "Artificial Earth Satellites" especially for this English edition.

Last, but not least, I gratefully acknowledge the helpfulness of Springer-Verlag, Heidelberg, where Prof. W. Beiglböck always gave every possible consideration to the translators' and my suggestions.

Irschenhausen *Günter D. Roth*
Summer 1994

Translators' Preface

It is a pleasure to present this work, which has been well received in German-speaking countries through four editions, to the English-speaking reader. We feel that this is a unique publication in that it contains valuable material that cannot easily—if at all—be found elsewhere. We are grateful to the authors for reading through the English version of the text, and for responding promptly (for the most part) to our queries. Several authors have supplied us, on their own initiative or at our suggestion, with revised and updated manuscripts and with supplementary English references. We have striven to achieve a translation of *Handbuch für Sternfreunde* which accurately presents the qualitative and quantitative scientific principles contained within each chapter while maintaining the flavor of the original German text. Where appropriate, we have inserted footnotes to clarify material which may have a different meaning and/or application in English-speaking countries from that in Germany.

When the first English edition of this work, *Astronomy: A Handbook* (translated by the late A. Beer), appeared in 1975, it contained 21 chapters. This new edition is over twice the length and contains 28 authored chapters in three volumes. At Springer's request, we have devised a new title, *Compendium of Practical Astronomy*, to more accurately reflect the broad spectrum of topics and the vast body of information contained within these pages. It should be noted that, while much of this information is directed toward the "amateur," it is equally applicable to the professional astronomer or physicist who teaches at a small college and is searching for suitable astronomical projects to give to his or her students.

The *Compendium of Practical Astronomy* is structured somewhat differently from its German counterpart. The former consists of three volumes, the latter two. Volume 1 is essentially the same as the corresponding volume in the *Handbuch für Sternfreunde*, but the chapters have been reordered in a sequence which, we feel, presents the topics in more homogeneous groups. In addition, "Astronomy Education and Instructional Aids" (H.J. Augensen) has been added as the last chapter. Volume 2 contains chapters covering the Earth and Solar System, including "The Terrestrial Atmosphere and Its Effects on Astronomical Observations" (F. Schmeidler), which appeared in Vol. 1 of the *Handbuch*. Also, "Artificial Earth Satellites" (R. Kresken) replaces the corresponding chapter by Petri in the German edition. Volume 3 is devoted to topics of stellar, galactic, and extragalactic astronomy. In the fourth German edition, these latter two volumes appear as one very thick second volume.

At the end of each volume, we have included a "Supplemental Reading List" for each of the chapters in that volume. We have prepared a new appendix, "Educational Resources in Astronomy" (Appendix A in Vol. 1), as a supplement to Chap. 12, and have also updated some tables, primarily those which constitute Appendix B, "Astronomical Data," in Vol. 3. Recognizing the fact that many readers will want to utilize computer techniques in the various reduction procedures presented in this *Compendium*, it is our hope that the references on computers and programmable calculators given in Sect. 12.4.4 of Vol. 1 will prove helpful.

The superb guidance provided by Prof. W. Beiglböck and the able assistance of his secretary Ms. S. Landgraf (Springer-Verlag, Heidelberg) has been invaluable in the completion of this project. We are also indebted to Mark Seymour at Springer for his thorough proofreading of the entire manuscript, and to Dr. Fred Orthlieb of Swarthmore College for his help in translating several technical terms. Finally, we gratefully acknowledge the unwavering assistance of Computing Services at Widener University, in particular Barry Poulson, James Connalen, John Neary, Kim Stalford, Lynn Pollack, and David Walls, who provided expert advice on the preparation of this entire manuscript in TeX.

One of us (WDH) still cherishes his acquaintance with A. Güttler, W. Jahn, R. Kühn, R. Müller, and K. Schütte—the now-deceased authors of the first edition, whose enthusiasm helped Günter Roth's project off to a good start over 30 years ago.

Swarthmore/Chester   *Wulff D. Heintz*
June 1994   *Harry J. Augensen*

Preface to the Fourth German Edition

The ability to employ objective techniques to appreciate the wide variety of cosmic phenomena quantitatively is not restricted to professional astronomers. It is the principal aim of the *Handbuch für Sternfreunde* to provide the astronomically interested public—amateur observers as well as teachers—with instruction and guidance in practical astronomical activities. This goal has remained unchanged since the first edition, which appeared in 1960.

What has changed is the technical content and organizational structure in various areas. Larger and more effective telescopes are currently within the reach of non-professionals. The professional accessories used in photography, photometry, and spectroscopy are now being operated by amateurs; schools and private observatories own electronic equipment. Thus equipped, the amateur can now engage in observational tasks ranging from photoelectric photometry of planets and variable stars to studies of high-resolution photographs of distant galaxies.

These developments are reflected in every chapter of the present work. The presentation of new tools, techniques, and tasks has required a significant expansion, so that the *Handbuch für Sternfreunde* now appears in two volumes.

Volume 1 provides the technical basis for astronomical observations and measurements with amateur equipment. These include fundamental methods for recording and processing light intensities in photography, photometry, and spectroscopy. The optical range is augmented by radio-astronomical observations, whose instrumental basics are described. Also included within this volume are instructions on how to organize astronomical observations and subsequently process the results by mathematical methods, a guide to the literature, and a brief history of astronomy.

The following chapters in Vol. 1 have been completely rewritten or newly added for the fourth edition: "Optical Telescopes and Instrumentation" (H. Nicklas), "Telescope Mountings, Drives, and Electrical Equipment" (H.G. Ziegler), "Astrophotography" (B. Koch, N. Sommer), "Principles of Photometry" (H.W. Duerbeck, M. Hoffmann), and "Historical Exploration of Modern Astronomy" (G.D. Roth).

Volume 2 presents the objects of astronomical study in detail, commenting on the execution and evaluation of observational tasks. The topics covered include, among others, the various Solar System bodies, the stars, the

Milky Way, and extragalactic objects. This volume contains an expanded section of tables, general literature references, and the cumulative subject index for both volumes. Also included as an appendix is the contribution "Instructional Aids in Astronomy" (A. Kunert).

The following chapters in Vol. 2 have been completely rewritten or newly added for the fourth edition: "The Sun" (R. Beck et al.), "Lunar Eclipses" (H. Haupt), "Noctilucent Clouds, Polar Aurorae, and the Zodiacal Light" (C. Leinert), "Stars" (T. Neckel), "Variable Stars and Novae" (H. Drechsel, T. Herczeg), "The Milky Way and the Objects Composing It" (T. Neckel), and "Extragalactic Objects" (J.V. Feitzinger).

I also wish to thank all the authors on this occasion for their successful collaboration. Welcomed as new contributors are: Dr. R. Beck and his coworkers V. Gericke, H. Hilbrecht, C.H. Jahn, E. Junker, K. Reinsch, and P. Völker (the "Sun" Working Group of the Vereinigung der Sternfreunde); Dr. H. Drechsel (Bamberg), Dr. H. Duerbeck (Münster), Prof. J.V. Feitzinger (Bochum), Prof. H. Haupt (Graz), Prof. T.J. Herczeg (Norman, OK), Dr. M. Hoffmann, B. Koch (Düsseldorf), Dr. C. Leinert (Heidelberg), Dr. T. Neckel (Bochum), Dr. H. Nicklas (Göttingen), and N. Sommer (Düsseldorf).

The preparation of the fourth edition was aided by the valuable advice given by Prof. F. Schmeidler (Munich) and Dr. H.J. Staude, managing editor of the magazine *Sterne und Weltraum*. Dr. W. Kruschel (Konstanz) supported the preparation of the tables in Vol. 2 by providing updated material. Photographs and tables were made available by C. Albrecht (Freiburg), H. Haug and coworkers at the Wilhelm-Foerster-Sternwarte (Berlin), and J. Meeus (Erps-Kwerps, Belgium).

As the representative of Springer-Verlag, Prof. W. Beiglböck has attended to this large project with care and provided numerous suggestions. The somewhat tedious task of preparing the manuscripts for printing was in the capable hands of Mrs. C. Pendl, to whom the editor and the authors are very grateful.

Irschenhausen *Günter D. Roth*
Summer 1989

Contents of Volume 2

13	**The Sun** by R. Beck, V. Gericke, H. Hilbrecht, C.H. Jahn, E. Junker, K. Reinsch, P. Völker, Fachgruppe Sonne der Vereinigung der Sternfreunde e.V.	1
13.1	Introduction	1
13.2	Observations of the Sun	1
13.2.1	Site Selection	1
13.2.2	Observing Conditions	4
13.2.3	Records	5
13.3	Sunspots	7
13.3.1	Development and Classification of Sunspots	7
13.3.2	Light Bridges	13
13.3.3	The Wilson Effect	14
13.3.4	Numerical Expression of Spot Activity	15
13.3.5	Sunspot Cycles	21
13.4	Photospheric Faculae	25
13.4.1	Structure and Appearance	25
13.4.2	Classification	26
13.4.3	Measurement of Faculae Activity	28
13.4.4	Polar Faculae	29
13.5	Chromospheric Faculae	30
13.6	Prominences and Filaments	31
13.6.1	Introduction	31
13.6.2	Prominences and Filaments	34
13.6.3	Classification and Types	34
13.6.4	Recording	35
13.6.5	Relative Numbers and Profile Areas of Prominences	37
13.6.6	Long-Term Observing Programs	38
13.6.7	Short-Term Observing Programs	40
13.7	Flares	44
13.7.1	Introduction	44
13.7.2	Classification	44
13.7.3	Recording of Data	46

13.7.4	Long-Term Observing Programs	47
13.7.5	Short-Term Observing Programs	47
13.8	Other Chromospheric Phenomena	50
13.9	Position Measurements of Solar Phenomena	50
13.9.1	Targets of Position Measurements	50
13.9.2	Heliographic Coordinates	57
13.9.3	Methods of Position Measurement	59
13.9.4	Calculation of Heliographic Positions	63
13.10	Solar Photography	67
13.10.1	Introduction	67
13.10.2	Suitability of Observing Instruments	68
13.10.3	Options for Light Reduction	69
13.10.4	The Most Suitable Camera	71
13.10.5	Comments on Photographic Materials	72
13.10.6	White-Light Photography	74
13.10.7	Photography in Narrow Spectral Ranges	75
13.10.8	Double Exposures as a Method of Measuring Positions	79
13.10.9	Observing Programs	79
13.11	Conclusion	80
	References	80
14	**Observations of Total Solar Eclipses** *by W. Petri*	85
14.1	Photography of the Solar Corona	85
14.1.1	The Coronal Continuum	85
14.1.2	The Structure of the Corona	85
14.1.3	Processing	86
14.1.4	Photography with the Telescope	86
14.1.5	Exposure Times	87
14.1.6	Amateur Photographs	87
14.2	Special Astronomical Programs	88
14.2.1	The Chromosphere	88
14.2.2	The Times of Contact	88
14.2.3	The Partial Phase	88
14.2.4	The Star Field	89
14.3	Special Terrestrial Programs	89
14.3.1	Brightness and Color of the Sky	89
14.3.2	Flying Shadows	89
14.3.3	Meteorological Observations	90
14.3.4	Biological Observations	91
14.3.5	The Ionosphere	91
14.4	The Observing Station	91
14.4.1	Devising the Program	91

14.4.2	Site Selection	92
14.4.3	Equipment	92
	References	93

15 The Moon
by G.D. Roth 95

15.1	Problems and Ideas for Lunar Observations	95
15.1.1	The Moon as a Test Object for Telescopic Work	95
15.1.2	Previous Studies and Space Missions	95
15.2	Conditions of Visibility	100
15.2.1	The Phases of the Moon	102
15.2.2	The Terminator	103
15.2.3	Libration	103
15.2.4	The Lunar Coordinate Grid	106
15.3	Lunar Formations	109
15.3.1	Maria	109
15.3.2	Terrae	109
15.3.3	On the History of Lunar Nomenclature	112
15.4	Observational Projects	113
15.4.1	Visual Observations	113
15.4.2	Photographic Observations	119
15.4.3	Photoelectric Observations	123
	References	127

16 Lunar Eclipses
by H. Haupt 131

16.1	Introduction	131
16.2	The Origin and Frequency of Lunar Eclipses	132
16.2.1	Principles	132
16.2.2	Historical Studies and the Saros Cycle	134
16.2.3	Canons: Statistics of Eclipses	135
16.3	Theory and Prediction of Lunar Eclipses	137
16.3.1	Geometric Theory	137
16.3.2	Photometric Theory of Lunar Eclipses	137
16.3.3	Prediction of Lunar Eclipses	139
16.4	Enlargement of the Earth's Shadow	144
16.4.1	General Considerations	144
16.4.2	Determination of the Enlargement of the Shadow by the Observation of the Transit of Craters Through the Terminator	145
16.4.3	Reduction of Measurements to Determine the Enlargement of the Shadow	146

16.5	Photometry of Lunar Eclipses	147
16.5.1	Principles	147
16.5.2	Global Photometry	149
16.5.3	Detail Photometry of Individual Features on the Moon	154
16.5.4	Photography of the Eclipsed Moon	155
16.6	Scientific Conclusions from Photometric and Astrometric Results	155
16.6.1	Cloudiness at the Earth's Terminator	156
16.6.2	Volcanic Eruptions and Meteoric Dust	156
16.6.3	The Ozone Problem	157
16.6.4	The Solar Influence: Luminescence on the Moon	157
	References	158

17	**Occultations of Stars by the Moon** *by W.D. Heintz*	159
17.1	The Moon as an Astronomical Clock	159
17.2	Predictions	161
17.3	Timing of the Contacts	162
17.4	Grazing Occultations	165
17.5	Occultations of Planets	166
17.6	Photoelectric Registration	167
	References	168

18	**Artificial Earth Satellites** *by R. Kresken*	169
18.1	Introduction	169
18.1.1	The Population of Satellites in Space	169
18.1.2	International Designations	170
18.2	Satellite Orbits	170
18.2.1	Undisturbed Motion	170
18.2.2	Orbit Perturbations	171
18.2.3	Classes of Orbits	174
18.3	Conditions of Visibility	176
18.3.1	Influence of the Inclination	177
18.3.2	Accurate Predictions	177
18.3.3	Simplified Predictions	178
18.3.4	Geostationary Satellites	181
18.3.5	The Brightness Behavior of Satellites	184
18.4	Optical Observations	186
18.4.1	Accurate Positional Observations	186
18.4.2	Photographic Observations	187
18.4.3	Light Variations	188

18.4.4	Changes in Color	189
18.5	Professional Observing Techniques	189
18.5.1	Photographic Techniques	189
18.5.2	Laser Techniques	190
18.5.3	Radar	190
	References	190
19	**Observations of the Planets**	
	by G.D. Roth	193
19.1	The Purpose of and Tasks for Planetary Observations	193
19.1.1	The Amateur Observer and the Planets	193
19.1.2	Observational Tasks	193
19.2	Observing Equipment	195
19.2.1	The Telescope	195
19.2.2	Accessories	196
19.3	Visibility of the Planets	197
19.3.1	Apparent Diameter, Phase, and Oblateness	197
19.3.2	Atmospheric and Environmental Influences	199
19.3.3	Personal Qualities	200
19.4	The Representation of Planetary Observations	201
19.4.1	Drawings	201
19.4.2	Photographs	202
19.4.3	Maps and Planispheres	204
19.5	The Planets	206
19.5.1	Mercury	206
19.5.2	Venus	208
19.5.3	Mars	215
19.5.4	Minor Planets (Asteroids)	224
19.5.5	Jupiter	231
19.5.6	Saturn	245
19.5.7	Uranus	251
19.5.8	Neptune	254
19.5.9	Pluto	255
	References	256
20	**Comets**	
	by R. Häfner	261
20.1	The Nature of Comets	261
20.2	Searching for Comets	265
20.3	Determining the Positions	267
20.3.1	Visual Observations	267
20.3.2	Photographic Observations	269

20.4	Studies of Structure	271
20.4.1	Visual Studies	271
20.4.2	Photographic Studies	273
20.5	Special Techniques	275
20.5.1	Photometry	275
20.5.2	Polarimetry	277
20.5.3	Spectroscopy	278
	References	280

21 Meteors and Bolides
by F. Schmeidler . . . 283

21.1	General Information on Meteors	283
21.2	Methods of Meteor Observation	284
21.2.1	Visual Observations	284
21.2.2	Photographic Observations	285
21.2.3	Radio Astronomical Observations	285
21.2.4	Observation by Television	286
21.3	Special Aspects of Observations	286
21.3.1	Meteors	286
21.3.2	Bolides	288
21.4	Orbit Determinations of Meteoroids	289
21.4.1	The Path Within the Atmosphere	290
21.4.2	The Orbit in Space	291
	References	292

22 Noctilucent Clouds, Polar Aurorae, and the Zodiacal Light
by Ch. Leinert . . . 295

22.1	Noctilucent Clouds	295
22.1.1	Early Observations	295
22.1.2	Characteristic Features	296
22.1.3	Classification of Types	296
22.1.4	Visibility	298
22.1.5	Their Origin	301
22.1.6	The Larger Picture	303
22.1.7	Future Observations	303
22.1.8	Comment on the Literature	303
22.2	Aurorae	304
22.2.1	Explanation	304
22.2.2	Apparent Shapes	307
22.2.3	Spectrum, Color, and Brightness	308
22.2.4	Observability	311
22.2.5	The Auroral Oval and the Geometry of the Magnetosphere	313
22.2.6	Origin	315

22.2.7	Accompanying Phenomena	316
22.2.8	Photographs	317
22.2.9	Comment on the Literature	317
22.3	The Zodiacal Light	317
22.3.1	The Brightness Distribution	319
22.3.2	The Spatial Distribution of the Interplanetary Dust Cloud	320
22.3.3	Properties of Interplanetary Dust Particles	323
22.3.4	Lifetime	325
22.3.5	The Origin of the Zodiacal Light	326
22.3.6	Comment on the Literature	329
	References	329
23	**The Terrestrial Atmosphere and Its Effects on Astronomical Observations** by *F. Schmeidler*	331
23.1	General Remarks on the Atmosphere	331
23.2	Weather-Dependent Phenomena	331
23.2.1	Assessment of Weather Patterns	331
23.2.2	Atmospheric Turbulence and Scintillation	336
23.2.3	Halos, Rainbows, and Other Optical Phenomena	337
23.3	Permanent Atmospheric Phenomena	338
23.3.1	Refraction	339
23.3.2	Extinction	340
23.3.3	Twilight	342
23.3.4	The Brightness of the Night Sky	343
23.3.5	The Polarization of Sky Light	344
23.3.6	The Apparent Shape of the Celestial Sphere	344
23.4	Site Selection for Astronomical Observations	345
	References	346

Supplemental Reading List for Vol. 2 349

Index 357

Contents of Volume 1

1. **Introduction to Astronomical Literature and Nomenclature**
   by W.D. Heintz

2. **Fundamentals of Spherical Astronomy**
   by F. Schmeidler

3. **Applied Mathematics and Error Theory**
   by F. Schmeidler

4. **Optical Telescopes and Instrumentation**
   by H. Nicklas

5. **Telescope Mountings, Drives, and Electrical Equipment**
   by H.G. Ziegler

6. **Astrophotography**
   by B. Koch and N. Sommer

7. **Fundamentals of Spectral Analysis**
   by R. Häfner

8. **Principles of Photometry**
   by H.W. Duerbeck and M. Hoffmann

9. **Fundamentals of Radio Astronomy**
   by W.J. Altenhoff

10. **Modern Sundials**
    by F. Schmeidler

11. **An Historical Exploration of Modern Astronomy**
    by G.D. Roth

12. **Astronomy Education and Instructional Aids**
    by H.J. Augensen

Appendix A: Educational Resources in Astronomy

Supplemental Reading List for Vol. 1

Index

Contents of Volume 3

24 The Stars
by T. Neckel

25 Variable Stars
by T.J. Herczeg and H. Drechsel

26 Binary Stars
by W.D. Heintz

27 The Milky Way Galaxy and the Objects Composing It
by T. Neckel

28 Extragalactic Objects
by J.V. Feitzinger

Appendix B: Astronomical Data

Supplemental Reading List for Vol. 3

Index

13 The Sun

R. Beck, V. Gericke, H. Hilbrecht, C.H. Jahn, E. Junker, K. Reinsch,
P. Völker, Fachgruppe Sonne der Vereinigung der Sternfreunde e.V.

13.1 Introduction

No observer can avoid being fascinated by the Sun, whose appearance changes from day to day, sometimes from hour to hour. On no day in the future will the Sun look exactly as it does today. A wide variety of phenomena are within reach of small telescopes, including those located in urban sites. A welcome feature of solar observations is that they can be pursued without losing nighttime rest. The enormous surface brightness of the Sun requires special equipment for observation and photography—*never use a telescope without sufficient filtering!*

Amateur solar observation has a long tradition in many countries. The first German inter-regional collaboration started in 1917, when the GEDELIA Society founded its solar section, named DARGESO (Deutsche Arbeitsgemeinschaft für Sonnenbeobachtung). The GEDELIA had ceased activities by the 1920s, but the DARGESO continued until 1965. In 1969, solar observers re-founded the Working Group on the Sun (Vereinigung der Sternfreunde e.V.) and a newsletter *Sonne* in 1977, whose editing team also took over the administration of the Working Group and, more importantly, promoted cooperation in various observational projects. The British Astronomical Association also has a working group on the Sun, and the AAVSO and the ALPO have a solar division which monitor sunspot numbers and flares.

The time-demanding monitoring of the Sun puts the amateur in a position of being able to increase the body of existing material, and to continue long-standing series of observations. The evaluation of these may even produce new scientific results.

13.2 Observations of the Sun *(H. Hilbrecht)*

13.2.1 Site Selection

The location of the observing site has a great influence on the quality of viewing, as observations made under inferior conditions cannot be improved even when extreme care is exercised and precision equipment used. Less-experienced observers frequently spend substantial sums of money on instruments and accessories, but this expense either is not justified or is vitiated by the inferior quality of the observing site. This applies also to sites of public observatories and stations located away from major cities, assuming a choice of sites is available at all.

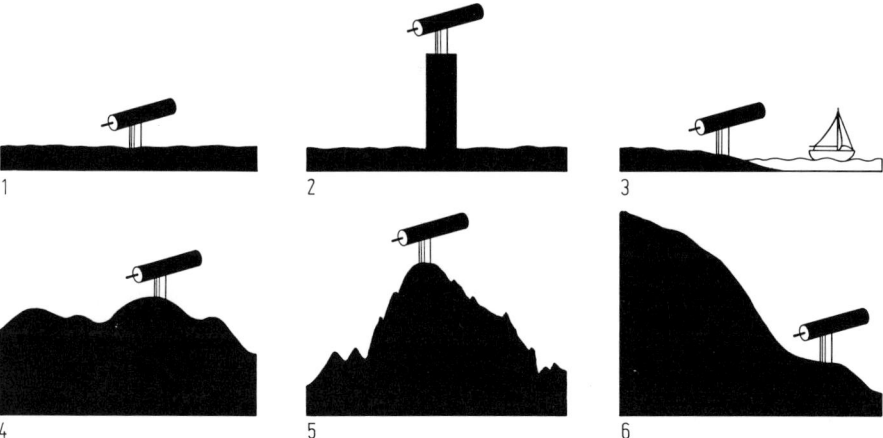

Fig. 13.1. Various solar telescope sites. 1. Telescope near the ground on a plain; 2. telescope on a free tower about 20 m above a plain; 3. telescope several meters above ground on the northern side of a lake; 4. telescope in highland on a hill exceeding 500 m and surrounded by equally high hills; 5. telescope on a mountain top about 1000 m high; 6. telescope about 50 m high on a southern or southwestern slope of a higher mountain facing a plain. Adapted from Kiepenheuer [13.2] and Müller [13.3].

The experiences of professional observers have been summarized by Kiepenheuer [13.2] and interpreted by Müller [13.3] for amateur astronomers. Figure 13.1 shows some typical sites of solar telescopes in Europe. Basically, the line of sight should lead over terrain which warms and cools homogeneously. Very unfavorable are built-up areas (houses, streets, etc.) as they heat and cool rapidly and, moreover, the heated air from smokestacks contributes to the notorious "twinkling" phenomenon, particularly in winter. Similarly disadvantageous are farm fields because of their low heat capacity; variations in the solar irradiation causes rapid changes in the ground temperature.

These radiative temperature changes refer particularly to the lowest air layers, for instance Site 1 in Fig. 13.1, where the observing conditions are good only in the early morning. On a tower 10 or 20 m above ground, a substantial fraction of the ground turbulence is eliminated and the conditions improve markedly (Site 2). A balcony on a high-rise can also be used as a tower as long as the objective extends beyond the hot air rising at the wall. An effective heat absorber placed near the telescope best prevents the interfering and annoying temperature variations. The best telescope site therefore should be at the northern side of a lake. Large continuous forest areas also often substantially diminish the turbulence by evaporative processes (Site 3). Site 4 is located in highlands where, independent of forests, the ascending air forms turbulence and causes the conditions to deteriorate already early in the morning. Site 5 shows an isolated peak in the mountains where the conditions are very good in the early morning, but may diminish quite rapidly with increasing insolation. Site 6 is an interesting and too-little considered possibility. Ascending air generates smooth, uphill winds at a place at least 50 m above the plain or valley on a southern or southwestern slope, a situation which is conducive to good observing. Such favorable conditions easily compensate the reduced view to the north.

Much turbulence originates in the immediate vicinity of the telescope, and therefore the observing area should be carefully shaped in order to obviate the heating. Roads should therefore approach the observing site from the north. The vicinity should contain continuous growth; walkways and parking areas should be more distant. Existing sites can often be much improved toward eliminating sources of heat turbulence. Obviously not all possibilities can be discussed here, and the observer must pay some attention to the particular features of the site. Bushes and trees should be planted abundantly and densely in the garden, as large lawns alone do not possess sufficient heat capacity for stability. Walkways and flat roofs should be white-gravelled, if possible, or even covered with grass or ivy growth.

Observing huts with rolling or folding roofs are of great advantage as ordinary domes store heat and exchange it with the outside air through the slit, thus generating turbulence. Leaving the dome door open for ventilation nevertheless can improve conditions to a surprising degree. The outside house walls, roof, and inside walls surrounding the instrument (observatory, balcony, attic) are best painted titanium white, which minimizes the heating of the walls. It is particularly important to cool tar roofs by spraying them with water, but this is only a temporary solution as the water heats quickly and evaporates. Floors, mounts, piers, chairs, and other such items also heat and should be painted white. The closer a heat source is to the telescope, the more it affects the observations. Of course, there should be no heat source at the site, as the heat exchange with the surrounding air may persist for hours, particularly in winter. Cold feet can be avoided by standing on a wooden platform or sheet of white styrofoam.

Most turbulence arises at and in the telescope. The best and most expensive solution is a *vacuum* telescope; the air is "simply" evacuated from an airtight tube. The experienced telescope maker will take into account the deformation of tube and optical parts by the pressure gradient. A broad ring placed around the tube in the vicinity of the objective is useful during observations, as it automatically shadows the telescope and also increases contrast on the projection screen. The dewcap should be removed in order that the objective may be cooled by air breezes. A complete covering of the telescope during observation every 20 to 30 minutes for about 10 minutes has proven advantageous. In any case, it is worthwhile to consider and attempt improvements of the site. Holes in the tube can, but need not, contribute to the outflow of warm air and to an improved image quality.

While inferior conditions hamper the visual observer by increasing the waiting time for those brief, but precious moments of "good seeing" (see Fig. 13.10), the photographer suffers also through high expenses for needless supplies. What can be saved in photo-materials over the years should be weighed against the expenses for simple constructional improvements. None of the hints given will work miracles, but each of them can effect small improvements which, taken together, can lead to optimal results.

13.2.2 Observing Conditions

Of the atmospheric effects interfering with telescopic observations, one distinguishes directional scintillation or "image motion" from "blurring," which makes the image unsharp or diffuse. The temperature differences and turbulence eddies in the atmosphere change the optical properties of air as if a lens were changing the focal length several times each second. In obtaining data on atmospheric conditions, which are important to judge the quality of a solar observation, the Kiepenheuer scale [13.2] (modified for amateur purposes by Dreyhsig and Reinsch) has proven useful. The image motion (M) measures the scintillation, the sharpness (S) the amount of "blurring."

Modified Kiepenheuer Scale

Image Motion M
1. No noticeable image motion on the disk nor at the solar limb.
2. Image motion of $\leq 2''$ seen at the limb, not on the disk.
3. Image motion $\leq 4''$ distinctly visible at the limb and on the disk; "boiling" limb.
4. Image motion $\leq 8''$ almost prohibits distinction of umbra and penumbra, with strongly "boiling" limb.
5. Image motion $> 8''$ reaches spot diameters, with strongly "boiling" limb.

Sharpness S
1. Granulation easily visible; fine structure in penumbra visible.
2. Granulation and penumbra easily seen but with very little fine structure; umbra-penumbra and penumbra-photosphere boundaries are sharp.
3. Granulation visible only in traces, but surface structures are easily found when moving the solar image; umbra and penumbra well-separated without fine structure; boundary to photosphere difficult.
4. Granulation not visible; umbra and penumbra separated only in large spots; boundary to photosphere diffuse.
5. Granulation not visible; umbra and penumbra not distinguished even in large spots.

For some purposes, it may be useful also to record the deviation from the average conditions, for example, on long-term sunspot counts (sunspot numbers), since here the deviations from a series are more important than its absolute quality. Using the following scale, turbulence, sharpness, transparency, and other factors affecting the

Quality Q	Description
Excellent	Reserved for days when details are observed unusually clearly.
Good	Average visibility of solar surface details corresponding to the conditions for that observer.
Fair	Substandard seeing but not yet considerable interference.
Poor	Considerable interference strongly limiting the value of the observation.
Worthless	Conditions so poor that evaluation of the observation is not useful.

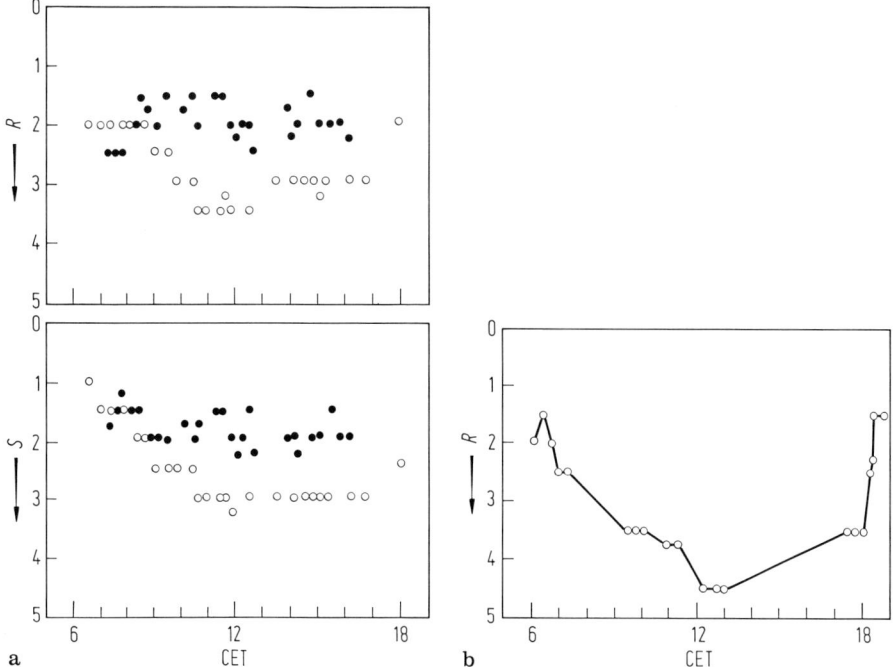

Fig. 13.2a, b. Daily variation of image motion and sharpness of the solar image. **a** From Kiepenheuer [13.2] from observations made on 1962 August 14. *Open circles*: Schauinsland Observatory, Black Forest; *Closed circles*: Upper Rhine plane (tower telescope); M and S observed in the plane from the tower telescope are about constant. The Schauinsland records show distinctly better observing conditions in the early morning and evening hours, and worst around noon. **b** Daily variation in image motion observed by Brandt [13.4] on 1967 August 22 at the Schauinsland Observatory. Excellent conditions were recorded for only about 1 hour each before and after noon. The times of these minima shift depending on the seasonal variation in sunrise and sunset.

image quality are considered. "G" (for good) refers to the average conditions at a site; note that "G" for one observer who works at a particular site may correspond to a "P" (for poor) for another who works at a more favorable site.

Figure 13.2 shows the variation of M and S over one day, the diurnal change of turbulence. Usually, conditions are best one to two hours after sunrise or before sunset, but the data from the Rhine valley provide a counterexample. Thus, it is worthwhile to study local conditions. Observations at noontime are preferred by most solar observers, but these observations are generally made under the poorest conditions.

13.2.3 Records

The previously used observing book has been replaced in solar observations by single recording sheets which can be more easily archived, evaluated, and copied. A general observing form is shown in Fig. 13.3. Special observations are recorded using other

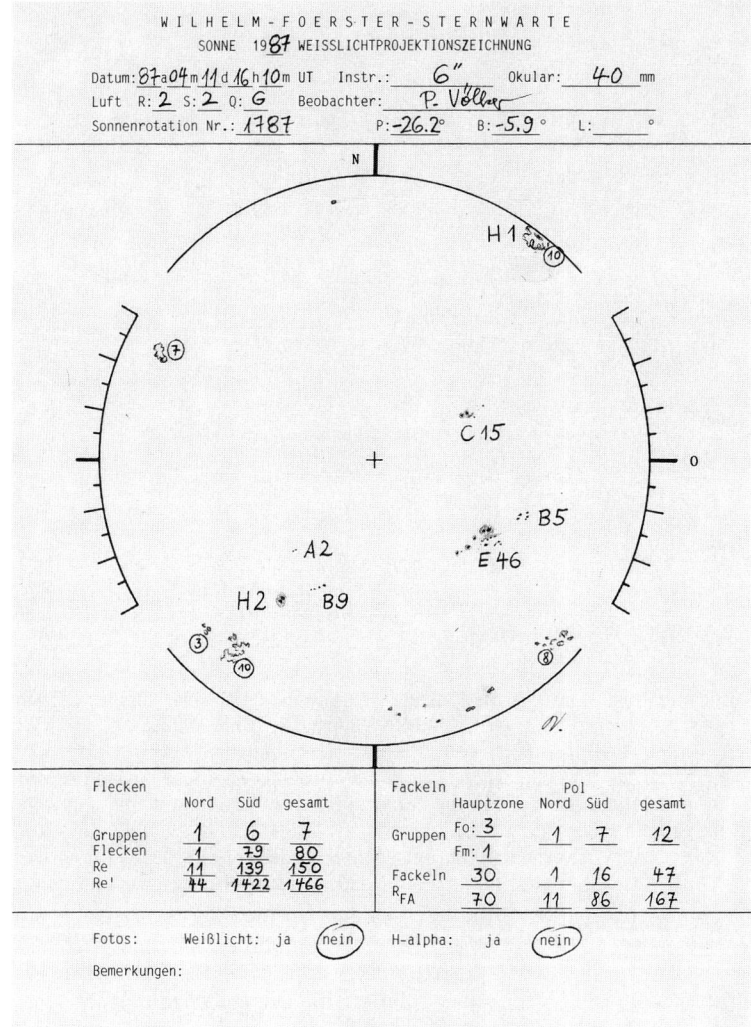

Fig. 13.3. Actual record form for observation of the Sun (in German). It contains all relevant data for amateurs and also permits the detection of changes in the graphed sunspot groups in the circle over the course of time.

forms. For many observing programs, forms designed by working groups can be obtained from the Wilhelm-Foerster-Sternwarte in Berlin or any national solar section. As an example, the record form for sunspot numbers is pictured in the section on solar activity (Fig. 13.3).

13.3 Sunspots *(R. Beck and K. Reinsch)*

13.3.1 Development and Classification of Sunspots

Sunspots are extended, strongly magnetic regions on the Sun which, owing to their reduced temperatures (3000–3500 K) radiate less visible light than the undisturbed photosphere (5800 K). The sizes of spots ranges from 2000 to over 100 000 km. (An angle of 1″ corresponds to a length of about 700 km on the Sun; 1′ is about 43 000 km.) Spots with angular diameters under 10″ are called *pores*. Larger spots usually consist of an *umbra* and *penumbra* (Fig. 13.4).

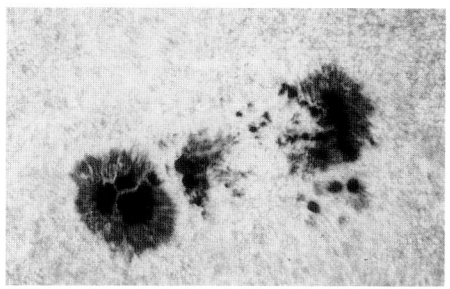

Fig. 13.4. Sunspot group on 1986 April 27, photographed by C.H. Jahn using a 200/3250 mm refractor with 12.5-mm eyepiece, solar prism, and green filter.

The *umbra* is the dark core of a spot. In direct viewing (e.g., using an objective filter) with low turbulence, brightness differences can be seen in the umbra. Larger telescopes show bright spots with a diameter of only about 500 km but almost as bright as the photosphere. Diffraction in the Earth's atmosphere and at the objective makes them appear larger than they really are. Their lifetime is 15 to 30 minutes.

The *penumbra* surrounds the umbra, and consists of bright and dark filaments connected radially to the umbra. The width of filaments is about 0.3″ or about 200 km, their duration about 2 hours. Changes and divisions can be observed within just a few minutes.

The ratio of radii of penumbra to umbra depends on the development of the sunspot and on solar activity in general. At sunspot maximum, the umbra is on the average larger relative to the penumbra than at minimum. The increase of the average magnetic field strength in sunspots from minimum to maximum is the presumed cause.

Sunspots usually occur in groups. Large groups often show a clustering of spots around two primary spots, which are termed *bipolar*. The occurrence of a spot group is linked to other phenomena tied to solar activity such as *faculae, prominences*, and *flares*, all of which are caused by magnetic fields. The development of these phenomena is of different speeds and also different in each spot region. The following example describes the general development of an activity center; see Newton [13.5], Bray and Loughhead [13.6], Wilson [13.7], and McIntosh [13.8] for more detailed information.

Day 1: A bundle of magnetic field fibers reaches the photosphere. When the magnetic flux density exceeds 0.1 tesla (= 1000 gauss), *faculae* appear.

Day 2: The first small sunspot appears at the western edge of the faculae region. This region increases in size and brightness. The magnetic field bundle continues to ascend and to increase in flux density.

Day 3: One or more spots appear at the eastern edge of the faculae region, with magnetic polarity reversed from that of the first spot. The area occupied by magnetic field and faculae continues to grow.

Day 4: Smaller spots merge into large ones. The western spot of the group forms a penumbra. The faculae region surrounds the spot group, but remains compact. The magnetic field distinctly shows bipolar structure. The first flares are observed. Small filaments near the western spot are not yet stable.

Days 5–13: On day 5, the eastern spot also forms a penumbra. Then numerous small spots between the main spot appear until the group reaches greatest extension. The brightness of the faculae region continues to grow, as does the area occupied by the magnetic field. Flare activity reaches maximum.

Days 14–30: All spots except the western primary spot disappear. The faculae region is very extended but starts to shrink. Flare activity diminishes. Magnetic flux density is maximum, but the area occupied decreases. A stable filament about 50 000 km long points in the direction of the western spot.

Days 31–60: The western spot shrinks and disappears. The faculae region diminishes in brightness and splits into smaller sub-areas. The magnetic field weakens and becomes irregular. The filament increases in length by $\sim 10^5$ km per solar rotation, dividing the active region into two parts.

Days 61–100: Chromospheric faculae disappear, photospheric faculae dissolve gradually. The filament reaches largest extension, almost parallel with the solar equator.

Days 101–250: No more faculae are found. The filament gradually dissolves together with the magnetic field.

While the sunspot group in this example has a lifetime of only 60 days, the magnetic field, which is the cause of solar activity, can be traced for about 250 days. A bundle of magnetic field lines from deeper zones reaches into the photosphere, and there expands (owing to the lesser pressure) into the form of an arch. The two points of penetration in the photosphere mark the two magnetic poles of the spot group. The details of the developments of spots, such as the formation of sharp boundaries between umbra, penumbra, and photosphere have not yet been clarified. Also, the origin of solar magnetic fields is still much under discussion (Schüssler [13.9], Giovanelli [13.10]).

Sunspots are short-lived phenomena. 90% of all groups disappear after 10 days or less, 50% after only 2 days. The most stable groups, however, have lifetimes of several months. As of this writing, the longest-lived group was observed from 1979 March 15 until 1979 August 03 (Fritz, Treutner, and Vogt [13.11]). The average lifetime of a group, about 10 days, varies periodically with the *long cycle* (cf. Sect. 13.3.5).

Only in about 10% of all groups does the eastern spot live longer, while in 40% the western spot dominates. Also, the area of the western spot averages 40% larger than that of the eastern spot.

As a bipolar spot group develops, the angle between the line joining the main spots (the axis) and the latitude circles shows a typical change: with increasing spot area, the angle decreases, reaches a minimum at the time of maximal development, and then increases as the number of spots in the group diminishes. The mean inclination of the axis also depends on the heliographic latitude of the group (Bendel [13.12]).

The typical evolution of a large group as described led Waldmeier in 1947 to introduce the following sunspot classes (Fig. 13.5):

A: Single spot or group of spots without penumbrae and without bipolar structure.
B: Spot group without penumbrae in bipolar structure.
C: Bipolar group, one of the two main spots has a penumbra.
D: Bipolar group, both main spots having penumbrae, and at least one spot shows some structure. The maximum extension of the group under 10° on the Sun (about 1.2×10^5 km).
E: Large bipolar group, the main spots are surrounded by penumbrae and have a complex structure. Between the main spots there are numerous smaller spots. Extension of the group at least 10°.
F: Very large bipolar or complex spot group, extension at least 15° (or 1.8×10^5 km).
G: Large bipolar group without small spots between main spots. Extension at least 10°.
H: Unipolar spot with penumbra, diameter greater than 2.5° (about 3.0×10^4 km).
J: Unipolar spot with penumbra, diameter under 2.5°.

Only about 2% of the spot groups pass through all Waldmeier classes. Most groups reach maximum extension in one of the early classes and then regress through classes C, G, or J. The largest number of sunspots is usually reached by the end of the first third of the lifetime of the group. This time asymmetry in the development increases with

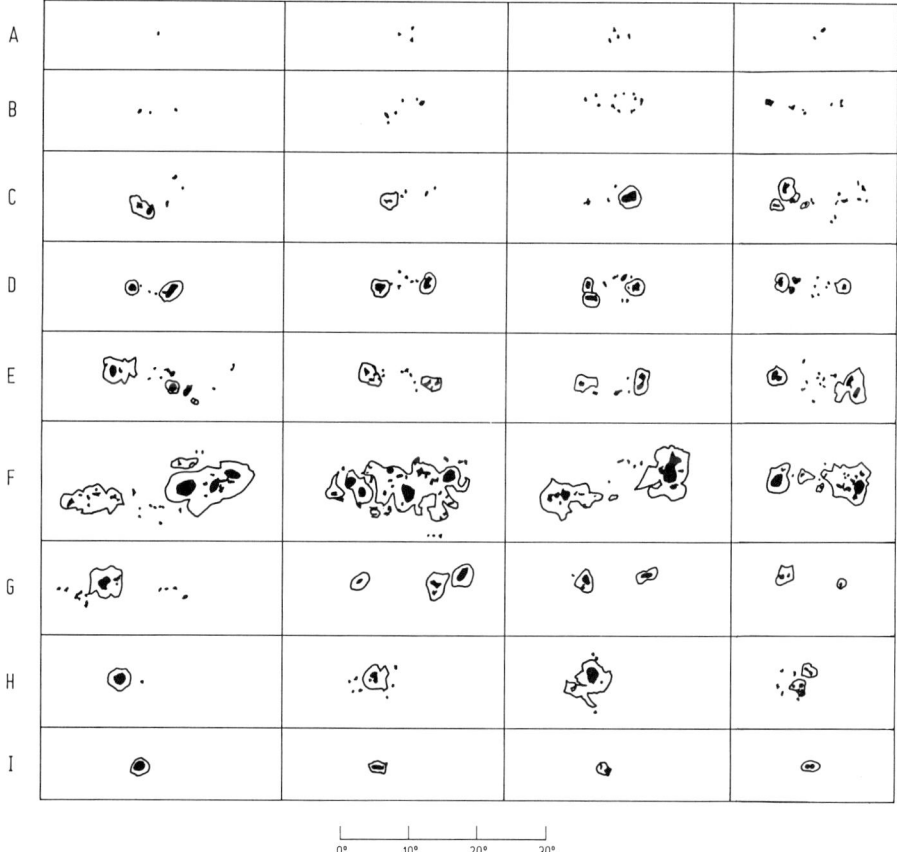

Fig. 13.5. Waldmeier classification scheme for sunspot groups.

the maximum class of development, in surprising analogy with one of the Waldmeier rules of sunspot cycles (see Sect. 13.3.5).

A useful task for observers is to count the frequencies of spot groups in the separate Waldmeier classes, and to note the change of that distribution during the course of an activity cycle. Various classes of spots reach greatest abundance at different times in the cycle (Künzel [13.13]).

For complex structured groups, the Waldmeier scheme does not suffice for classification. The specialists have thus modified and extended it, so that it is now known under the name *McIntosh Classification* [13.14] (Fig. 13.6).

The *first letter* (upper case) corresponds to the Waldmeier class but with classes G and J omitted. The class G, depending on extension, is counted among classes E or F; class J merges with H, which then contains all unipolar groups with penumbrae. If a magnetogram is not used, unipolar groups are characterized by the largest distance between two spots (or between edge of the penumbra of the main spot and any other spot of the group) not exceeding $3°$. When new large spots appear in the vicinity of an H-spot, this generally indicates the formation of a new bipolar group which should be treated as a group in its own right. When the diameter exceeds $5°$, it can be considered certain that both magnetic polarities exist within the penumbra (bipolar group), and the group is to be classified as Dkc, Ekc, or Fkc.

The *second letter* (lower case) in the McIntosh scheme expresses the appearance of the penumbra of the largest spot in the group:

x: no penumbra (corresponds to Waldmeier class A or B).
r: rudimentary penumbra partially surrounds the largest spot. This penumbra is incomplete, granular rather than filamentary, brighter than a mature penumbra, and extends as little as 3 arcseconds (2200 km) from the spot umbra. Rudimentary penumbrae may be either in a stage of formation or dissolution.
s: small, symmetric (corresponds to Waldmeier class J). The largest spot has a mature, dark filamentary penumbra of circular or elliptical shape with little irregularity to the border. There is either a single umbra, or a compact cluster of umbrae, mimicking the symmetry of the penumbra. The north–south diameter across the penumbra is $\leq 2.5°$.
a: small, asymmetric. Penumbra of the largest spot is irregular in outline and the multiple umbrae within it are separated. The north–south diameter of the penumbra is $\leq 2.5°$.
h: large, symmetric (corresponds to Waldmeier class H). Same structure as type "s," but the north–south diameter of the penumbra is $> 2.5°$. The area, therefore, must be ≥ 250 millionths of the solar hemisphere.
k: large, asymmetric. Same structure as type "a," but with north–south diameter $> 2.5°$, and surface area ≥ 250 millionths of the solar hemisphere. This type sometimes contains spots of opposite polarity, and may indicate potential for proton flares.

The *third letter* (lower case) distinguishes the distribution of spots within a group:

x: undefined for unipolar groups (classes A and H).
o: open. Few, if any, spots between the leader and follower. Interior spots of very small size. Classes E and F of the *open* category are equivalent to Waldmeier class G.
i: intermediate. Numerous spots lie between the leading and following portions of the group, but none of them possesses a mature penumbra.
c: compact. The area between leading and following ends of the spot group is populated with many strong spots, with at least one interior spot possessing a mature penumbra. In the extreme case of a compact distribution, the entire spot group is enveloped in one continuous penumbral area.

The second and third letters of the McIntosh code contain information on the magnetic field structure in the group. Statistical data using this classification thus

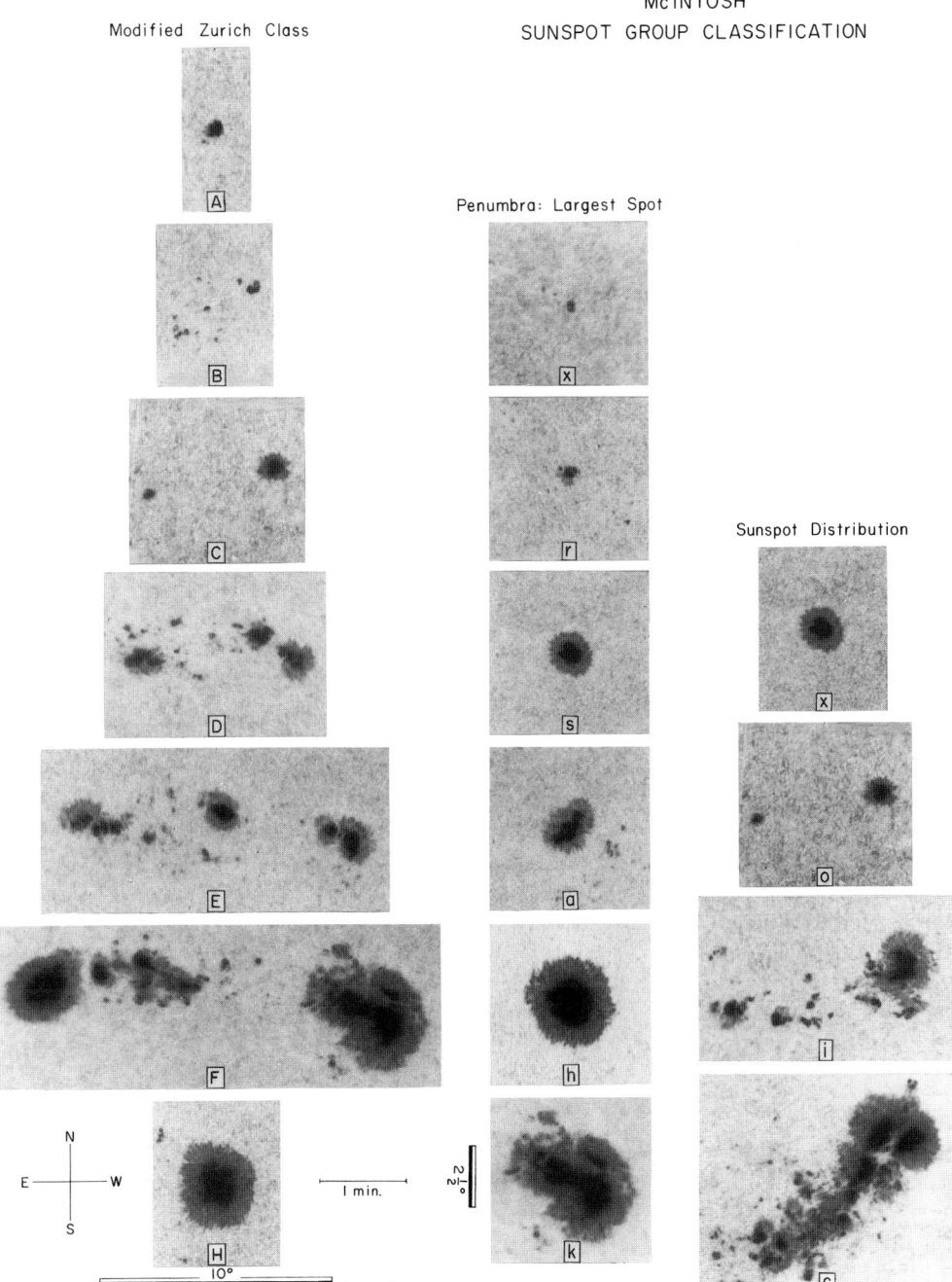

Fig. 13.6. McIntosh classification scheme for sunspot groups.

gain in significance. The McIntosh scheme is also recommended for use by amateur observers.

Another relevant quantity which describes the evolutionary state of a sunspot group is the *area* A_i occupied by the spots. It is expressed in 10^{-6} of the visible hemisphere of the Sun (MH), considering the geometrical forshortening in the vicinity of the limb:

$$A_i = 10^6 \times \frac{A_i'}{2\pi R^2 \cos \rho_i}, \tag{13.1}$$

where A_i' is the area of the spot group in the solar image, R is the radius of the solar image, and ρ_i the angular distance of the center of the image, given in good approximation by (cf. Sect. 13.9.4):

$$\rho_i = \arcsin(r_i/R), \tag{13.2}$$

where r_i is the linear distance of the center of the spot group from the central point of the image. The *area number* A is the sum over the A_i of all groups. On the average, the number A is about 17 times larger than the *Wolf number* $\mathfrak{R}$ (see Sect. 13.3.4) (Waldmeier [13.15]).

The area of a spot is intimately related to the magnetic field strength, and thus the development of the spot area reflects the evolution of the field. There is a direct connection between maximum area of the group and its lifetime. The faster the area of the group grows in the early development, the higher the maximum will be; the same also applies for the *number* of spots in the group. After maximum development, the number of spots drops rapidly, but the area is determined by the long-lived western main spot, which decreases rather slowly. Also, the magnetic field diminishes only slowly.

The measuring of spot areas can be done most expediently by graphing the solar image precisely on a firmly mounted and shaded projection screen, and then taking measurements with transparent millimeter paper. A stable mount, precise tracking, and a moderate-size telescope are required for good area measurements. The accuracy can be improved using photography, where the negative of the solar image is usually enlarged and projected onto millimeter paper in the darkroom. Solar photographs of adequate quality may be obtained at the primary focus of a 2-inch (5-cm) refractor with a good solar filter (see Sect. 13.10). The focal length of the objective and magnifier should be fairly long in order to reduce image distortion. The demands on time and material are low as copies of the negatives are not needed.

Every feature of a spot group (class, inclination of axis, spot number, area, size and brightness of the associated faculae, number and type of light bridges, frequency of flares, etc.; cf. later sections of this chapter) can be graphed against time as a curve of development characterized by the following parameters:

(a) the time interval (lifetime of the spot group),
(b) the height of maximum,
(c) the area under the curve,
(d) mean slopes of ascending and descending branches.

Statistical processing should start with these parameters. Long-term series of observations permit the study of the following questions:

(a) Do the parameters of the development curves change within the 11-year activity cycle? If so, which ones?
(b) What connections exist between parameters of the mean development curve of a feature?
(c) What connections exist between the mean development curves of different features?

During the lifetime of a large group of spots, rapid changes in their appearance occur. Spots appear, others vanish, new penumbrae form, light bridges materialize, and spots change relative positions (see Sect. 13.9). These changes can be observed within hours or days. Small spots sometimes show evolution within minutes.

As the eye is subject to illusions, photographs of rapid changes are desirable (see Sect. 13.10). Caution is also advised here because atmospheric turbulence may feign structural changes. The best material from which to study rapid developments are series of photographs taken at intervals of 10 minutes to 1 hour—the ideal observing program for weekend observers!

13.3.2 Light Bridges

Nearly all sunspots contain bright areas resembling the photosphere. Since they extend usually as narrow tongues into the spots, they are referred to as *light bridges*. Their formation is characterized, first, by the appearance of indentations on opposite sides of the spot. These indentations develop by moving toward each other until connecting to form a bridge. The spot appears divided at this point (Fig. 13.4). Apart from these "classical" light bridges, bright islands are noted particularly in older, stable sunspots (classes H and J; cf. Fig. 13.5), which, with time, may break out into the photosphere.

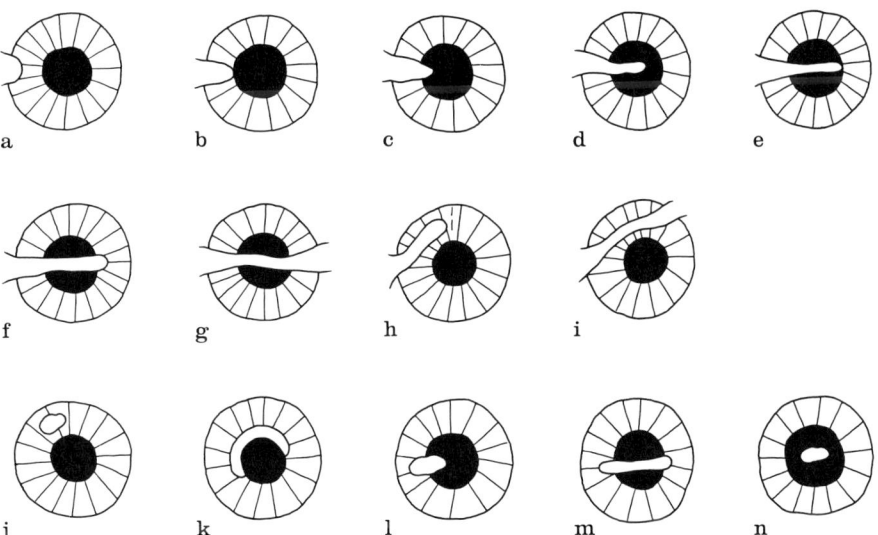

Fig. 13.7a–n. Scheme for assigning types to light bridges. After Hilbrecht [13.1].

Light bridges are closely connected with the general development of sunspot groups, which can be only incompletely understood without observing the bridges. Since 1978, the classification by Hilbrecht ([13.1], pp 401) has been applied, extended in Fig. 13.7 by the Type "n" introduced by Seebörger-Weichselbaum.

When studying the light bridge development with this encoding, it is found that the early and late phases of spots (classes C, D, H, and J) show, per area and number of spots, more light bridges than active groups do (classes E, F, and G). In active groups bridges penetrate from the photosphere far into the spot and evolve rapidly, but otherwise mostly slow changes are observed. In H- and J-groups, in contrast to C- and D-groups, "classical" light bridges are less pronounced than islands, as the bridges delineate primarily irregularities in spot contours.

From long-term series of observations, amateur observers can obtain and process valuable data on light bridges. Since 1977, the VdS Working Group on the Sun has collected amateur observations and introduced observers to the subject. A description (with references) of observing programs with results from professional observatories was published by Hilbrecht [13.1].

13.3.3 The Wilson Effect

In 1769, Alexander Wilson discovered the following solar phenomenon: sunspots which display a symmetrical shape at the center of the solar disk appear deformed as they approach the western limb, such that half of the penumbra directed toward the center narrows and may disappear, while the other half nearly retains its shape (Fig. 13.8). This feature is appropriately called the *Wilson effect*. Wilson explained it by assuming each sunspot to be a cavity of truncated conical shape in the solar surface and which therefore generates the observed effect when seen in perspective (Fig. 13.9). His calculations of depths gave values of several 10^3 km.

The models recently developed by solar physicists resemble the cone model of Wilson. Yet the Sun has no solid surface; it is gaseous with continuously decreasing density outward. The umbra is more transparent than the penumbra, and the latter more transparent than the photosphere. Hence, the light from the umbra comes from lower, deeper layers than that of penumbra and photosphere. The optically measured depths range between 500 and 1000 km, where the depth increases with the size of the spots (Wilson [13.16]).

Study of the Wilson effect requires an accurate definition in order to quantitatively measure how pronounced it is. Assigning ranges according to an often recommended point scale has been found to be unsatisfactory. It is more precise and significant to use the ratio of apparent width of the penumbra determined on a projection screen or on photographs (Jahn and Reinsch in [13.1], pp 393). The "depth" of the spot is found from the change of penumbra width during the motion toward the western limb (or from the eastern limb toward the center). The purpose of measurements is to determine the "depth" as depending on size and class of spot, and on phase of the activity cycle.

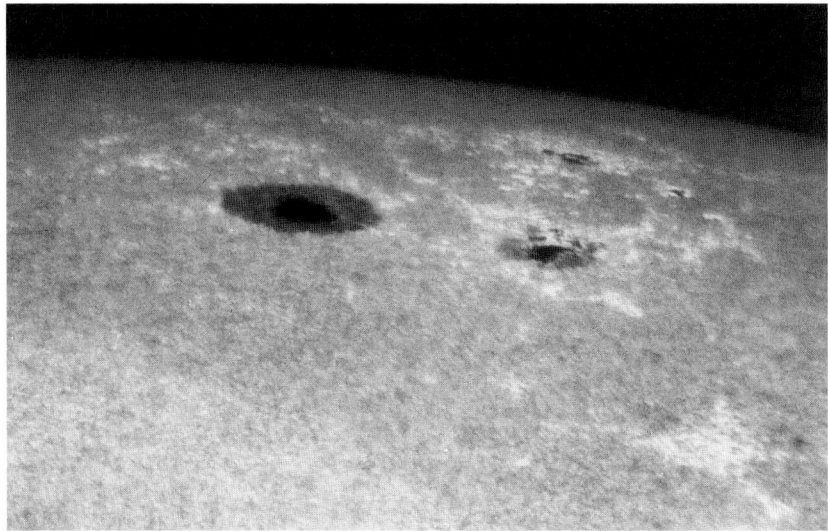

Fig. 13.8. Wilson effect in a sunspot and photospheric faculae, photographed on 1985 June 04 by C.H. Jahn using a 200/3250 mm refractor with 12.5-mm eyepiece, solar prism, and green filter.

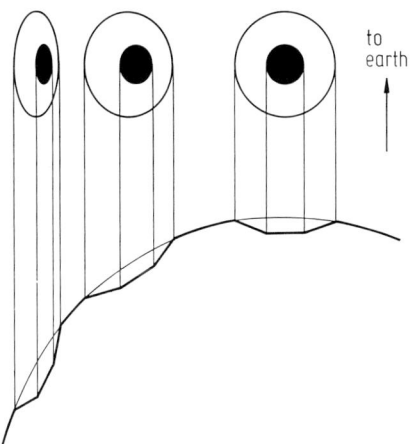

Fig. 13.9. The cause of the Wilson effect.

13.3.4 Numerical Expression of Spot Activity

The degree of sunspot activity is generally described by a single quantity which has been internationally accepted: the *Wolf sunspot number*. It was introduced by Rudolf Wolf in 1848 and is the basis of many investigations of solar-terrestrial relations. Determining Wolf numbers is a favorite amateur observing program as it can easily be performed using small telescopes.

The Wolf sunspot number includes the number of observed single spots s as well as that of entire spot groups g, as the appearance of spots in groups is an important feature of activity. Wolf multiplied the number g of groups with a factor 10 in order to express the fact that the appearance of a new group is weighted 10 times higher than a new spot within an existing group. An isolated spot is redarded as a separate group. The Wolf number $\mathfrak{R}$ is combined from spot and group numbers according to the relation

$$\mathfrak{R} = 10g + s. \tag{13.3}$$

Example: What is the Wolf sunspot number $\mathfrak{R}$ from the observing data in Fig. 13.3? The following groups are observed: an A-group with 2 spots, a B-group with 5 spots, a B-group with 9 spots, a C-group with 15 spots, an E-group with 46 spots, an H-group with 1 spots, and an H-group with 2 spots,

Thus, there are 7 groups with a total of 80 spots. The Wolf number is therefore

$$\mathfrak{R} = 10 \times 7 + 80 = 150.$$

The ordering of spots into groups may be difficult during times of high activity when several groups are closely adjacent or show complex structures not immediately fitting into Waldmeier's classification scheme (Sect. 13.3.1). In this case, the following rules (Künzel [13.17]) may be of help:

1. Spots within an area covering up to $5° \times 5°$ are counted as *one* group if no bipolar structure is noted. Bipolar groups, however, may extend to lengths of $20°$ or more.
2. Single spots separated in heliographic longitude by up to $15°$ are counted as one group when they are the remnants of a larger, formerly contiguous group.
3. A bipolar clustering of spots is considered one group when its western part has the same or a lesser latitude than the eastern part. The average inclination of the axis is $1°$ to $2°$ at latitudes $\pm 10°$, and about $4°$ at latitudes $\pm 30°$.
4. An isolated spot is counted as one group.

There may also be problems in counting the separate spots. Since the development is a smooth transition from granulation-free photosphere to pore to spot, the exact moment of the birth of a spot cannot be given. Under very good atmospheric conditions, the Sun's surface appears bespattered with small A-groups, but these are mostly enlargements of the inter-granular area with a lifetime of some minutes. These should *not* be counted in the Wolf number. For the statistics of numbers, Waldmeier defines a sunspot as having a diameter of at least $3''$ and a minimum lifetime of 30 minutes.

Very large sunspots often show several umbrae in a joint penumbra. Here, every umbra is counted as a spot when completely separated from other umbrae. Thickening of penumbral filaments are not counted as spots. Numerous circumstances of observation influence whether or not spots are visible and recorded, and thus the Wolf number:

- Atmospheric conditions (air turbulence, sharpness, wind, clouds, haze, etc.).
- Instrument and observing methods (aperture, focal length, optical quality, magnification, filter, mounting and its stability).
- Observer (vision quality, care in observing, experience, physical and psychological condition).

In order to intercompare the daily Wolf sunspot numbers obtained under various conditions, all values must be reduced to a standard scale. The observations made for over a century at the Zürich Observatory using its Fraunhofer refractor with an aperture 8 cm, focal length 110 cm, and magnification 64× were used as international standards up until the end of 1980. In 1981, the collection, evaluation, and publication of international Wolf numbers $\mathfrak{R}_{\text{int}}$ was transferred to the Sunspot Index Data Center (SIDC) in Brussels, Belgium, and since then these values have been used as the standards; they are published, for instance, in *Sky & Telescope* and other magazines.

The reduction factor k must be found for each observer and each instrument from extended simultaneous observations in order to convert the individual Wolf numbers to the international scale. This factor k is found from the ratio of average Wolf numbers over a certain period of time (at least one year) between the standards and the observed series:

$$k = \overline{\mathfrak{R}_{\text{int}}}/\overline{\mathfrak{R}}. \tag{13.4}$$

Of course, the averages $\overline{\mathfrak{R}_{\text{int}}}$ and $\overline{\mathfrak{R}}$ are formed only over days covered by simultaneous observations.

The value of the k-factor says *nothing* about the quality of a series, but the scatter of k-values from day to day or from year to year indicates the long-term homogeneity of observed series. Every observer should try, after an initial trial period, to maintain the observing method once chosen and to keep the observing conditions as constant as possible. Once the k-factor of an observer is determined, his/her observations are reduced to the international scale by the relation:

$$\mathfrak{R}_{\text{int}} = k(10g + s) = k\,\mathfrak{R}. \tag{13.5}$$

The Wolf number is a simply determined measure of activity, but in practice it has its disadvantages; for instance, the sizes of individual spots is not considered. A tiny A-spot contributes as much to the number as a giant H-spot of diameter 50 000 km.

Several measures of solar activity deviating from Wolf's method have been suggested to remedy these drawbacks, and are mentioned in the following discussion.

A physically meaningful expression of solar activity as observed in white light should be closely connected with magnetic field strength on the Sun. This could be expressed by the area of a sunspot. As for stable spots, the maximum field strength at the center increases with the area of the spot. Measuring areas, however, is much more demanding than simply recording Wolf numbers (cf. Sect. 13.3.1).

The new sunspot number (or, more correctly, new area number) $\mathfrak{R}'$ after Beck [13.18] purports to combine the advantages of the area number with the simple counting of the Wolf number. It uses the knowledge that for each group type of the Waldmeier classification there is an average ratio of spot area to spot number. When the observer determines the spot number and the Waldmeier class of the ith group, he/she can estimate the area of the group by multiplying the spot number s_i with a weight factor G_i representing the average area/number ratio for the group type. The new area number $\mathfrak{R}'$ is obtained by adding the area numbers of all groups i:

$$\mathfrak{R}' = G_1 s_1 + G_2 s_2 + G_3 s_3 + \cdots. \tag{13.6}$$

The following weight factors G_i for the Waldmeier classes are used:

Waldmeier class =	A	B	C	D	E	F	G	H	J
Weight factor G_i =	4	4	8	18	25	36	50	44	37

The highest weights occur in the late classes G, H, and J, with a much-diminished number of spots but still a large area covered by the main spots.

Example: What is the new area number $\Re'$ for the observation in Fig. 13.3?
Calculation yields

$$\Re' = 4 \times 2 + 4 \times 5 + 4 \times 9 + 8 \times 15 + 25 \times 46 + 44 \times 1 + 44 \times 2 = 1466.$$

The new area number $\Re'$ has been determined by amateur astronomers since early 1977 and has been compared with Wolf sunspot numbers and with area numbers (Fig. 13.10).

In 1976, the Astronomische Arbeitsgemeinschaft Paderborn proposed a different Wolf number, the *Paderborn sunspot number* S_{PB} defined as [13.19]:

$$S_{PB} = Gr + Grsp + Grs + Esp + Es. \tag{13.7}$$

where Gr = group (i.e., the number of phenomena with at least two umbrae, also a penumbra with two umbra but no separate spots), Grsp = total of all spots with

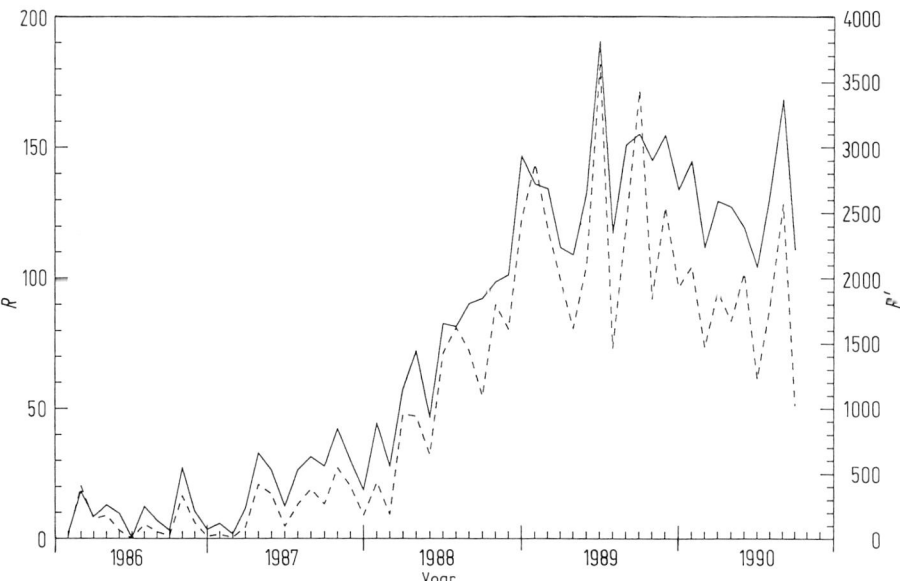

Fig. 13.10. Comparison of Wolf spot number $\Re$ (*solid line*) and new area number $\Re'$ (*dashed line*). After observations by the Working Group *Sonne* made during the period 1986–1990.

penumbra within groups, Grs = total of all spots without penumbrae in groups, Esp = number of individual spots with penumbrae outside groups, Es = number of individual spots without penumbrae outside groups. The Paderborn number, in contrast to the Wolf number, indicates the fraction of spots with penumbrae and the distribution of spots with respect to groups and singles. The individual terms of the sum are actually more important for the evaluation than is the total S_{PB}. The weight factor 10 for the number of groups is omitted and thus S_{PB} corresponds approximately to the number of spots in the Wolf count.

Another sunspot number, one which considers the formation of penumbrae as signs of high solar activity, was suggested by Pettis [13.20]:

$$SN = 10p + s, \qquad (13.8)$$

where p is the number of spots with penumbrae and s that of spots without penumbrae. A penumbra with more than one umbra within it is counted as $p = 1$.

Considerations similar to those leading to the definition of the Beck area number have also led the Norwegian Astronomical Society (NAS) to introduce *classification values* CV [13.21]. The CV numbers give larger and more complex groups higher weight than small spots. The system is based on the McIntosh Spot Classification (Sect. 13.3.1) which admits 60 possible classes. These are arranged in a sequence giving the highest weight (60) to an F-group with large, symmetric penumbrae and with complex spot distribution. A single A-spot is assigned a CV-weight of 1. The CV-values of all groups are added and thus form the measure of spot activity. Notwithstanding the somewhat arbitrary definition of weights for the spot classes, the CV system has the advantage that spot counts and area measurements may be dispensed with; it needs merely the classification of all groups in the McIntosh scheme.

The most readily determined spot value is the number A of spots visible to the unaided eye [13.22]. (This observation requires eye protection.) It permits a connection between present observations and records from pre-telescopic times. Keller [13.23] has performed naked-eye sunspot observations from the Zürich Observatory for over 10 years, and claims good agreement with the Wolf numbers.

All of the aforementioned measures of solar activity have in common that they are all subject to strong external influences (e.g., observing conditions and selection effects) and random effects (e.g., the invisibility of the other side of the Sun), and hence their daily values are essentially meaningless. It is only from the *average* of daily numbers over one *apparent* rotation period of 27 days that a fairly dependable description of solar activity can be obtained.

A solitary observer cannot determine the spot activity without experiencing gaps due to weather and personal commitments. Thus, in order to achieve continuous supervision of solar activity and follow any long-term patterns over years and decades, it is advisable to cooperate with several observers distributed in locations as widely separated as possible. Such observing networks exist in several countries; the oldest amateur group is the solar division within the American Association of Variable Star Observers (AAVSO), which publishes their results monthly in the *Solar Bulletin* and in *Sky & Telescope*. In Germany, a network of observers of Wolf numbers was founded in 1977, and today counts over 130 contributers worldwide and is assisted by

SUNSPOT NUMBERS IN MONTH															19	

Observer (Name, Address):

Instrument/Method:
O Refractor Ø= _____ mm, f= _____ mm
O Reflector Magnification: _____ x
O Binocular: _____ x _____ mm

geogr. Longitude: Latitude:
O visual; O Projection, Ø: ____ mm; O photogr.

Day	Time UT hh:m	Motion M 1-5	Sharpn S 1-5	Quality Q	North g_n	North f_n	South g_s	South f_s	total g g_n+g_s	total f f_n+f_s	total R ≥10g+f	Beck index R^I	Pettis p	Pettis s	Pettis SN ≥10p+s	Classif. value CV	naked eye A
1																	
2																	
3																	
4																	
5																	
6																	
7																	
8																	
9																	
10																	
11																	
12																	
13																	
14																	
15																	
16																	
17																	
18																	
19																	
20																	
21																	
22																	
23																	
24																	
25																	
26																	
27																	
28																	
29																	
30																	
31																	
Sum:																	
Number:																	
Mean:																	

Quality: 1=E (xcellent), 2=G (ood), 3=F (air), 4=P (oor), 5=W (orthless); Motion, Sharpness: modified Kiepenheuer scale, values: 1 (no image motion), 2, 3, 4 or 5 (motion>8)

Fig. 13.11. Form for recording Wolf number observations.

the Working Group on the Sun of the Vereinigung der Sternfreunde.[1] Results and lists of participants are published quarterly in the circular *Sonne*. Other observer groups report the new area number $\mathfrak{R}'$, Paderborn number S_{PB}, the Pettis number SN, and the number *A* of spots observed with the naked eye.

A sample list to record observations of Wolf numbers follows in Fig. 13.11. Every observation should include, in addition to the number of groups *G*, the number of spots *s* (or *f* in Fig. 13.11), the Wolf number $\mathfrak{R}$ and the new area number $\mathfrak{R}'$, the time (UT), the image motion *M*, the sharpness *S*, and the quality *Q* of the solar image (Sect. 13.2.2). Division of activity into the northern and southern hemispheres of the Sun (g_N, s_N, g_S, s_S) is useful in conjunction with position determinations of spots (cf. Sect. 13.9). The heading includes month and year as well as data on observer and instrument. Forms may be obtained from the amateur working groups mentioned (see also Appendix Sect. A.3 in Vol. 1).

13.3.5 Sunspot Cycles

The face of the Sun changes from day to day, as individual sunspots appear and disappear unpredictably. Long-term studies of sunspots, however, reveal regular patterns, the most obvious being the cyclic variation of solar activity—noted as early as 1843 by the amateur astronomer Heinrich Schwabe—in which the appearance of sunspots varies with a mean period of 11.1 years. This *11-year sunspot cycle* can be traced back, by means of ancient naked-eye observations, to pre-Christian times. Since the beginning of regular telescopic observations in 1749, the sunspot cycles are consecutively numbered. Figure 13.12 shows the annual averages of Wolf numbers over the cycles 0 to 21. The current cycle is number 22, which began in 1986.

Underlying the phenomenon of solar activity is a *magnetic cycle* during which the magnetic polarity of the sunspot groups is reversed; its mean period of 22 years comprises exactly two visible spot cycles.

Individual sunspot cycles vary between 9.0 and 13.6 years in length, and the height of the maxima can also be quite different. The lowest maximum thus far was recorded for the 6th cycle, with a smoothed monthly mean of 48.7, while the 19th cycle reached a maximum height of 201.3. Often no spots can be found on the Sun for weeks during minimum, while at maximum often 10 to 20 spot groups may be seen simultaneously. The steeper the rise from minimum to maximum, the higher the maximum becomes. The ascent takes on the average 4.4 years, while the descent to the next minimum averages 6.5 years in duration. Only the very shallow low maxima display any symmetry with respect to time. These rules regarding the pattern of sunspot cycles were formulated by Waldmeier [13.24] and can be used to predict solar activity. Bendel and Staps [13.25] revised the coefficients of the Waldmeier rules by including the cycles numbered 19 through 21, and by applying a new smoothing method (*P17-averages*; see below).

---

[1] Contact address: Peter Völker, c/o Wilhelm-Foerster-Sternwarte, Munsterdamm 90, D-12169 Berlin, Germany.

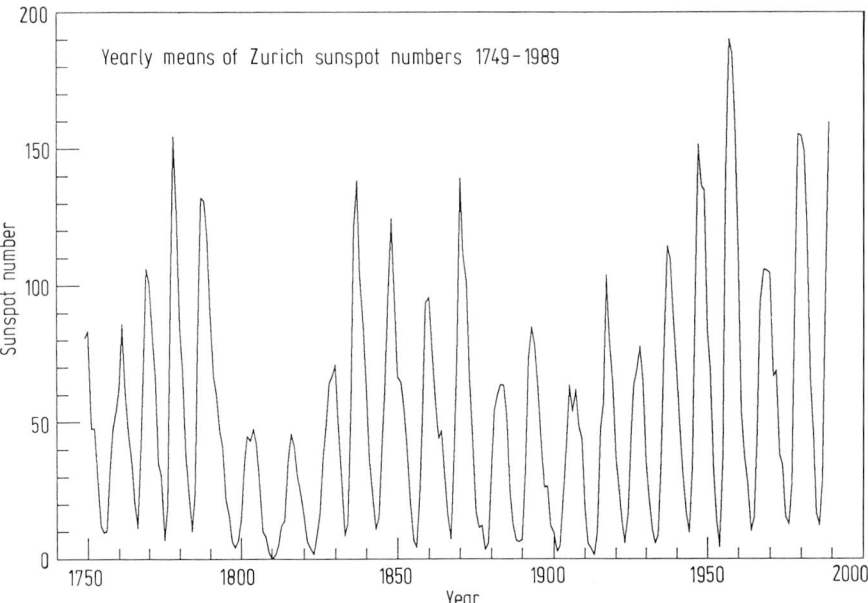

Fig. 13.12. Annual means of Wolf relative numbers for 1749–1989 (Cycles 0–22).

In addition to the 11-year cycle, a longer spot cycle of about 80 years, as inferred from the heights of the maxima, is suspected (Gleissberg [13.26]). However, the limited number of the long cycles observed so far is not sufficient to conclusively prove its existence or to determine its duration.

The distribution of solar activity between northern and southern hemispheres also shows asymmetries which are potentially connected with the Gleissberg cycle. Observers able to determine the positions of spots (see Sect. 13.9) should therefore record spots for the two hemispheres separately.

The monthly averages of Wolf numbers are subject to too much scatter to permit the exact fixing of epochs of sunspot minima and maxima. A smoothing of the curve is needed to find the instants of time where solar activity was lowest and highest. Smoothing is usually performed by the calculation of 13-month means (*A13-averages*): The means of 13 consecutive months are added, counting the months 2 through 12 twice, and then the sum is divided by 24. This yields the smoothed mean for the 7th month included. The procedure is continued for the next month by shifting the 13 averages included by one. The calculated epochs and heights of maxima and minima of cycles 0 to 22 are given in Table 13.1.

One serious drawback of this method is that the number 13 (months) is not an integer multiple of the solar rotation. It is therefore preferable to average over an interval of either 9 months (10 rotations) or 17 months (19 rotations). The discontinuous weighting function of the 13 months is also a disadvantage. An alternative method which is better suited to obtain a smooth curve and yet includes fairly short-period

Table 13.1. Epochs of maxima and minima of solar activity

Cycle No.	Year of Minimum	Lowest Smoothed Monthly Relative Number	Year of Maximum	Highest Smoothed Monthly Relative Number	Interval of Ascent (years)	Interval of Descent (years)
0	1745.0	—	1750.3	92.6	5.3	4.9
1	1755.2	8.4	1761.5	86.5	6.3	5.0
2	1766.5	11.2	1769.7	115.8	3.2	5.8
3	1775.5	7.2	1778.4	158.5	2.9	6.3
4	1784.7	9.5	1788.1	141.2	3.4	10.2
5	1798.3	3.2	1805.2	49.2	6.9	5.4
6	1810.6	0.0	1816.4	48.7	5.8	6.9
7	1823.3	0.1	1829.9	71.7	6.6	4.0
8	1833.9	7.3	1837.2	146.9	3.3	6.3
9	1843.5	10.5	1848.1	131.6	4.6	7.9
10	1856.0	3.2	1860.1	97.9	4.1	7.1
11	1867.2	5.2	1870.6	140.5	3.4	8.3
12	1878.9	2.2	1883.9	74.6	5.0	5.7
13	1889.6	5.0	1894.1	87.9	4.5	7.6
14	1901.7	2.6	1907.0	64.2	5.3	6.4
15	1913.6	1.5	1917.6	105.4	4.0	6.0
16	1923.6	5.6	1928.4	78.1	4.8	5.4
17	1933.8	3.4	1937.4	119.2	3.6	6.8
18	1944.2	7.7	1947.5	151.8	3.3	6.8
19	1954.3	3.4	1958.2	201.3	3.9	6.6
20	1964.8	9.6	1968.9	110.6	4.1	7.6
21	1976.5	12.2	1979.9	164.5	3.4	6.8
22	1986.7	12.3	1989.6	157.9	2.9	

variations is to use smoothed weighting functions which assign the highest weight to the mean month of the period to be smoothed and continuously decreasing weight to the adjacent months. The *P17-averaging* proposed by Karkoschka [13.27], where the monthly averages of Wolf numbers over a 17-month interval are included and are weighted with a polynomial of the form $(1 - x^2)^3$, has proven advantageous. Figure 13.13 compares the smoothing procedures A13 and P17 for monthly means of cycle 21.

Example: Calculate smoothed monthly means from the table below according to the "A13" and "P17" prescriptions.

Observed SIDC monthly means for 1985 and 1986.

1985		1985		1986		1986	
Jan:	16.5	Jul:	30.8	Jan:	2.3	Jul:	17.8
Feb:	16.1	Aug:	10.4	Feb:	23.6	Aug:	7.4
Mar:	11.9	Sep:	3.9	Mar:	15.7	Sep:	3.9
Apr:	16.1	Oct:	18.5	Apr:	20.4	Oct:	35.7
May:	27.4	Nov:	16.6	May:	13.1	Nov:	14.7
Jun:	24.2	Dec:	17.2	Jun:	0.8	Dec:	6.4

The A13-mean for December 1985 is calculated from the averages from June 1985 to June 1986, counting the 11 inner numbers double:

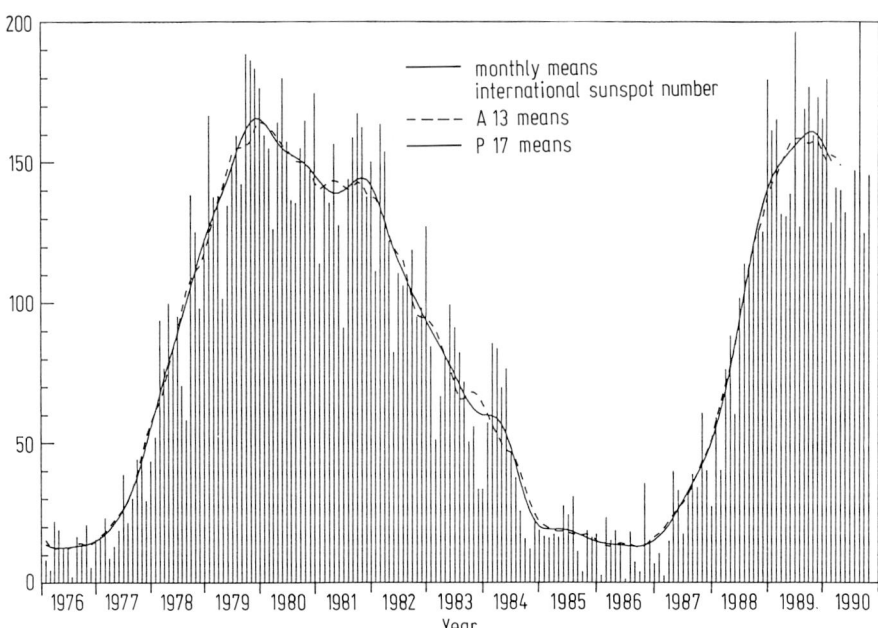

Fig. 13.13 Comparison of monthly means of international Wolf numbers during the 21st and 22nd cycles, smoothed by different methods. *Vertical lines*: unsmoothed monthly means; *Solid curve*: P17-averages; *Dashed curve*: A13-averages. It is seen that the P17 values follow the run of solar activity more accurately than do those of A13.

$$\overline{\Re}_{A13}(\text{Dec. 1985}) = [24.2 + 2 \times (30.8 + 10.4 + 3.9 + 18.5 + 16.6 + 17.2 +$$
$$2.3 + 23.6 + 15.7 + 20.4 + 13.1) + 0.8]/24$$
$$= 15.4.$$

The P17-mean for December 1985 is computed from the months April 1985 through August 1986 with the aid of the weighting factors in Table 13.2.

$$\overline{\Re}_{P17}(\text{Dec. 1985}) = [0.009 \times 16.1 + 0.062 \times 27.4 + 0.171 \times 24.2 \ldots$$
$$0.963 \times 16.6 + 1.0 \times 17.2 + 0.963 \times 2.3 + \ldots$$
$$0.009 \times 7.4]/8.226$$
$$= 15.0.$$

Table 13.2. Weighting factors for the P17-means calculated from the formula $[1 - (M/9)^2]^3$.

M	Coeff.	M	Coeff.	M	Coeff.
±8	0.009	±5	0.330	±2	0.859
±7	0.062	±4	0.517	±1	0.963
±6	0.171	±3	0.702	±0	1.000

The "P17"-average calculated for the month $M = 0$ includes mean for that month and also those of the 8 preceding ($M = -8 \ldots -1$) and the 8 following months ($M = +1 \ldots +8$) with the tabulated weights. The weighted monthly means added are divided by the sum of the coefficients ($= 8.226$).

The determination of minimum and maximum epochs will depend on the averaging procedure used. The difference between methods is maximally half a year.

According to predictions, the next sunspot minimum is scheduled for 1996 (Schatten [13.28]), 1997 (Bracewell [13.29]), or 1998 (Wilson et al. [13.30], Malde [13.31]).

13.4 Photospheric Faculae *(V. Gericke)*

13.4.1 Structure and Appearance

Bright veins or streaks of light can be observed almost every day on the surface of the Sun. These phenomena are the *photospheric faculae* which form over the entire solar surface (cf. Sect. 13.5), although in white light they are generally visible only at the limb. Their observability is thus very limited (Fig. 13.14; cf. also Fig. 13.8).

When the Sun is observed through monochromatic filters, such as in the Calcium H- and K-lines or in the hydrogen Hα-line, faculae can also been seen outside the disk. These *chromospheric faculae*, which represent the protrusion of photospheric faculae into the chromosphere, are to be described in Sect. 13.5.

Faculae consist of *aligned mottles* 5000 to 10 000 km wide and up to 50 000 km long. They in turn are composed of oval-shaped *coarse mottles* with diameters of about 5000 km. The coarse mottles are made up of the *faculae granules*, about 1000 km in size.

Faculae—and almost all solar surface phenomena—are spawned by magnetic fields, and characterize a region of enhanced activity in which sunspots can also appear. Faculae not only presage sunspots but they also outlive them, often by several weeks. The average lifetime of photospheric faculae is 90 days. Their integrated area is substantially larger than that of the associated spot group.

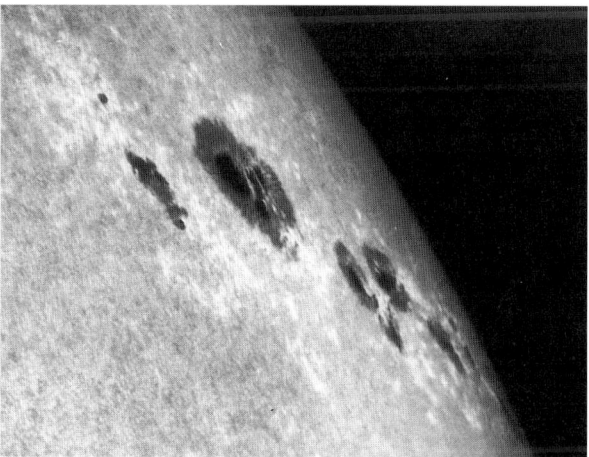

Fig. 13.14. Photospheric faculae photographed by C.H. Jahn on 1984 April 21 using a 200/3250 mm refractor with 12.5 mm eyepiece, solar prism, and green filter.

The temperature in faculae is several hundred (degrees) Kelvin higher than that of the quiet, undisturbed photosphere. The brightness contrast to the photosphere is about 10% at a heliocentric angle $\rho = 72°$. The magnetic flux density in the brighter faculae is between 0.1 to 0.2 T (T stands for the unit tesla). The physics and evolution of faculae has been described in detail by De Jager [13.32].

The direct observation of faculae, using a telescope with a suitable filter of course, is preferred over the projection method because of the better visibility of the faculae. Solar photography will be dealt with in Sect. 13.10.

13.4.2 Classification

There does not exist an analogy with the Waldmeier classification code of sunspots (Sect. 13.3.1) for the classification of photospheric faculae. Two practiced methods to be discussed in the following paragraphs relate to the structure or to the area of faculae.

Classification according to faculae structure goes back to a suggestion by Gericke [13.33]. Five types are distinguished and coded with lower-case letters (cf. Fig. 13.15):

Type a: streaky, net-like structure
Type b: large contiguous areas
Type c: subdivided areas
Type d: point faculae
Type e: cluster of point faculae

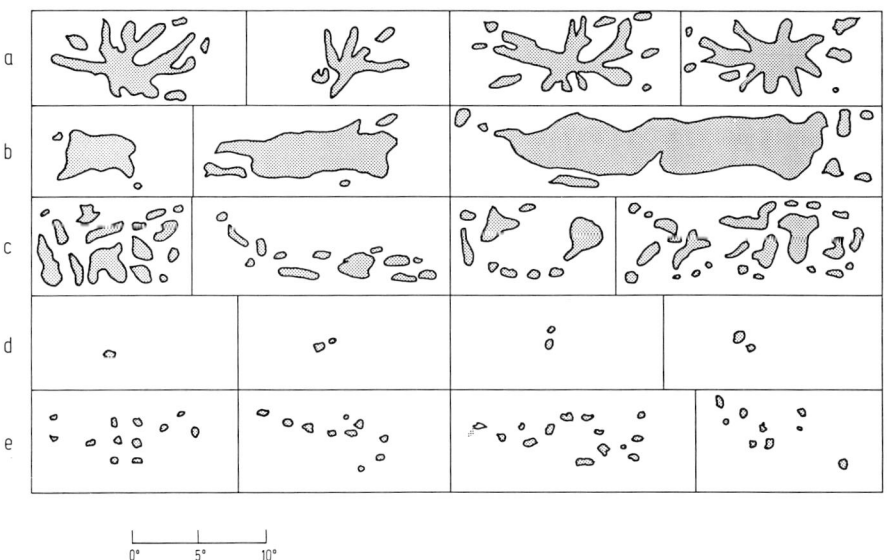

Fig. 13.15. Schematic examples for classifying faculae.

The area of a faculae region should be at least 80% contiguous to be counted as Type b. Type d includes maximally two small separate faculae; Type e includes more than two. Just as the sunspot classification in the Waldmeier code sometimes causes problems, so too the classification of a faculae region is not always unambiguous. Statistical results on the frequency of the different types have been given by Gericke [13.34]. Similar, but less differentiated schemes were proposed by Reble [13.35,36] and Wadsworth [13.37].

Classification according to the area of a faculae region goes according to a 10-step scale estimating the size of the area in square degrees of heliographic coordinates. In this case, the projection method is more helpful than direct observation, as the projection template with a comparison surface can be prepared and used. The scale contains the following 10 steps (the area in square degrees):

0 (0–1), 1 (2–3), 2 (4–6), 3 (7–12), 4 (13–20), 5 (21–30), 6 (31–45), 7 (46–60), 8 (61–75), 9 (> 75)

Figure 13.16 shows the pattern for a simplified 4-step scale.

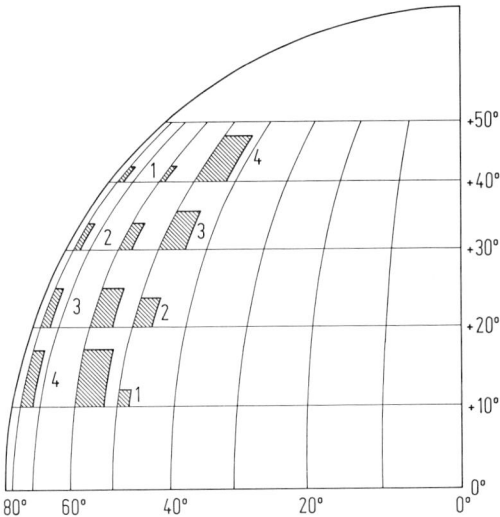

Fig. 13.16. A solar projection grid.

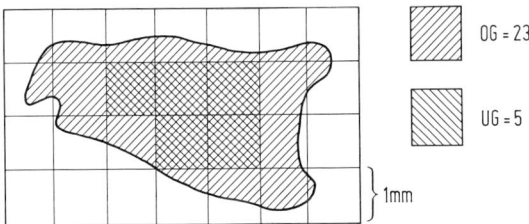

Fig. 13.17. Example of area measurements.

Better accuracy can be achieved if the areas are measured from sketches or photographs overlaid with graph paper. In this case, the actual measuring is performed not at the telescope but at the desk: transparent graph paper is placed over the photograph or sketch and millimeters are counted with a magnifying glass. The total of all millimeter squares contained entirely within the area gives a lower limit (LL). All squares which lie partly within the region are added to obtain an upper limit (UL).

The area F is found from the relation

$$F = \frac{(UL + LL)}{4\pi R^2 \cos \rho}, \qquad (13.9)$$

where again ρ is the heliocentric angle. The factor $\cos \rho$ corrects for foreshortening and R is the radius of the solar image (in mm).

13.4.3 Measurement of Faculae Activity

Three different quantities for expressing faculae activity have been proposed for use by amateur astronomers. The *number* of faculae F_{tot}, i.e., the total sum of all areas of faculae observed on the Sun, was suggested for use by Gericke in 1978. For its determination, the areas separated according to faculae with sunspots (F_w) and without sunspots (F_o) are counted on each day, and the sum is the total number of faculae:

$$F_{tot} = F_o + F_w, \qquad (13.10)$$

One should always record not only F_{tot} but also F_o and F_w because F_{tot} gives only a coarse survey of the activity in the course of time. The monthly rotational period or annual means can be formed from the observations if faculae also follow the 11-year activity cycle [13.38]. Recording both F_o and F_w also affords the chance to study the ratio F_w/F_o as a function of time or of total activity F_{tot}. Figure 13.18 shows an example of processing. The observations were supplied by participants in the observing network of the Working Group *Sonne* of the Vereinigung der Sternfreunde.

The second quantity is the faculae number RFA, defined by Völker in 1971 in analogy to the sunspot number (Sect. 13.3.4). As with the latter, the number of faculae areas F_{tot} is multiplied by 10 and the individual appearances of faculae FE added:

$$RFA = 10F_{tot} + FE, \qquad (13.11)$$

where F_{tot} is already known from the previous calculation. The multiplication factor 10 serves the same purpose as for the sunspot number, although here the value 10 is arbitrary. FE is the sum of individually appearing faculae, corresponding to the sum of individual spots. The individual appearance of faculae is the number of separate streaks, partial surfaces, and fragments recognized with the telescope used. The possibilities for processing are similarly diverse as for sunspot numbers, but they also result in the same problems as occur in evaluating sunspot activity. The faculae number is well-correlated in time with faculae areas ([13.1], pp 432).

This connection leads directly to the third conceivable quantity, the *area of faculae*, the measurement of which has been previously explained. In order to use this quantity as a measure of activity, the faculae areas of individual regions on one day are added.

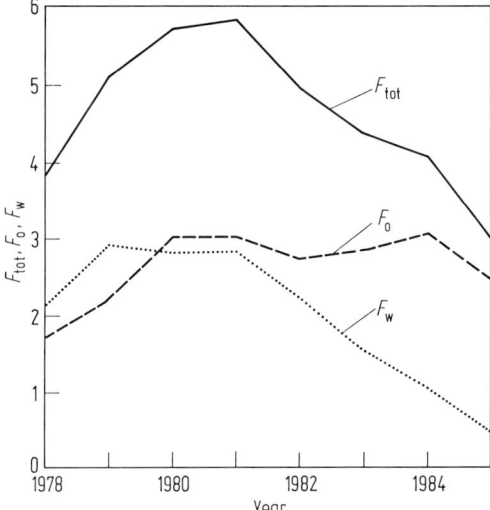

Fig. 13.18. Development of faculae activity.

Which quantity should be used when observing? The number F_{tot} is most widespread and easiest to find. It can be found using any small amateur telescope, for instance, a 60-mm refractor. The faculae number RFA is better determined at larger telescopes than at smaller ones as it is easier to notice the individual appearances of faculae FE with telescopes of higher resolution. The area number requires a very stable projection screen or photographic equipment. Thus the choice of the quantity depends on size and quality of the telescope available. For users of larger instruments, such as refractors with apertures 100 mm and over, it is possible to study the correlation of the quantities.

13.4.4 Polar Faculae

It is widely known that sunspot activity occurs not uniformly across the entire solar surface, but in two zones parallel to the equator and extending to about ±45° of heliographic latitude. The same holds true for the faculae which can be observed daily on the Sun. Nevertheless, there do exist *polar faculae* in regions of high heliographic latitudes (Fig. 13.19).

Polar faculae are smaller than the main-zone faculae; their average diameter is around 2300 km. Their shape is pointlike or oval. Their lifetimes range from a few minutes to some hours, but the decisive difference from the main-zone faculae lies in their activity cycle. When spots and faculae of the main zone are at minimum, the polar faculae have their maximum activity, and vice versa. The latter thus go through the activity cycle in opposite phase.

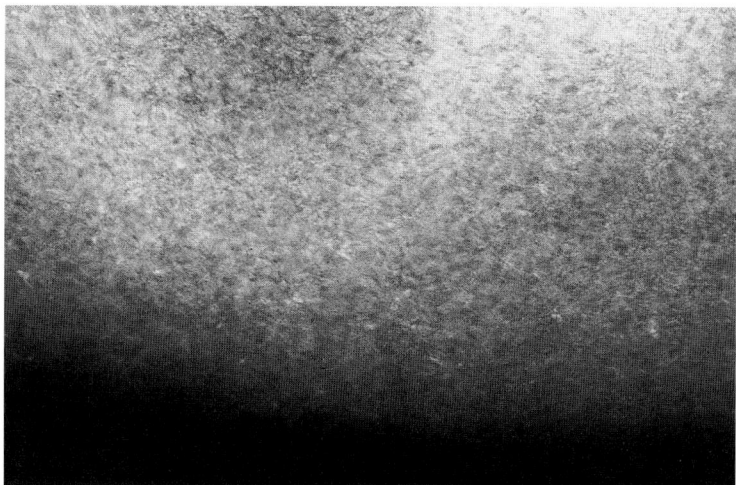

Fig. 13.19. Polar faculae, photographed by C.H. Jahn on 1986 July 30 with a 200/3250 mm refractor equipped with a 12.5-mm eyepiece, solar prism, and green filter.

As the polar faculae are all very tiny, observations of them should be made using large telescopes. Only the largest and brightest of these phenomena can be reached with a small telescope, and this can lead to erroneous and/or incomplete results.

To determine the activity in polar faculae, the number of these faculae are counted on the northern and southern hemispheres of the Sun. All faculae in latitudes higher than $\pm 50°$ are counted as polar faculae. In any event, the recording of activity must be separated from that of main-zone faculae; in no case should the number of polar faculae be added to that of the main zone! Otherwise, the inverse cyclic behavior would vitiate the faculae statistics. The relative position of the solar equator changes periodically during the course of a year; thus at certain seasons, more of the northern and at others more of the southern hemisphere is visible. This fact should be allowed for in the determination of the polar faculae activity. One possibility is to observe the southern hemisphere from March 15 of one year to April 15 of the next, and similarly the northern hemisphere from August 15 to September 15.

Position measurements of polar faculae can show the distribution of activity with latitude, and observations made over several hours on one day yield data on the lifetime of the faculae. Indeed, the polar faculae provide an interesting province of study for observers during the time of minimum solar activity [13.39,40].

13.5 Chromospheric Faculae

When the Sun is observed in the monochromatic light of Hα or the Ca II K-line, faculae are observed over the entire solar disk (see Sect. 13.6); these are *chromospheric faculae* (see Fig. 13.21 a,b in Sect. 13.6.1), an extension of photospheric faculae (Sect. 13.4) into the chromosphere. A definitive explanation of their precise connection with the features of photospheric faculae is still lacking [13.41,42]. The smallest resolved faculae granules in the photosphere are in the subarcsecond realm, in the chromosphere several arcseconds, and in the higher layers (the *transition region*) approach the arcminute realm (linear size about 15 000 km).

The observation of chromospheric faculae or *plages* (from the French *plage faculaire*) is of great interest for several reasons.

Statistical data on plages can be obtained in analogy to the procedures given in Sect. 13.4 (cf. also [13.1], pp 513). Section 13.3.1 has indicated that the plage phenomenon heralds the appearance of sunspots. Therefore, if plages can be observed over the entire solar disk, then it is possible to predict the positions of new activity centers which, given sufficient magnetic strength, will soon manifest themselves in the form of other phenomena such as spots and prominences. The observation of plages is also important because their general arrangement agrees rather well with the magnetogram structures of the regions observed (Mangis [13.43], p 20). Plages can also be followed by means of frame-by-frame cinematography, in which the plages with all their changes can be seen in motion over the entire disk. The necessary tools and techniques are detailed on p 185 in [13.1].

Waldmeier has called attention to one characteristic in particular of plages. He noted that while all other chromospheric phenomena (spicules, flares, prominences, etc.) exhibit growth and decay on short time scales, the same is not true for the plages. He states that a sudden disappearance of plage regions has never been observed, but that with each rotation these areas appear more extended and diffuse until they finally no longer stand out from the chromospheric structure. The observer should note: do not confuse plages with flares, and vice versa, although there may be connections and interactions in some cases.

13.6 Prominences and Filaments *(P. Völker)*

13.6.1 Introduction

Prominences and *filaments* are, as are the plages described in Sect. 13.5, phenomena which occur in the solar *chromosphere*. The term "chromosphere" ("chroma" is Greek for "color") was coined by Lockyer and Frankland in 1869, who noted its distinct pinkish-red hue (due to the hydrogen Hα emission line). The chromospheric layer lies immediately above the photosphere, and its expanse varies strongly with place and time; the layer thickness can be given only very roughly as 2000 km. *Spicules* (Sect. 13.8) are part of the chromosphere, but can rise 8000–10 000 km above the photosphere. The temperature variation through the chromosphere is expressed in Fig. 13.20.

The gas density is about 10^{-8} g cm^{-3} at the level of the photosphere, and around 10^{-12} g cm^{-3} in the chromosphere; the outlying *corona* is much thinner still, with continued density decrease outward. The chromosphere is outshone by the enormous

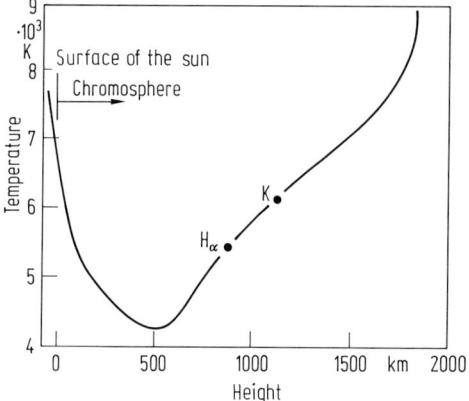

Fig. 13.20. Temperature variation within the chromosphere.

brightness of the photosphere, and can, unless special filters are used, be observed only during the brief moment of total solar eclipse when the lunar limb has completely covered the photosphere but not yet the chromosphere.

The continuous monitoring of chromospheric activity requires special instruments and filters (see Chap. 4).

In 1930 Lyot designed a device known as a *coronograph*, which creates an artificial solar eclipse with a conic diaphragm located in the light path of a refractor. This instrument was intended to monitor the inner corona, but unless it is installed at a high-altitude site (i.e., on a very high mountain), the scattered light from Earth's atmosphere will overwhelm the delicate coronal features.

The *prominence telescope* is a modified form of the coronograph which was suggested to solar observers by Nögel [13.44,45]. This instrument uses additional optical parts which "fools" the scattered light. The construction is described in detail by Nemec [13.46] and explained in Chap. 4 of this *Compendium*.

Hanisch [13.47] and recently Lille ([13.1], pp 96) described a *prominence attachment* based on the principle of the prominence telescope, but with the length reduced to 20–25 cm. It can be inserted into the eyepiece tube of any refractor. Richter [13.48] and Dobrzewski ([13.1], p 102) introduced an even smaller construction with a *prominence eyepiece*, where all relevant optical parts are accomodated within one eyepiece. It is to be noted that all of the instruments mentioned above permit the observation of chromospheric phenomena only at the solar limb.

The significance of the layer above the photosphere was recognized in the 19th century, and subsequently attempts were made to design an instrument which would permit observations of it on the solar disk. Around the last turn of the century, Deslandres (France) and Hale (USA) [13.49] independently succeeded with the construction of a *spectroheliograph*. It permitted only photographic observation, because a composite image of the Sun (a *spectroheliogram*) was composed by scanning the solar disk in one discrete spectral "line," a process which required several minutes. Some years later in 1925, however, Hale followed with his invention of the *spectrohelioscope* [13.50], which permitted direct visual observations of the chromosphere. The spectrohelioscope has been reintroduced to amateur observers by Veio ([13.1], p 89, [13.51,52,53]).

In the early 1930s, Lyot presented the basic concept of the *polarization-interference filter* [13.54,55], which is employed in all monochromators in current use. One spectral line is filtered out by multiple interference of several birefracting calcite crystals (calcium carbonate) in series in a complex, but ingenious arrangement. It is now called a Lyot filter, and its construction is described in Chap. 4. Modern technology permits a bandwidth as low as 0.1 Å. Interference filters from well-known manufacturers can be quite expensive, sometimes running into six digits. Only professional and large educational institutions can afford this special accessory to solar observations.

For the benefit of less affluent users, Woods developed the so-called *DayStar filter* (a Fabry-Perot etalon filter) [13.56], which has since become recognized for its high quality. These filters are full-fledged polarization-interference filters, but a technological trick makes them affordable: the very expensive plane-parallel calcite crystals are replaced by specially vapor-coated foils, called *etalons*.

Solar observers intending to monitor chromospheric events on the disk thus need a spectrohelioscope or an interference filter (Lyot or DayStar). Of course both can also detect activity out to the solar limb.

Long series of observations have shown that the most interesting chromospheric features appear in the hydrogen Hα-line ($\lambda = 6563$ Å $= 656.3$ nm) and also in the Ca II K-line ($\lambda = 3934$ Å $= 393.4$ nm); these differ also in depth of the particular layer of the chromosphere which they show (cf. Fig. 13.20). Figures 13.21 a and 13.21 b are photographs taken in Hα and Ca II, respectively.

The brightest hydrogen line, Hα, gives a visually fascinating picture when viewed through a Lyot filter. The Ca II K-line, on the other hand, lies so far into the violet spectral range where the sensitivity of the eye is low that only photographic observations are useful.

 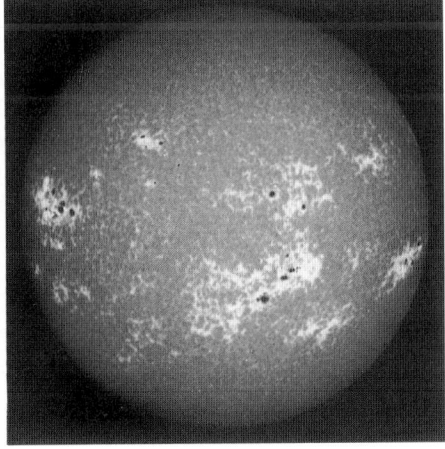

a   b

Fig. 13.21a, b. The Sun in monochromatic light; **a** in Hα light, photographed 1978 May 04 by W. Paech; **b** in Ca II light, photographed 1981 September 06 by G. Appelt.

Suggestions for processing observations have thus far been available only in the technical literature, most likely because small-telescope solar work is a rather young subject. Nevertheless, some systematic methods of analyzing observing records on chromospheric activity can be achieved; the first concrete suggestions appeared in a series of articles by Völker ([13.1], pp 471, and [13.57]).

13.6.2 Prominences and Filaments

Prominences and filaments are manifestations of the same phenomenon: when viewed at the solar limb in front of the dark background, it appears bright and is called a *prominence*, but when seen as a dark structure silhouetted against the bright solar disk, it is termed a *filament*. With a Lyot filter, the transition from one to the other in the vicinity of the limb can be directly visually observed, but less easily with photography as disk and limb require different exposure times (Sect. 13.10).

A prominence occurs when solar material is ejected from the Sun or falls back onto it. It is always directly connected with the magnetic field of an activity center, and even provides a visible representation of the magnetic field lines. Its density is $\sim 10^{-11}$ g cm^{-3} and its temperature $\sim$ 7000 K.

Prominences may be broadly subdivided into *quiescent* and *active prominences*, which in general is correlated with the age of the feature in the life of an activity center (see Sect. 13.3.1). Prominences are only about 5000 km wide, but differ greatly in height and length, ranging from the size of spicules ($\sim 10^4$ km) up to an extent of several 10^5 km.

13.6.3 Classification and Types

Scientists have repeatedly presented prominence classification schemes (e.g., Tandberg-Hanssen [13.58]). This and other sources are compiled on p 501 in [13.1]. Their drawback is that they rely chiefly on physical parameters, which are not easily obtainable, or require extensive series of observations in order to classify the prominence in question. A simple scheme as that of Waldmeier for spots (Sect. 13.3.1) has not been found.

From his own observations, Völker [13.59] in 1969 derived a classification scheme which is suitable for amateurs; it permits every visible prominence to be immediately (i.e., at the moment of observation) assigned a type according to its appearance. This scheme is shown in Fig. 13.22. It distinguishes with a letter the bar-shaped, arch-shaped, and area prominences, even those which, due to the effects of foreshortening (which produces an apparent visual "density") or the resolution limit of the instrument, cannot be further resolved. The Völker classification scheme also considers the relative apparent size and the frequent case of a prominence detached from (i.e., "hovering" above) the solar limb.

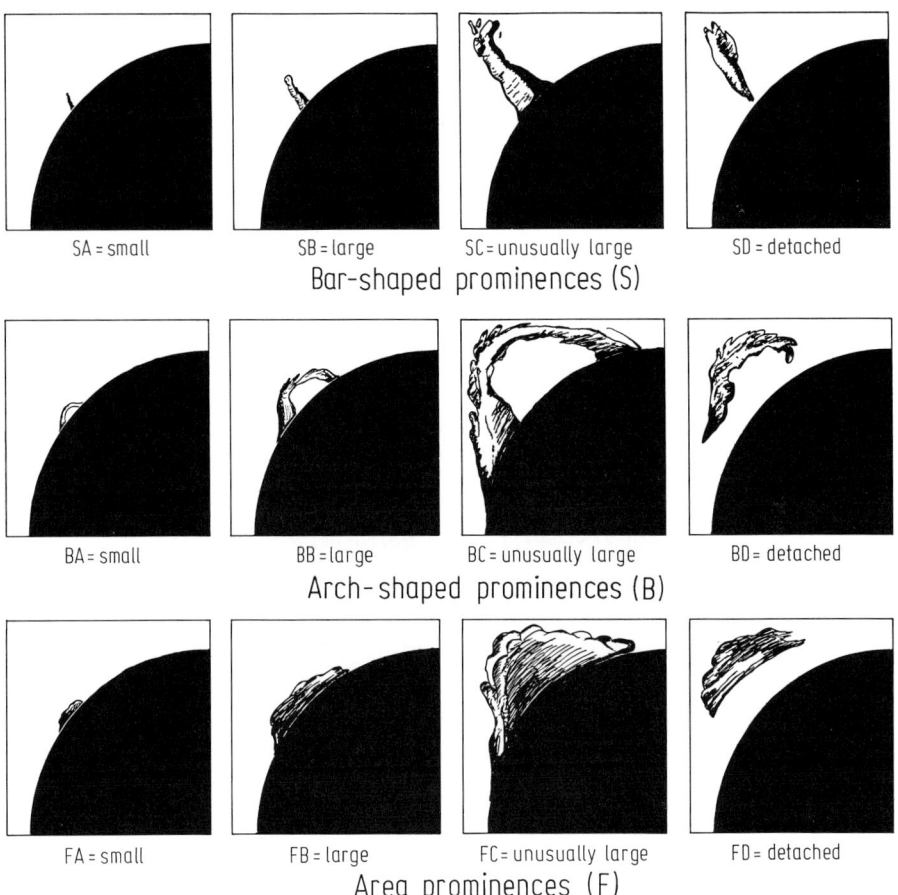

Fig. 13.22. Prominence types. After Völker [13.59].

13.6.4 Recording

The recording of visual observations of prominences is handled in a fashion analogous to that of sunspots and plages (see Sect. 13.2.3). Stencils have been found to be useful, as is shown in an example given in Fig. 13.23. The solar diameter should not exceed 9 cm, so a standard sheet of paper provides enough space to write down the individual types and to sketch the sizes of prominences as accurately as possible. For limb observations, the interior of the "eclipsed" Sun is colored black, but is left white if surface features are being sketched also.

Observational data such as time, instrument, etc. are recorded in the usual manner. The air quality is noted according to the Kiepenheuer scale (Sect. 13.2.2). Wedel ([13.1], p 498) expanded it by including a four-step scale which also considers the image contrast. In instruments with conic diaphragms, the effects of the weather are considerable; even under clear skies, a too-bright sky background can make an

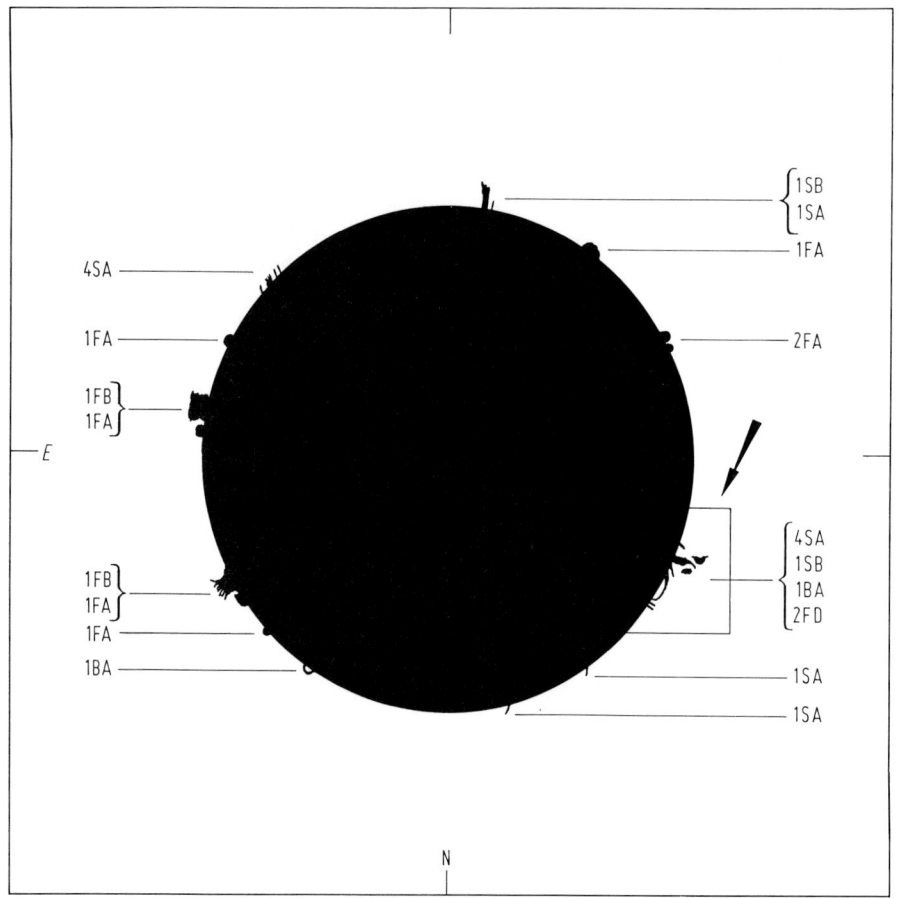

Fig. 13.23 An observing record from 1970 June 10.

Table 13.3. Scale for rating atmospheric conditions when observing prominences. After Wedel [13.1], p 498.

Value	Sky Background
1	Very dark background, prominences exceptionally distinct;
2	Dark background, prominences distinct;
3	Slightly brightened background, but prominences still recognizable;
4	Bright background, prominences seen only with difficulty.

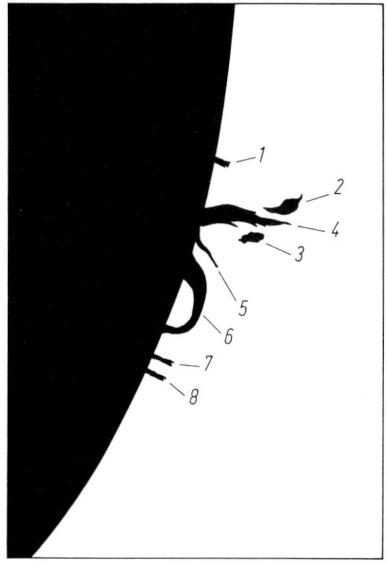

1,5,7 and 8 = 4 small bar-shaped prominences = 4 SA
4 = 1 large bar-shaped prominence = 1 SB
6 = 1 small arch-shaped prominence = 1 BA
2 and 3 = 2 detached area prominences = 2 FD
There are thus a total of 8 prominences in this group

Fig. 13.24. Naming of individual prominence features within an area (excerpted from Fig. 13.23).

observation useless (Table 13.3). This scale has been found to be quite useful in practice.

The observation of 1970 June 10 (Fig. 13.23) now serves to draw the following conclusions:

1. Prominences occur in local concentrations just as sunspots do in groups. Such prominence areas are aptly termed "Herde" (hearths) in German because of their flame-like appearance. The grouping of all features and phenomena of the active Sun is explained by their belonging to an active region (Sect. 13.3.1).
2. Prominences occur not only in the main spot zone but also at the poles. Comments will follow later.
3. Within prominence areas, individual features can be distinguished with the aid of the Völker scheme. An example is given in Fig. 13.24.

13.6.5 Relative Numbers and Profile Areas of Prominences

Based upon the obvious similarity between sunspots (groups/single spots) and prominences (areas/individual features), Völker [13.59] defined the *prominence number* $\Re_P$ as a measure of prominence activity. It is expressed by the relation

$$\Re_P = 10H + E, \tag{13.12}$$

where H (for "Herde") stands for prominence areas and E (for "einzel") for individual features. The observation in Fig. 13.23 shows 12 areas with 26 features, which yields

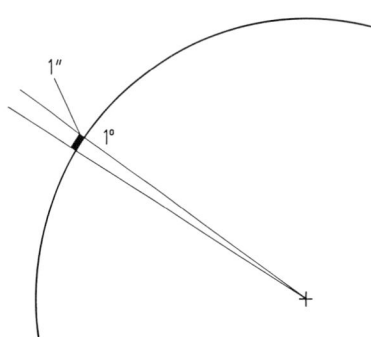

Fig. 13.25. Unit of prominence area.

a prominence number $\Re_P = 10 \times 12 + 26 = 146$. It is completely analogous to the Wolf sunspot number with respect to statistics, averaging, etc. (Sect. 13.3.4).

Another method for recording prominence activity is to determine their area. Since the 1920s, scientists have done this in the following manner:

1. All prominences with sizes over about $20''$ are measured with a prominence area defined as unity of $1°$ (heliocentric) multiplied by $1''$ (geocentric) (Fig. 13.25).
2. The sum of the units is the daily prominence profile area.

The method will not be detailed here but can be found in [13.60].

For amateur observers, Völker ([13.57], pp 121 and 153) devised a simplified method for determining prominence areas. Transparent graph paper is overlayed on the photographs or projected negatives, and the number of square millimeters filled with prominences are counted. The sum gives the area $A(\text{mm}^2)$. This method requires an always-constant diameter of the images to be measured.

Owners of Lyot filters can, using the above procedures, prepare statistical analyses of filament activity.:

- The prominence number $\Re_P$ is replaced with the *filament number* $\Re_F = 10H + E$. The areas H are the activity centers and the individual features E are the details noted in them (lines, knots, etc.), and they are counted.
- The prominence area number A (mm^2) is analogously employed for all filaments on the solar disk.

These methods for amateurs were introduced by Völker ([13.57], p 153).

Professional solar observers determine the following quantities [13.61]:

$n=$ number of quiescent filaments visible on the day of observation (daily number), and
$l=$ total apparent length expressed in solar diameters.

13.6.6 Long-Term Observing Programs

Once statistics have been prepared according to one of the methods described, the prominence activity can be compared with that of sunspots and plages. Attention is to be given to the activity not being synchronous in the main spot zone and in the polar zone.

- The activity and distribution of the prominences in the main zone correspond well to that of the spots, but the prominences stay about 10° farther off the equator than the spots, and their minimum occurs one year before that of the spots.
- The prominences of the polar zone above ±50° appear about 2 years before the spot minimum, but their activity increases after the spot minimum while they migrate further poleward. The maximum of polar prominences occurs about 2 years before sunspot maximum. Shortly after the latter transpires, the prominence activity vanishes. Asymmetries between northern and southern hemispheres have been observed in the intensity of activity and also the different times of occurrence of maximum activity.

In a statistical comparison with sunspots or plages, it is advisable to use only the main-zone prominences; otherwise the curves may show strong deviations from one another.

A comparison between polar prominence and polar faculae activity may be useful. Waldmeier published the run of the latitude distribution of prominences up to 1980 in the *Astronomische Mitteilungen der Eidgenössischen Sternwarte Zürich* [13.62]. Stetter [13.63] provided a method by which the solar observer using the Völker prominence number $\mathfrak{R}_P$ can obtain similar results (Fig. 13.26).

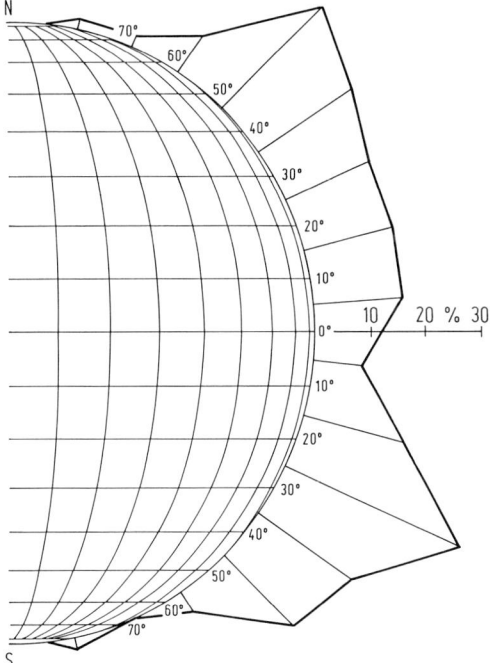

Fig. 13.26. Prominence activity in 1985; $\mathfrak{R}_P$ is graphed against heliographic latitude and normalized to percent. Observed by H. Stetter.

Besides $\mathfrak{R}_P$, of course, the positions of prominences must be known, and they are found analogously to that of spots and plages (Sect. 13.9).

The possibilities just described for recording chromospheric activity require, as is the case for spots and plages, long series of observations and, if possible, coordination of observations with other observers in order to obtain meaningful statistics.

13.6.7 Short-Term Observing Programs

Prominences and filaments also show interesting short-term variations, some of which are well worth following. The often bizarre shapes sometimes perform fantastic movements with ballet-like grace, as has been documented in cinematographic time-lapse films.

Amateur astronomers can also impressively record such events in series of photographs (Fig. 13.27a, b) or as a sketched presentation in phases (Fig. 13.28).

The development of long-lived filaments is well presented in Fig. 13.29.

It is obvious that velocities of moving prominence or filament details can be deduced from single phases of an observed series. Filaments should be scanned by using the *Doppler effect* (a line-shift in the Lyot filter), but this can lead to errors, as has been shown by theoretical studies made at the Kiepenheuer Institute in Freiburg; a detailed discussion can be found in [13.1] (pp 537) and [13.65].

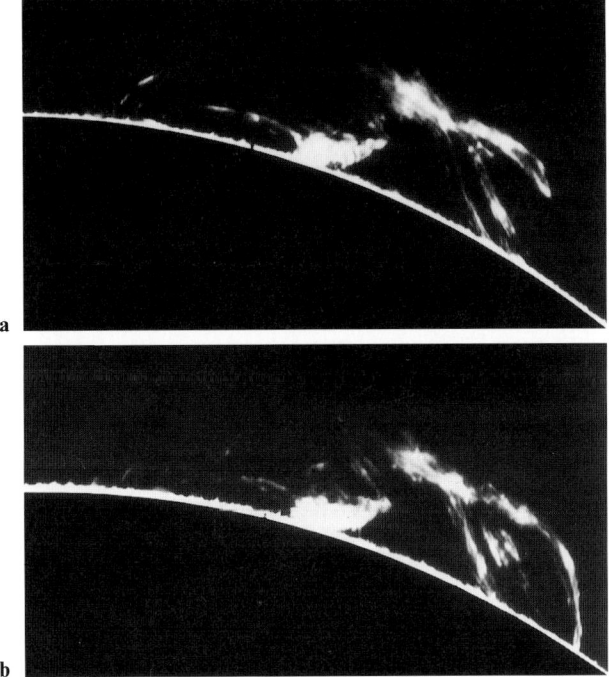

Fig. 13.27a, b. An active prominence, observed 1979 June 02 by W. Lille with a 125/1300 mm refractor: **a** 13.45 UT; **b** 14.00 UT.

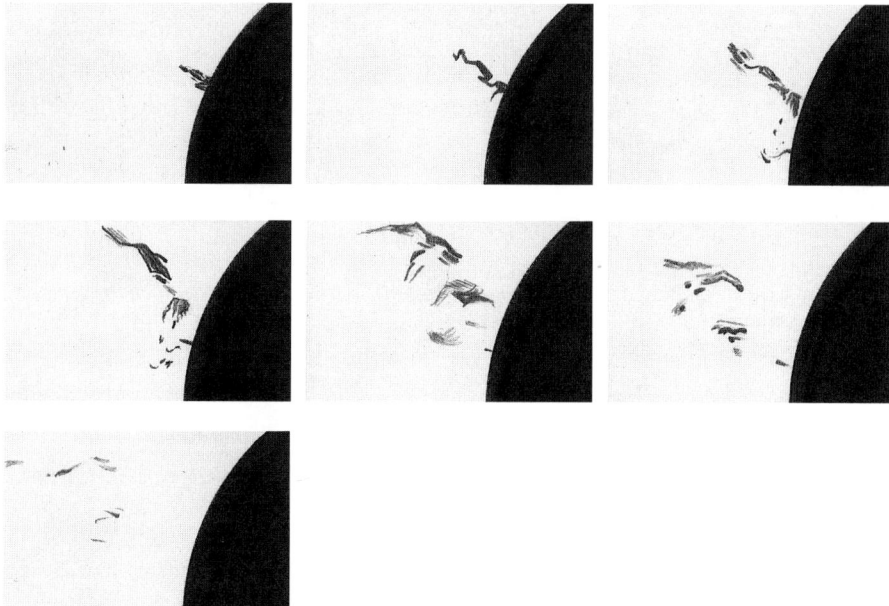

Fig. 13.28. Ascending and dissipating prominence (spray), observed 1969 May 24 (selected from 14 sketches) by P. Völker, Wilhelm-Foerster-Sternwarte, Berlin. 13.30 UT: beginning of active phase; 14.05 UT: complete dissipation. The phenomenon was 6×10^5 km high at maximum, with no trace left at the solar limb!

Prominences present a problem in that absolute velocities cannot be found owing to the perspective foreshortening, which may cause part of the feature to apparently move in a curved path. But amateurs can readily determine radial velocities (with respect to the solar surface), i.e., projected into the image plane of the observer. All that is required is a precisely sketched sequence of a moving feature similar to that shown in Fig. 13.28, but it is more accurate and preferable to use photography. From the abundance of details, two or three knots or clouds which can be unambiguously followed are selected. These are measured and their height above the solar limb calculated.

This study includes the following relations (Fig. 13.30):

$$h = \left(\frac{d}{R}\right) R_\odot, \tag{13.13}$$

where h is the height of a knot in km above the solar limb,
$R_\odot$ is the radius of the Sun in km,
R is the radius of the measured image of the Sun in mm,
D is the distance of the knot in mm from the center of Sun (in the image),
$d = D - R$ is the height of the knot above the solar limb. The radial velocity v_r is

$$v_r = \frac{h}{t}. \tag{13.14}$$

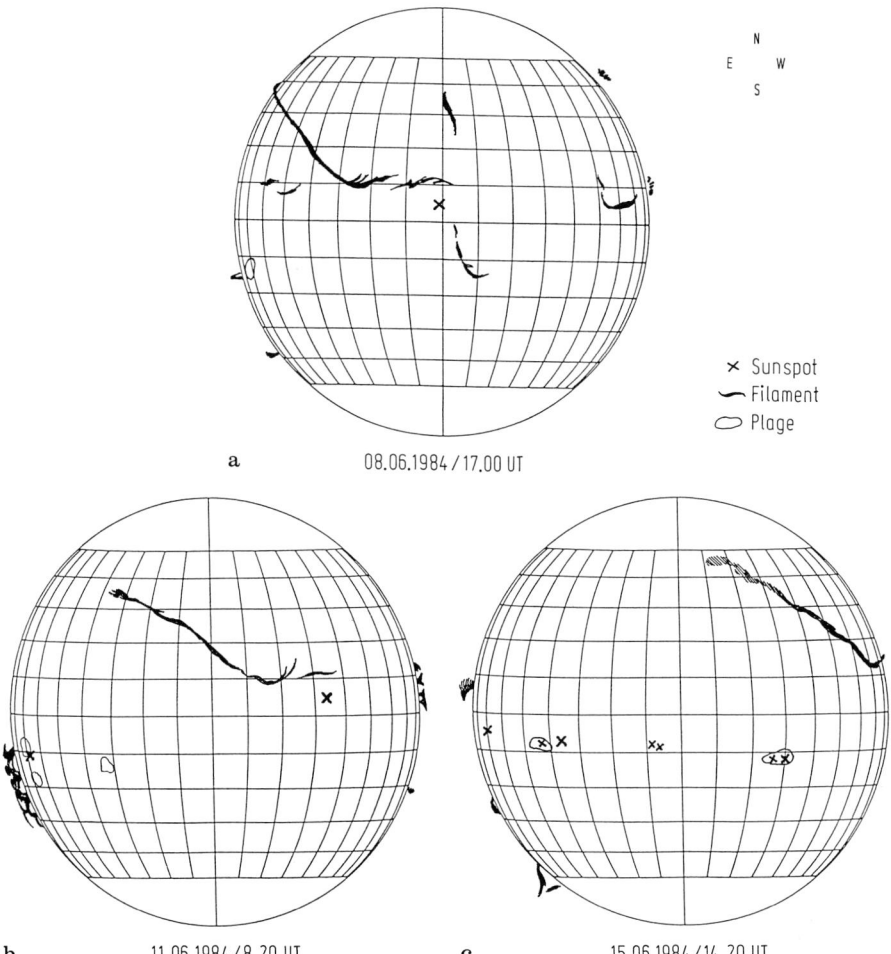

Fig. 13.29 a–c. A giant filament, observed by I. Glitsch, Switzerland, and shown in excerpts of a phase sequence covering 21 days (1984 June 02 – June 22). Owing to the height of the filament, it can be followed as a prominence to the rear side of the Sun [13.64].

If the knot moves away from the solar surface, v_r is given as positive (height increases); if toward the surface, v_r is negative (height decreases).

The values thus calculated permit various conclusions to be drawn. For two examples, Fig. 13.31 shows motion curves of the measured knots, and Fig. 13.32 the height-time graph, which directly indicates sudden changes in speed in active prominences.

Besides following, representing, and evaluating events of chromospheric activity, there are additional possibilities which cannot be mentioned here due to space constraints. (Most of these are usually reserved only for very experienced and well equipped solar observers.) Some of these include:

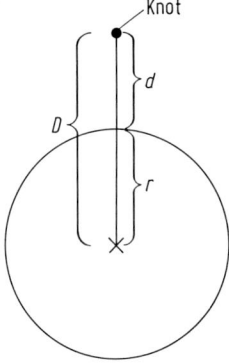

Fig. 13.30. Measurements of knots against the solar disk. (See text for an explanation.)

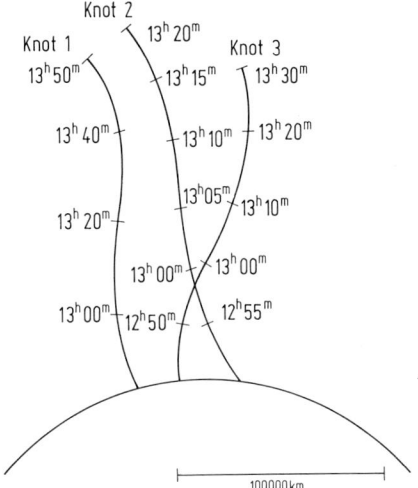

Fig. 13.31. Motion curves of measured knots.

- Spectroscopy [13.1], pp 50, and Chap. 7;
- Photometry (estimates, measures, equidensitometry/ photographic or with electronic image-processing: light curves of chromospheric details, microdensitometry) [13.1], pp 161 and pp 175, and Chap. 8;
- Cinematography (on spectroscopic material or Super-8 film) [13.1], pp 187;
- Videography (video monitoring of chromosphere or recording of active features, e.g., for educational use in classroom and lectures) [13.1], pp 206.

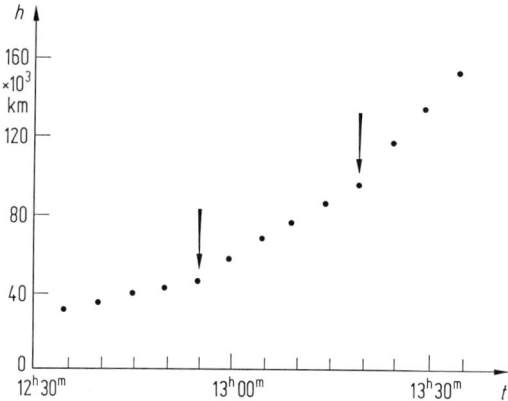

Fig. 13.32. Height-time diagram.

13.7 Flares *(P. Völker)*

13.7.1 Introduction

Flares were once believed to involve ejections of physical matter similar to that of prominences. It is known today that flares are only radiation bursts, and that those flares visible in monochromatic light constitute only the secondary result of an instability in the coronal region above the chromosphere. The physical background is well presented by Giovanelli [13.10].

Flares occur in complex sunspot groups and can, although they are only radiation bursts, strongly excite (prominences: *surges* (Fig. 13.33 a–c) and *sprays* (Fig. 13.28)) or change (e.g., the disparition brusque phase of a filament) the surrounding material.

Various forms of flares are distinguished: *plage flares*, *two-ribbon flares*, *homologous flares*, etc., whose explanation can be found in [13.1], [13.41], or [13.66]. A specialty is the extremely rare *white-light flare* [13.67, 13.94], which is limited to very large sunspot groups and can be observed in integrated light. Here, small points with brightnesses greater than that of the photosphere constitute parts of a very strong flare viewed in monochromatic light.

13.7.2 Classification

Flares are divided into classes according to a quantity called their *importance*. Formerly, the code values were −1, 1, 2, 3, and 3+, but since 1966 January 01, the international code is S (for subflare), 1, 2, 3, and 4. Table 13.4 gives the pertinent quantities.

As Table 13.4 shows, flares are, compared with other active phenomena, very short-lived.

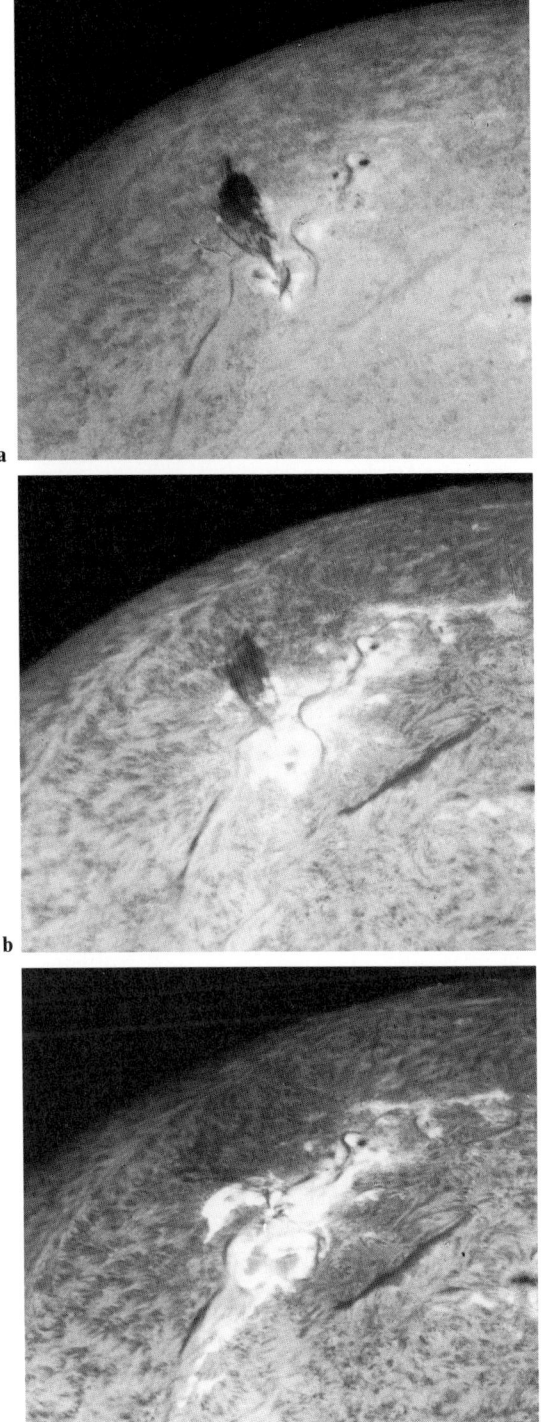

Fig. 13.33 a–c. Matter is ejected by a flare and returns to the Sun after the activity phase. Observed 1979 June 10 by G. Appelt, Kaufbeuren with a refractor (150/3000 mm) with DayStar filter 0.5 Å; **a** 8.12 UT; **b** 8.18 UT; **c** 8.44 UT.

Table 13.4. Classification and properties of solar flares.

Code	Degree	Seconds	Dimensions 10^{-6} vis. Hemisphere	Hα Line Width	Lifetime
S	< 2.06	< 600	< 100	1.5	< 4 min
1	2.06 – 5.15	600 – 1500	100 – 250	3.0	4 – 43 min
2	5.15 – 12.4	1500 – 3500	250 – 600	4.5	10 – 90 min
3	12.5 – 24.7	3500 – 7200	600 – 1200	8	20 – 155 min
4	> 24.7	> 7200	> 1200	15	0.9 – 7.2 hr

In addition to the importance class of a flare, the flare *intensity* is given in the steps f = faint, n = normal, and b = bright or brilliant. There is no quantitative definition. The total description of a flare is thus "Sf" for the smallest features and "4b" for the largest. It may be worth mentioning that the classification should apply to the time of maximum intensity, and not to that of greatest areal extension.

13.7.3 Recording of Data

It is virtually impossible to obtain sketches of flare phenomena, as their complex structure changes within a very short time. Therefore, the best approach is to take rapid sequences of photographs.

The following data should be recorded:

Dat: Date;
ph: photographic observations or series;
v: visual observations or series;
I: state of the studied heliogram (1 = very poor ... 5 = excellent);
TB F: Time of beginning of flare (UT);
TB O: Time of beginning of observation (UT);
TE F: Time of end of flare (UT);
TE O: Time of end of observation (UT);
TM: Time of maximum of flare (UT);
LAT: Heliographic latitude of center of phenomenon;
L: Heliographic longitude of center of phenomenon;
IMP: Importance and intensity steps according to Table 13.4;
O P: Partial observation;
O C: Complete observation;
T: UT of measurements;
AP: Maximum projected area in millionths of the solar disk;
AC: Corrected area in square degrees (if the heliocentric angle ρ is smaller than 65°);
F: Maximum intensity relative to the local undisturbed chromosphere;
R: Remarks according to international rules, as follows:

A: eruptive prominence, base at $\rho > 90°$;
B: probably the end of a more important flare;
C: was not visible 10 minutes earlier;
D: brilliant point;
E: two or more brilliant points;
F: several eruptive centers;

G: no spot visible in the neighborhood;
H: flare with high velocity dark surge;
I: very extensive active region;
J: plage with flare shows marked intensity variations;
K: several intensity maxima;
L: filaments show effects of sudden activation;
M: white-light flare;
N: continuous spectrum shows effects of polarization;
O: observations have been made in the Ca II lines H or K;
P: flare shows helium D in emission;
Q: flare shows Balmer continuum in emission;
R: marked asymmetry in Hα line;
S: brightening follows disappearance of filament (same position);
T: region active all day;
U: close and somewhat parallel bright filaments (II or Y shape);
V: occurrence of an explosive phase;
W: great increase in area after time of maximum intensity;
X: unusually wide Hα emission;
Y: onset of a system of loop-type prominences;
Z: major sunspot umbra covered by flare.

Two flare phenomena are considered as independent if the distance between the flare knots is over 3° or if their time of appearance differs by more than 5 minutes.

13.7.4 Long-Term Observing Programs

Two questions regarding the activity cycle are of special interest:

1. Does the total number of flares noted distinctly follow the cycle?
2. Which flare classes occur at the different phases of the cycle and with what frequency?

13.7.5 Short-Term Observing Programs

One still-important study involves the appearance of the fine structure of the chromosphere and the sunspots before a flare occurred. This so-called *pre-flare phase* is difficult to catch, since there is no way of knowing when and where a flare will occur. After a flare has occurred, however, the observer can backtrack to locate these structures on a photograph obtained previously during a routine survey.

When a flare occurs, there are numerous ways to describe the event. Figure 13.34 shows the variety of shapes or forms which can occur.

A *surge* produced by a flare is shown in Fig. 13.33. Such effects on solar material have often been observed and are worth studying. In addition to surges, there are also the *sprays*, the *flare loops*, and even the *Moreton Waves* and the previously mentioned *disparition-brusque phase* of filaments [13.1,41,66]. Changes in photospheric structures are also observed in the wake of flares.

Constructing the light curves of flares is an engaging project because, although some similarities can be recognized, each phenomenon follows a different course.

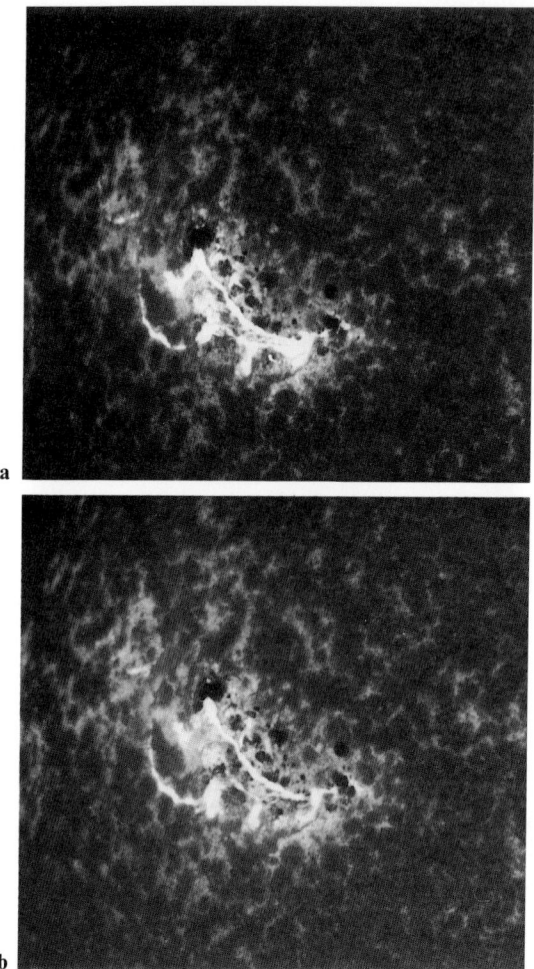

Fig. 13.34 a, b. Flare in the light of the Ca II K-line, photographed 1982 July 12 by G. Appelt, Kaufbeuren, using a refractor (125/3000 mm) with DayStar filter 1 Å. It is evident that dramatic changes have occurred in only 25 minutes.

There are two possible ways to present light curves, as is demonstrated in Figs. 13.35 and 13.36. Figure 13.35 graphs the line width of the flare (see Table 13.4) and Fig. 13.36 the intensity of the flare in the middle of the line relative to the intensity of the undisturbed chromosphere, versus time.

Although the curves in the two figures are somewhat different, both show that the flare quickly ascends from the pre-flare stage to maximum activity (*flash phase*) and then fades away slowly. There is justifiable interest in light curves of flares because significant deviations from the typical behavior have been observed. In one case, the flash phase may occur almost without pre-flare, and in another the brightness may initially decline in the pre-flare phase and then abruptly transit to the flash phase.

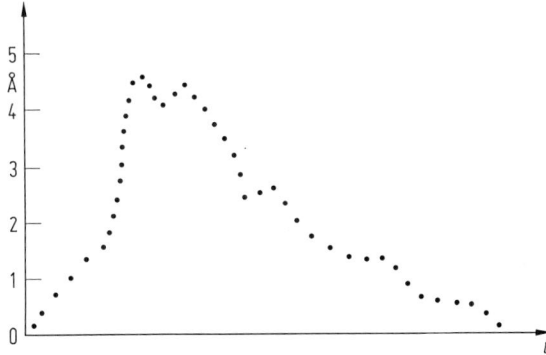

Fig. 13.35. Graph of the line width of a flare versus time.

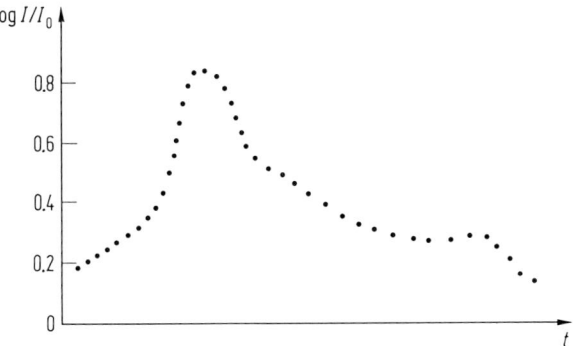

Fig. 13.36. Graph of the intensity of the flare in the middle of the line measured, relative to the intensity of the undisturbed chromosphere, versus time.

The many "knots" which are noticeable in flares can serve to derive data on motions. To that end, either the negatives are underexposed, or copies are made from normally exposed negatives which show only the central parts so that these "knots" can be more easily followed. The further evaluation then proceeds in analogy to the representation of movements of sunspots within a group (Sect. 13.9.1).

There are many other interesting studies on flares (see [13.1]), and a brief listing of some key subjects will be used to complete this discussion:

- Spectroscopy (see Chap. 7);
- Cinematography (time-lapse films);
- Videographs for flare monitoring, but again also for sharing with larger groups (instruction);
- Radio observations:
  (a) Monitoring of flares (called *bursts* with radio telescopes; see Chap. 9);
  (b) Interference with radio broadcasts (particularly shortwave) after large, optically observed flares (see [13.94]).

13.8 Other Chromospheric Phenomena *(P. Völker)*

The preceding three sections have dealt with the chromospheric features and events that are most evident and most accessible to the observer. These include plages, prominences/filaments, and flares.

The following section will reveal that the structure of the chromosphere is much more detailed, and presents a greater variety of forms than the photosphere. Again, for detailed discussions, reference should be made to the technical literature ([13.41,68–72]; a compilation from several sources is found in [13.1], pp 488).

Astronomers distinguish between (1) phenomena of the *quiet chromosphere* and (2) phenomena of the *active chromosphere*.

The best-known features of the quiet chromosphere are the following:

- *spicules* (see Fig. 13.27a, b), the jagged, flame-shaped structures directly above the solar limb. They reach heights of around 10^4 km and diameters of only $1''$, and have a lifetime of 5–10 minutes;
- *the chromospheric network* (Fig. 13.21 b), a large-scale, chromospheric "superstructure."

Also counted among these are structures which thus far are known only in the technical literature under the following terms: *bright mottles in the low chromosphere, dark bands at the base of the chromosphere, dark mottles, grains,* and *bright patterns.*

The best-known phenomenon of the active chromosphere may be the *superpenumbra*, where the fibrils continue from the photospheric penumbra into the chromosphere. Its appearance is similar to that of the white-light penumbra, but need not correspond with the photospheric fine structure of the penumbra.

Terms found in the specialized literature also include the *fine mottles* (bright and dark), *coarse mottles* (bright and dark, also called *rosettes* and *bushes*), *arch filaments* (also called *arch filament systems* or *field transition arches*), *moustaches* (also *bright points* or *Ellerman bombs*) and *filigree structure.*

13.9 Position Measurements of Solar Phenomena *(E. Junker)*

Determining the position of the previously mentioned solar features permits a detailed study of the properties of individual objects as well as those of the entire activity cycle of the Sun. The following sections will describe measuring techniques as well as the possibilities open to amateur astronomers for observing and processing solar position measures.

13.9.1 Targets of Position Measurements

Some of the results supplied by position measurements of features on the Sun will be mentioned first. For many observing programs, cooperation with other observers in a network is strongly recommended.

The *rotation of the Sun* is noticeable in the motion of active regions across the solar disk, observed on several consecutive days. A sunspot appearing at the eastern limb

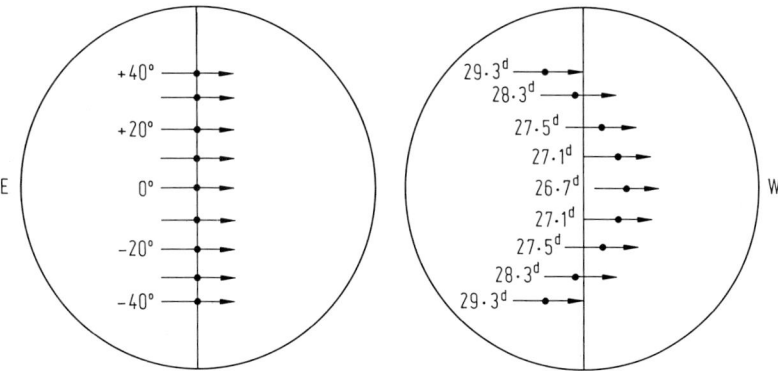

Fig. 13.37. An illustration of the differential rotation of the Sun using a symbolic race of sunspots. At the start (*left*), all spots are on the central meridian, a situation which never occurs in practice. One rotation later (*right*), at the "finish line," the latitude dependence of the duration of rotation is seen. The recorded synodic rotation times are derived from over 100 observations on sunspots, calcium clouds, prominences, and metallic lines (from spectroscopy).

of the Sun moves, as seen from Earth, by about $13°.2$ westward per day. Measuring this displacement of the spot on the Sun gives the *synodic solar rotation* as:

$$\frac{360°}{13°.2/d} = 27^d.3.$$

However, the gaseous sphere of the Sun rotates non-rigidly; the polar regions need over one week longer to complete one rotation than those near the equator. This *differential rotation* is portrayed in Fig. 13.37 using a symbolic race of spots and faculae located at various latitudes.

It is of interest to follow the movement of sunspots continuously, because in this way the average or even the variation of rotation with latitude can easily be found. The accurate determination of the law of differential rotation requires numerous accurate position measurements of stable spots at different latitudes (Zerm [13.73]).

The differential solar rotation is expressed by the relation

$$\omega(B) = a - b \sin^2 B, \tag{13.15}$$

where $\omega(B)$ is the angular velocity as depending on latitude B, a is the angular velocity at the equator, and b the coefficient of deceleration with B. The values for the coefficients a and b quoted in the literature have a substantial range:

$$13.8° \text{ day}^{-1} < a < 14.6° \text{ day}^{-1} \qquad 1.7° \text{ day}^{-1} < b < 3.4° \text{ day}^{-1}.$$

For the time interval 1874 to 1976 (Balthasar et al. [13.74]):

$$\begin{aligned} a &= 14.551(6)° \text{ day}^{-1}, \\ b &= 2.87(6)° \text{ day}^{-1}. \end{aligned} \tag{13.16}$$

The two constants depend on the type of spots used, the time interval within the cycle, the hemisphere observed, and perhaps other parameters not yet considered.

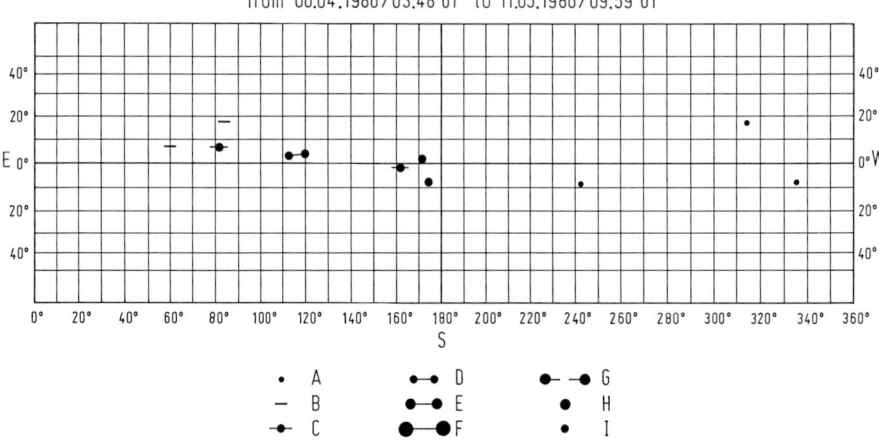

Fig. 13.38. Synoptic map of the Sun for synodic rotation No. 1774 in heliographic coordinates. Graphics by J. Hoell, E. Junker, G. Schwaab, and M. Schwab.

Studying the *distribution* of spots requires a compilation of many observations over months and years. Graphing solar charts in order to find the change in spot positions with time is also important. A *synoptic map* (Fig. 13.38) graphs mean positions of all events (spots, faculae, filaments, etc.) observed during one solar rotation. It can be, for instance (Fig. 13.39), the superposition of all spots observed within 27.3 days (0–360° solar longitude). The actual lifetime of the spots is not considered in the recording.

A synoptic map gives a variety of information: active zones are evident; mean latitudes of spots can be found from a frequency diagram; the north–south asymmetry of a cycle is seen; groups whose p- and f-spots (preceding and following) are on different hemispheres are noted; the beginning of a new cycle is shown by the appearance of spots at high latitudes, and the distribution of the spot group types can be studied; the overall number of groups is a quantitative measure of solar activity. By superimposing several synoptic maps onto an annual chart, the positions of activity centers are clearly revealed (Fig. 13.39).

A *position diagram* graphs only one spherical coordinate (heliographic longitude or latitude) against the time as abscissa. The spots are then recorded according to their time of observability and their position in the diagram. Connecting the points referring to the same group on various days reveals the actual movement of spots on the Sun. Thus, in contrast to the synoptic map, a position diagram investigates phenomena of motion.

The *butterfly diagram* (Fig. 13.40) illustrates the zonal migration of spots (*Spörer's law*). The mean latitude of spots or groups is graphed versus the time of their observation. This is a position diagram in heliographic latitude but with a much compressed time axis. The sunspots migrate from higher latitudes during the course of a cycle toward the solar equator. Even before the spots of the old cycle disappear near the equator at the time of minimum, the spots of the new cycle appear at high latitudes. This overlapping of cycles, the occurrence of spots in two zones symmetric with the

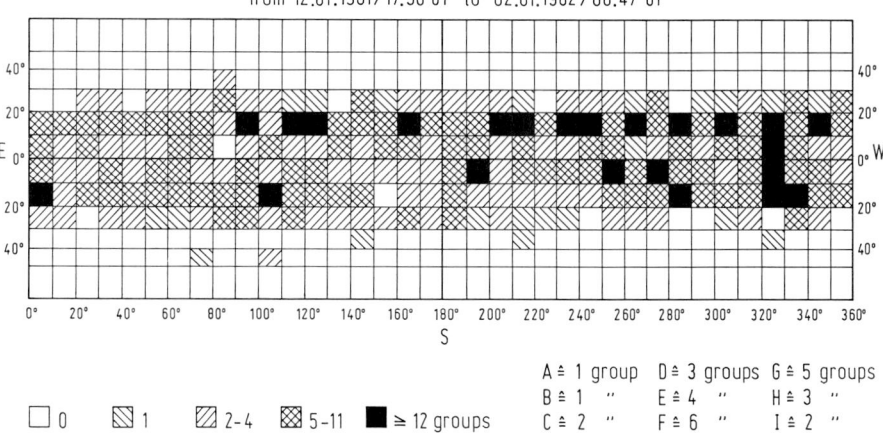

Fig. 13.39. Synoptic map for the year 1981, combined with the synoptic maps for rotation Nos. 1704–1716. Each spot group was assigned a number of regions according to its Waldmeier type (Sect. 13.3.1). These regions have then been distributed over the entire area in which the group appeared in fields of size $10° \times 10°$.

equator, the interval of maximum activity, the intensity of a cycle, and the differences between hemispheres can all be read from a butterfly diagram.

A more detailed analysis of the zonal migration reveals that there exists a second spot zone. The spot zone forming at high latitudes at the beginning of the cycle splits so that one component moves toward the pole instead of the equator. This secondary spot zone, however, is recognizable only during the ascending phase of a spot cycle until about the maximum. Any theory on the origin and development of sunspots must also account for the migration in latitude of the spots. The butterfly diagrams of cycles 11–20 (Fig. 13.41) are based on sunspot positions measured in Greenwich from 1874 to 1976.

The position measures are well suited for a systematic study of properties of sunspot groups. The *length* (i.e., maximum extension) of a spot group is given by the relation:

$$s = 12\,148\sqrt{(\Delta L \cos B)^2 + (\Delta B)^2}, \tag{13.17}$$

where s is the extension of the group in km, B the mean heliographic latitude of the group in degrees, ΔL the east–west extension in degrees, and ΔB the north–south extension in degrees. (1° of heliographic coordinates equals 12 148 km at the solar equator.)

The *inclination α of the group* is given by:

$$\alpha = \arctan\left(\frac{\Delta B}{\Delta L \cos B}\right), \tag{13.18}$$

where α is positive when the p-spot (preceding in the sense of solar rotation) has a lesser distance from the equator than the following f-spot. In most groups, α is

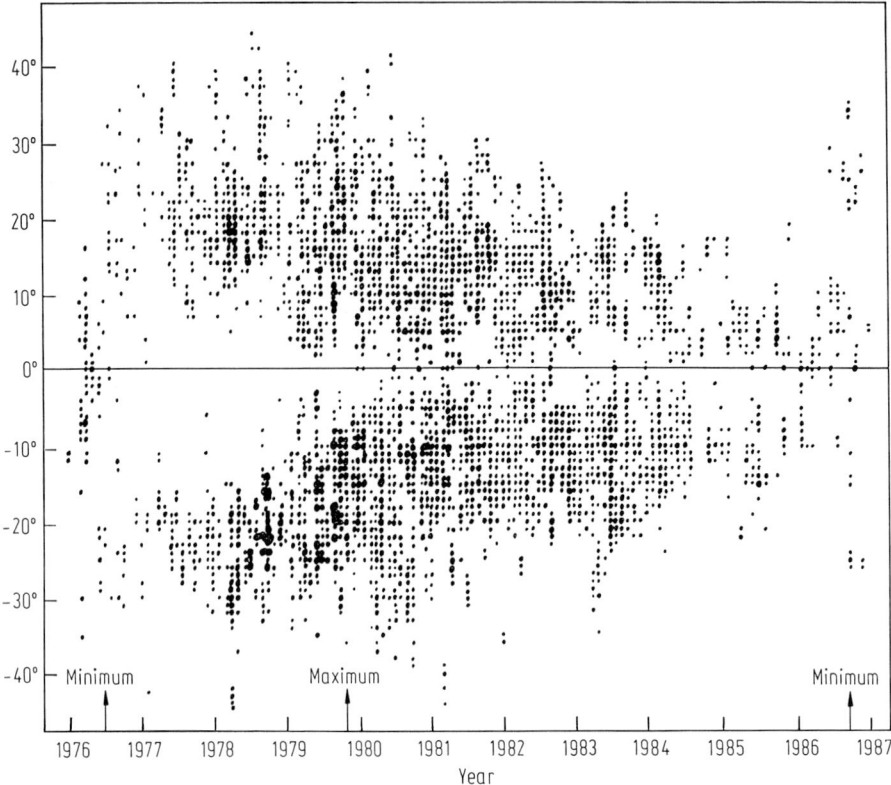

Fig. 13.40. Butterfly diagram for cycle No. 21, composed of 6701 individual positions from 1224 observing days, over the time interval 1976 January 17 to 1987 February 02. From Hammerschmidt [13.75].

positive, but the numbers depend on the latitude of the group and on the time within the cycle. The existing literature on this subject is conflicting.

Studies of *individual motions* within sunspot groups demand that extreme care has to be taken during the measuring process in order to obtain satisfactory accuracy in the positions, as the individual spots move usually only a few tenths of a degree per day. Relative measurements, however, are more precise than absolute measurements, and therefore the requisite precision can be reached (Reinsch in [13.1], pp 377; Beck in [13.1], pp 250; Jahn [13.77]; Pfister [13.78]). Pfister distinguishes several types of motions within groups: rotation of the line joining the umbrae when spots divide, shifts within a p-spot complex of a group, approach of spots of equal polarity, divergence motions, interpenetration of spot groups, and drifts in heliographic longitude or latitude. Drift motions, formation and dissipation of spots are all distinctly visible in the F-group of 1983 June. Figure 13.42 illustrates the development of that group. Mehltretter [13.79] observed substantial motions in spots entering the penumbra of a larger spot. It is not clear whether the motion of smaller spots near large penumbrae

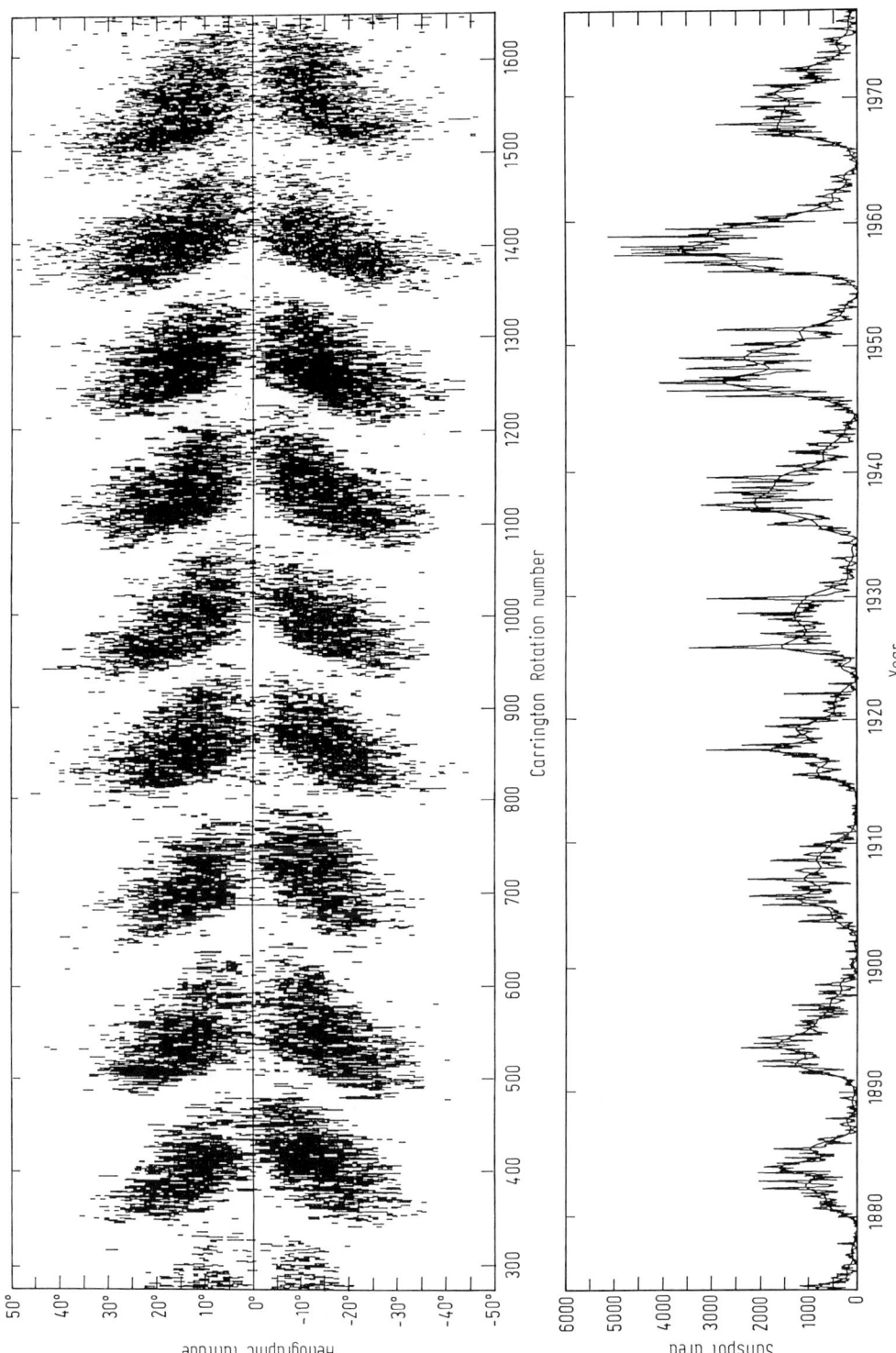

Fig. 13.41. *Top:* Butterfly diagrams for Cycles 11–20; *Bottom:* Mean daily sunspot area (in parts per million of visible solar hemisphere) per solar rotation. From Yallop and Hohenkerk [13.76].

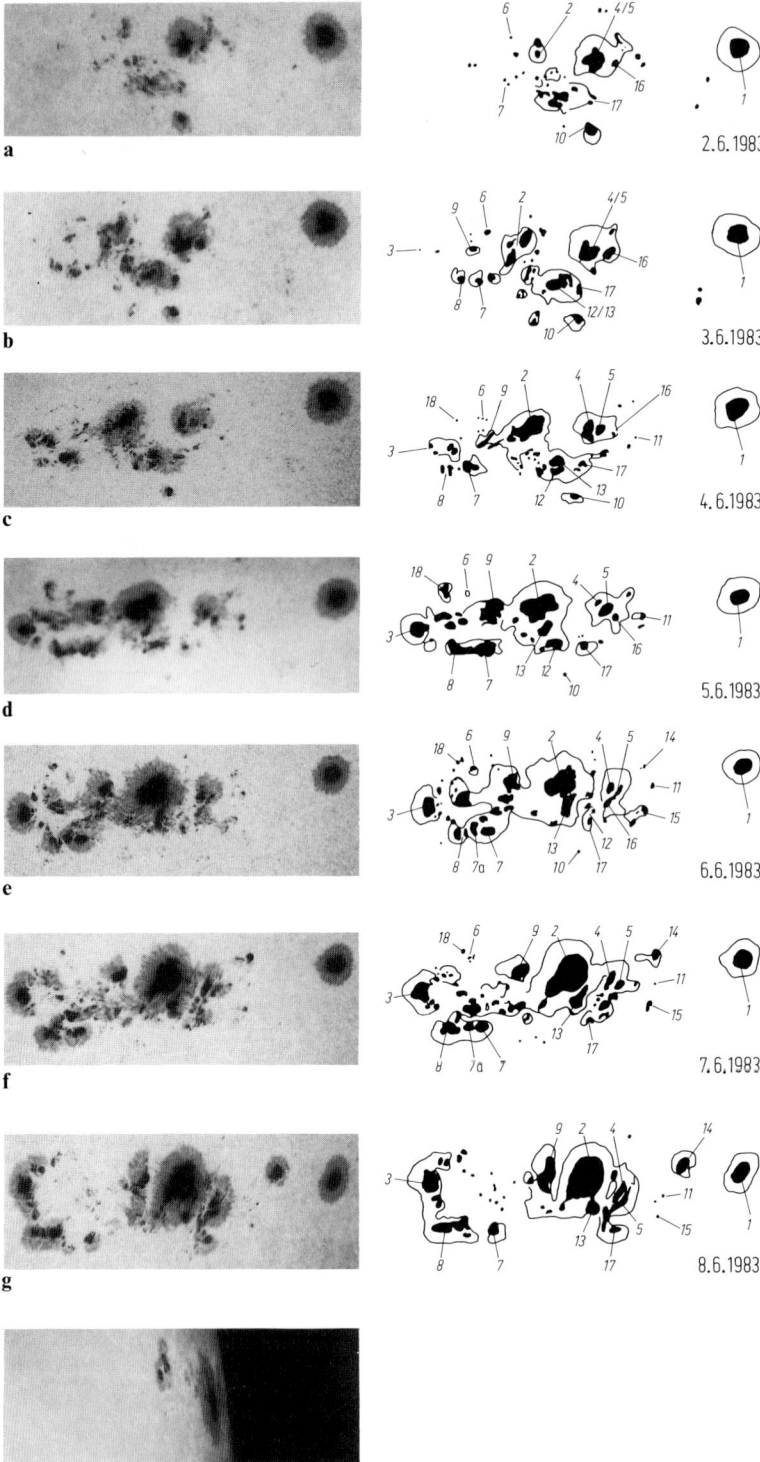

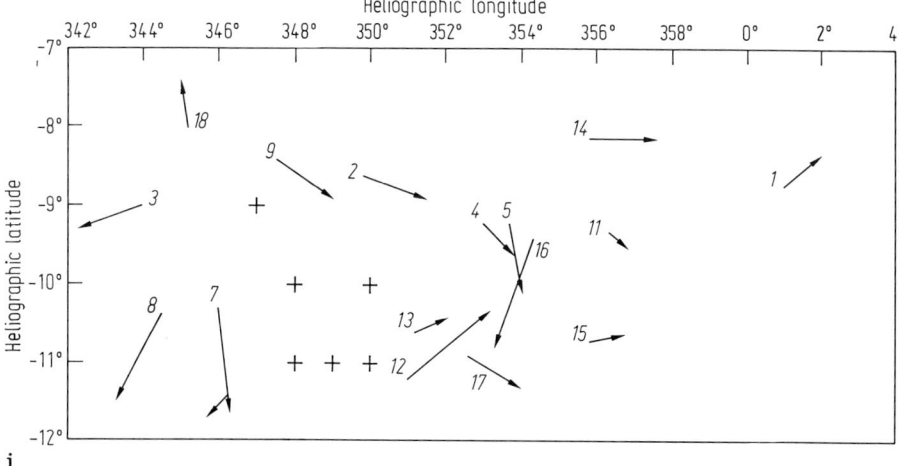

Fig. 13.42 i. Graph of motion of the F-group in 1983 June, showing the motion vectors over 4 days. Crosses indicate the position of the largest flares.

Fig. 13.42 a–h. *Left column*: Photographs of the F-group in 1983 June. All photos were taken on TP2415 by C.H. Jahn using the 200/3250 mm refractor (with 12.5-mm eyepiece, solar prism, green filter, and exposure time 1/1000 s) of the Astronomical Station of the Universität Hannover. Dates of exposures are (in UT): June 02 13^h17^m, June 03 09^h15^m, June 04 10^h00^m, June 05 13^h34^m, June 06 10^h43^m, June 07 06^h54^m, June 08 10^h51^m, June 11 08^h14^m. *Right column*: Sketches of the group. The spots marked have been used to determine their individual motions.

is accelerated or decelerated, or whether there is a connection between motions and flare frequency (Sect. 13.7.3), and in which evolutionary phase of a group the largest motions occur.

Photography is preferred over direct visual observations for these and similar studies of the Sun (Fig. 13.10) since interpretive errors which may strongly influence sketches are eliminated.

The methods of measuring solar positions are not limited to sunspots, but are applicable to plages, flares, prominences, filaments, and other phenomena of the solar photosphere and chromosphere. The polar faculae (Sect. 13.4.4) in latitudes over $\pm 50°$ should be given more attention during time of minimum (Brauckhoff, Delfs, and Stetter [13.80]).

13.9.2 Heliographic Coordinates

The system of the *heliographic reference frame* must first be defined. As in the case of the Earth, the *poles of the Sun* are the points where the rotation axis penetrates the surface. The North Pole is that point which, as seen from the Earth, points toward the northern sky. Each plane through the solar poles gives a *meridian* of the Sun. All planes at right angles to the solar axis intersect the surface in latitude circles,

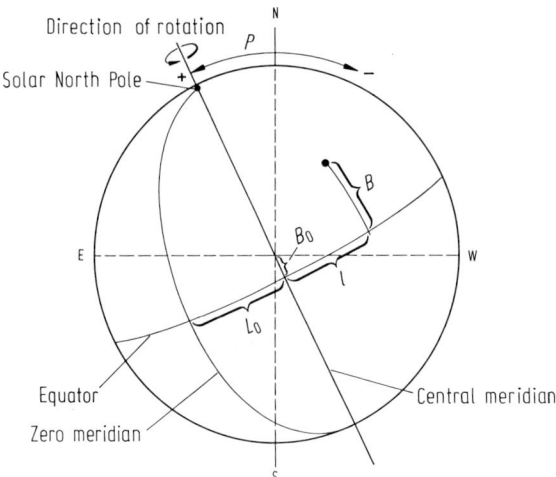

Fig. 13.43. The heliographic coordinate system (seen in perspective).

and the largest of them, which defines the plane through the center of the Sun, is called the *solar equator* (Figs. 13.43 and 13.47). The *heliographic latitudes B* are counted from the solar equator north (+) or south (−). Since there is no permanent distinguishable point on the Sun which could define the zero meridian—as Greenwich is on the Earth—the definition of *heliographic longitudes L* is more difficult. For each series of observations, the meridian is chosen from which longitudes are counted west (W) or east (E). The chosen meridians may be one of the following:

1. the *central meridian* at the moment of observation.
2. a *zero meridian*, internationally defined as that meridian which passed on 1854 January 01 at 12^h UT through the ascending node of the solar equator on the ecliptic. From this so-called *Carrington zero meridian*, heliographic longitude is counted toward west from 0° to 360°. The instant of transit of Carrington's zero meridian through the central meridian characterizes the beginning of a synodic solar rotation, whose average duration is defined to be $27^d.2753$ (at $B = \pm 16°$, about the latitude of maximum spot activity). The variable speed of the Earth in its elliptical orbit causes this number to vary between $27^d.20$ and $27^d.34$. At the beginning of a synodic solar rotation, all active regions lying west of the central meridian are counted to the old Carrington rotation. Rotation No. 1 began on 1853 November 09, and by 1990 December 19 exactly 1836 rotations (transits of the zero meridian through the central meridian) have elapsed.

It is not always useful to connect position measurements with the Carrington rotation cycle (cf. Sect. 13.9.1): Differential rotation, for instance, can of course be determined only by using the instantaneous central meridian, whereas synoptic maps use the international zero meridian.

The times of the beginning of the synodic solar rotation are given in the *Astronomical Almanac* and also in some magazines; for approximate values, see Appendix Table B.12 in Vol. 3.

In geophysics, a duration of the solar rotation of $27\overset{d}{.}0$ is used; the *Bartel rotation* No. 2001 began on 1979 December 14 at 0^h UT.

The rotation elements of the Sun given in the *Astronomical Almanac* are those published by Carrington in 1863: Ω = angle between the intersection point of the solar equatorial plane on the ecliptic and the vernal equinox of 1850 = $73°\!.67$; i = inclination of the solar equatorial plane against the ecliptic = $7°\!.25$. From spot group positions obtained in Greenwich from 1874 to 1976, Balthasar et al. [13.81] have derived small corrections to Carrington's numbers: $\Omega = 73°\!.75 \pm 0°\!.15$, $i = 7°\!.137 \pm 0°\!.017$. For amateur work, the old values in the *Almanac* are adequate.

The *apparent motion* of spots depends on the position of the Earth in its orbit. Since both the solar equator and the Earth's equator are inclined (by $i = 7°\!.25$ and $i = 23°\!.43$, respectively) against the ecliptic, sunspots as seen from the Earth usually move in ellipse sections and not in straight lines across the solar disk. They perform the latter only when the Earth is located in one of the two points of its orbit which intercept the plane of the solar equator. Figure 13.44 graphs the annual variations in position of solar axis and equator for the terrestrial observer. The heliographic latitude B_0 of the disk center M varies between $+7°\!.25$ and $-7°\!.25$. The *position angle P* (sometimes called P_0) of the solar axis against the north–south direction in the sky varies between $+26°\!.4$ and $-26°\!.4$ (+ indicates the tilt of axis is toward east, − toward west).

Figure 13.43 illustrates the relative quantities for a position measurement on the Sun. P, B_0, and L_0 are tabulated in almanacs and special periodicals for every day of the year, and must be interpolated for the specific time of observation.

The seasonal changes of the coordinate system preclude an immediate estimate as to whether a particular spot belongs to the northern or southern solar hemisphere, or a facula is or is not in the range of the polar faculae (Sect. 13.4.4). In order to make these simple distinctions, a sketch of the Sun including the solar axis and equator regarding B and P_0 is required.

13.9.3 Methods of Position Measurement

Irrespective of the particular method used to determine positions (to be described below), the *orientation of the solar image* must first be clarified: With the telescope drive switched off, a spot travels across the sky from east to west by about one solar diameter in 2 minutes. Moving the telescope toward the north celestial pole (i.e., toward higher declinations), the northern hemisphere of the Sun is last to disappear from the field of view.

It is important to use an equatorially mounted telescope. While an alt-azimuthal mount can provide a survey of the daily spot picture, the lack of tracking and the substantial change in the parallactic angle during the day makes it virtually impossible to obtain reasonably dependable position measurements. (The only exception would be position photographs taken with alt-azimuthally mounted telescopes having focal

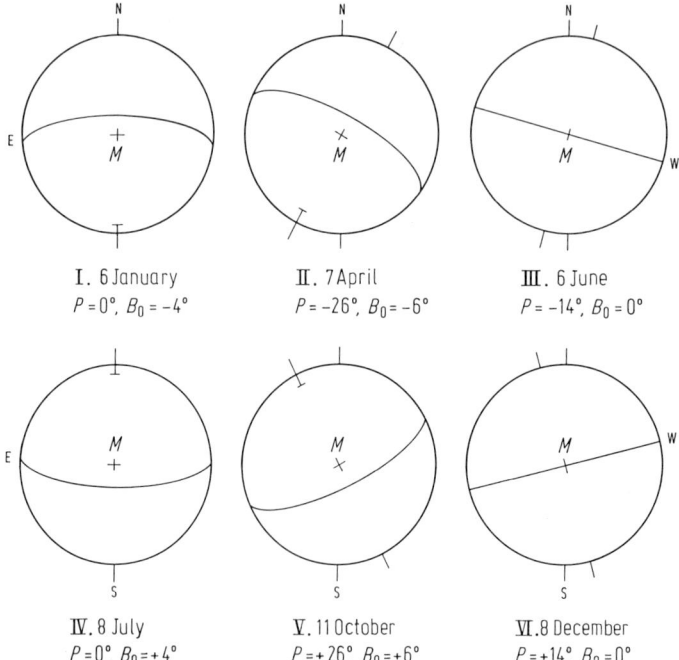

Fig. 13.44. Position of solar axis and apparent motion of spots in various seasons (schematic). *Graph I*: Around January 06, the position angle of the axis is $P = 0°$, and the rotation axis coincides with the north–south direction in the sky; B_0 is at this time about $4°$, the (imaginary) spots on the solar equator thus travel $4°$ above the center of the disk and across the central meridian. The south pole of the Sun is tilted toward Earth. *Graph II*: Around April 07, the position angle reaches its greatest western elongation, with $P = -26°.4$. The center of the disk has the heliographic latitude $B_0 = -6°$. *Graph III*: Around June 06, P diminishes to around $-14°$. As $B_0 = 0°$, sunspots travel in straight lines across the Sun, and the solar equator symmetrically halves the northern and southern parts of the visible disk. *Graphs IV–VI* show the corresponding situation when P and B_0 are positive.

lengths of less than 1.3 m.) Portable instruments require great care regarding the exact alignment of the equatorial mount. Permanent markings made on the ground will help to get the instrument into the same carefully checked alignment every day. But a control of this adjustment is still needed before each measurement is taken.

The *direct marking technique* is, owing to its simplicity, the most-used procedure. It requires a projection screen which is attached to the telescope very rigidly and precisely at right angles to the optical axis. Commercially available screens are usually far too unstable, and should therefore be strengthened. On the screen is placed a prepared sheet of white paper which contains only a circle of diameter 12 to 15 cm and a right-angle cross through its center, the latter serving to mark the north, south, east, and west directions.

With the telescope drive switched on, the positions of the individual sunspots and faculae are recorded using a light touch with a soft, well-sharpened pencil, without

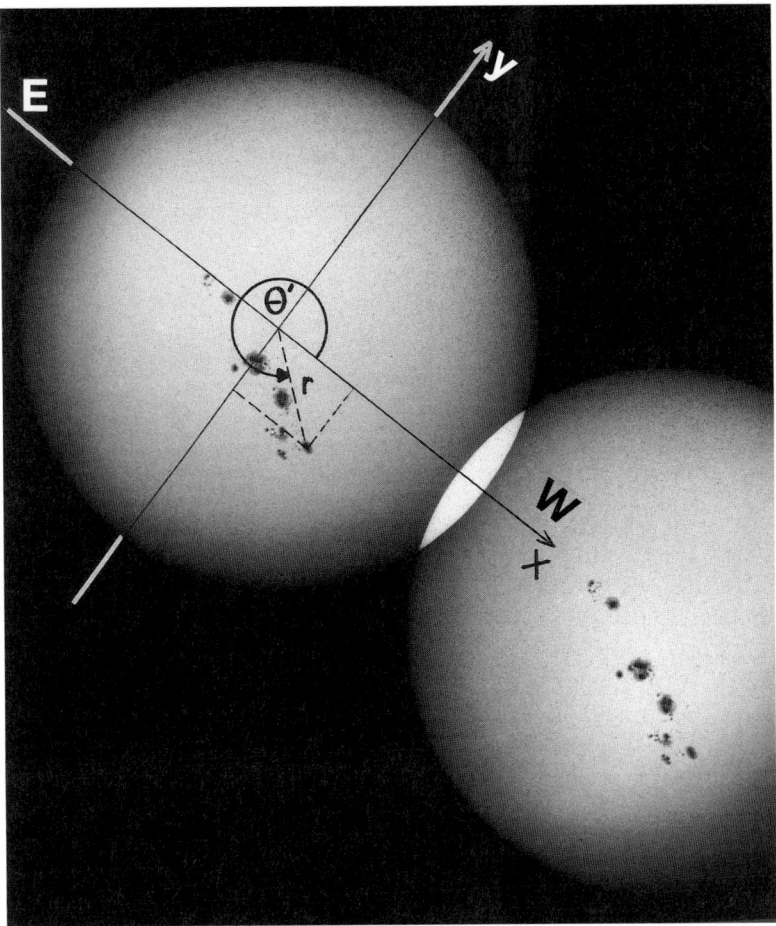

Fig. 13.45. Position photograph taken on 1984 April 29 by C.H. Jahn on Kodak TP2415 with the 200/3250 mm refractor of the Astronomical Station of the University of Hannover. Graphics by G. Schwaab and E. Junker, Bonn.

touching the screen. Position photography requires a double exposure of the focal image of the Sun at the primary focus (Fig. 13.45). One image of the complete disk will suffice. The camera must be held rigidly at the eyepiece tube, because the shutter must be cocked between the two exposures. Details on suitable materials and developer and on the technique of exposure can be found in Sect. 13.10 (and especially Sect. 13.10.8).

Quite satisfactory results can be obtained with very little effort using an ordinary camera with a multiple-exposure switch and commercial 35-mm or 110 film. All spot positions are thus recorded simultaneously in a short time. The atmospheric turbulence which has been "frozen" onto the film may, however, distort the recorded position.

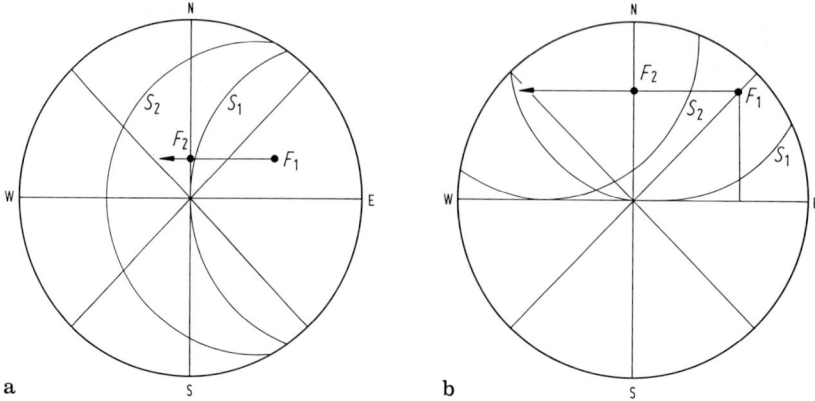

Fig. 13.46. a The Sun trails in an east–west direction across the grid; b the southern edge of the disk moves along the east-west direction. After Vogt ([13.1], pp 339).

This method gives the most precise positions, but it must be remembered that it is not necessary to obtain "pretty" pictures in order to determine the positions.

For the *trail method* (Vogt in [13.1], pp 339; Vogt [13.82]), a grid with the x-axis (east–west direction) and y-axis (north–south) and the diagonals in between, is drawn on a graph sheet (Fig. 13.46). After carefully adjusting the instrument in the east–west direction, the solar image is allowed to trail across the screen, and the time T_W between the transits of the western limb of the Sun (limb S_1, spot F_1 in Fig. 13.46 a) and of the spot (S_2, F_2 in Fig. 13.46 a) through the y-axis is stopped. The second step is to allow the southern limb of the Sun to trail along the x-axis. Here, the time T_S between the transits of the spots, for instance through the diagonal $y = x$ (S_1, F_1 in Fig. 13.46 b), and through the y-axis is stopped. The spot then has the following position on the projection screen:

$$x = T_W - R_t,$$
$$y = T_S - R_t, \tag{13.19}$$

where $R_t = \cos\delta\, \Gamma'/15$ is the radius of the Sun in units of time, Γ' its angular radius, and δ its declination. (cf. Sects. 13.9.1 and 13.9.4).

The method has also been used on fine grids. Also, the field distortion can be compensated by incorporating it into the grid.

Applying the trail method with a suitable *eyepiece micrometer* provides distortion-free measures, and thus no accessories, like a screen or camera, whose attachment makes high demands on the stability of the telescope mount, are needed.

The trail method gives the sunspot position not in a graph but rather in a numerical form which can be directly converted into positions (Sect. 13.9.4).

One should not expect an accuracy of better than ±1° using the methods described here. A stable instrument and the exercising of care on the part of the observer may improve the results.

With some improvements and corrections, e.g., using flat film which must be placed anew into the camera before every double-exposure, position photography can yield mean deviations of 0°.1 and less (Fritz, Treutner in [13.1], pp 350; Treutner [13.83]; Jahn [13.77]).

13.9.4 Calculation of Heliographic Positions

A special grid or template with coordinates (Fig. 13.47) will serve to provide an expeditious determination of heliographic positions. The correctly scaled template, when placed in proper orientation (position angle P and B_0) over sketches and photographs, permits the direct reading of heliographic latitude B and the longitude difference l against the central meridian of solar phenomena (Bendel [13.1], pp 355).

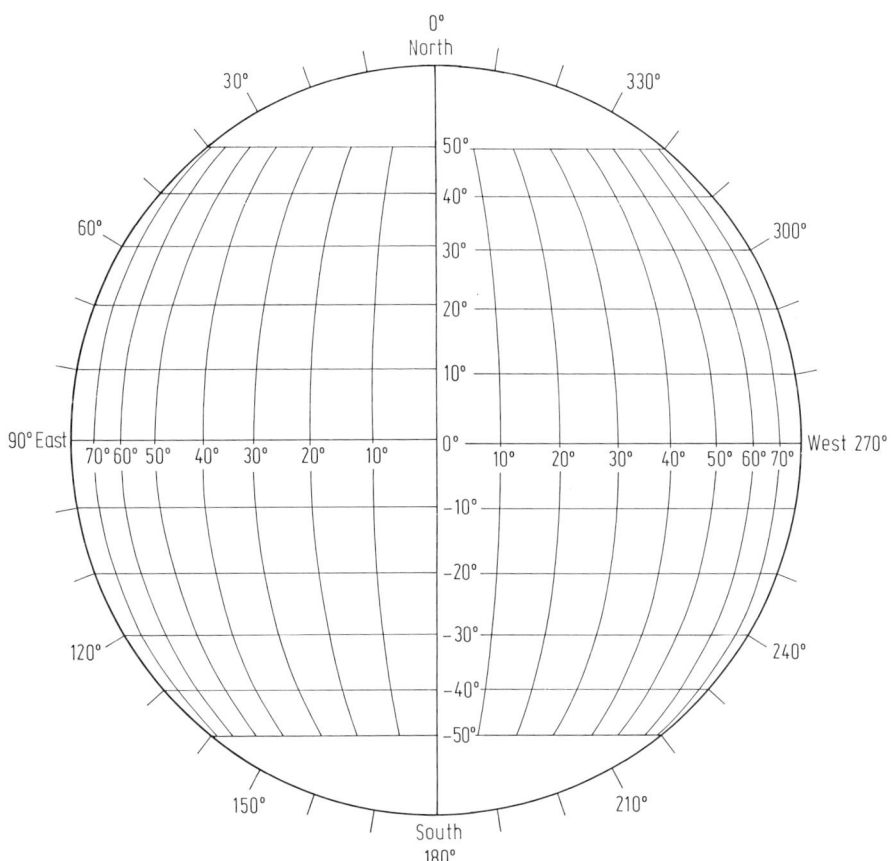

Fig. 13.47. Heliographic coordinate system with $B_0 = 0°$. Graphed along the limb of the Sun are position angles, which run from north via east from 0° to 360°.

Eight grids will suffice for the values $B_0 = 0, 1, 2, \ldots, 7°$. For negative B_0, the grids are rotated 180°.

The grids are calculated according the the formulae:

$$x = R \cos B \sin l,$$
$$y = R(\sin B \cos B_0 - \cos B \sin B_0 \cos l), \quad (13.20)$$

where x = distance in mm of longitude circles from the central meridian at a given latitude B, y = distance in mm of latitude circles from the equator ($B = 0°$), R = radius of stencil in mm.

For the special case $B_0 = 0°$, the coordinate grid is quickly calculated: $y = R \sin B$. As an example, for the coordinates $B = 40°$, $l = 20°$, the following distances from the grid center are calculated for $R = 50$ mm (cf. Tables 13.5 and 13.6 and Fig. 13.47):

$$x = 50 \text{ mm } \cos 40° \sin 20° = 13.10 \text{ mm},$$
$$y = 50 \text{ mm } \sin 40° = 32.14 \text{ mm}.$$

Using Tables 13.5 and 13.6, a grid with $10° \times 10°$ division (Fig. 13.47) can be easily constructed.[2]

Using only the grid for $B_0 = 0°$ (which, strictly speaking, holds for only two days of the year; cf. pictures III and IV in Fig. 13.44), but considering the true position angle P for the latitudes and longitudes—let them be called B' and l'—then the true B and l can be calculated:

$$\sin B = \cos B_0 \sin B' + \sin B_0 \cos B' \cos l',$$
$$\cot l = \cos B_0 \cot l' - \sin B_0 \tan B' / \sin l'. \quad (13.21)$$

The use of this latter method is expedited by employing a programmable calculator, which then gives the positions of the spot directly from the measured coordinates.

Because of the higher precision afforded, this numerical technique is preferable to a graphical method (Joppich [13.84]). Calculating coordinates from a sketch or photograph is simplified through the use of a pocket calculator or personal computer in the reduction. The latter in particular enables storage of the results and processes larger amounts of data.

The position of a spot on the Sun is recorded either in cartesian coordinates (x, y) or in polar coordinates (r, θ') (cf. Fig. 13.48).

Polar coordinates are calculated from cartesian ones via the relations

$$r = \sqrt{x^2 + y^2},$$
$$\theta' = \arctan(y/x), \quad (13.22)$$

where particular attention must be paid to the quadrant where the spot is located: θ' is counted from west over north; in quadrants 2 and 3, 180° is added, and in quadrant 4, 360°.

The angular distance ρ of the spot from the center of the disk is obtained with sufficient precision (approximately $0°.1-0°.2$) for most applications via

$$\sin \rho = \frac{r}{R}, \quad (13.23)$$

where R is the radius of the solar image. The angular distance of the spot from the solar east–west direction follows as

$$\theta = \theta' - P, \quad (13.24)$$

as Fig. 13.49 illustrates.

---

[2] A set of grids for the 8 different B_0-values with diameters 11 cm and 15 cm can be obtained from the VdS Supply Center, Munsterdamm 90, D-12169 Berlin, Germany.

Table 13.5. Distance (in mm) of the latitude circle of the equator for the grid $B_0 = 0°$ for a grid diameter of 100 mm (see Fig. 13.47).

B'	0°	10°	20°	30°	40°	50°
y (mm)	0.00	8.68	17.10	25.00	32.14	38.30

Table 13.6. Distance (in mm) of the longitude circles at various latitudes B off the central meridian for the grid $B_0 = 0°$ at a grid diameter of 100 mm (see Fig. 13.47).

l B	10°	20°	30°	40°	50°	60°	70°	80°
50°	5.58	10.99	16.07	20.66	24.62	27.83	30.20	31.65
40°	6.65	13.10	19.15	24.62	29.34	33.17	35.99	37.72
30°	7.52	14.81	21.65	27.83	33.17	37.50	40.68	42.64
20°	8.16	16.07	23.49	30.20	35.99	40.69	44.15	46.27
10°	8.55	16.84	24.62	31.65	37.72	42.64	46.27	48.49
0°	8.68	17.10	25.00	32.14	38.30	43.30	46.98	49.24

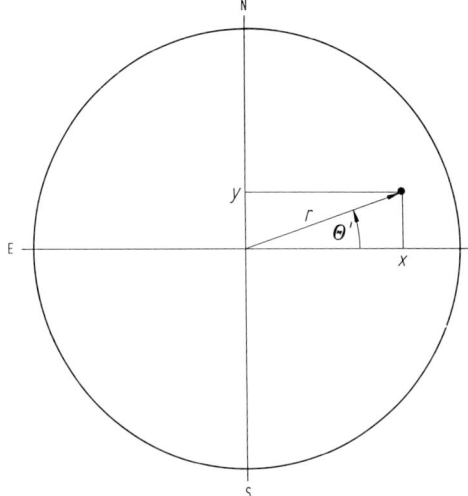

Fig. 13.48. Plane coordinate system for position measurements (polar and cartesian coordinates).

The heliographic latitude B and the longitude difference l of the spot off the central meridian are then found using the relations

$$\sin B = \cos \rho \sin B_0 + \sin \rho \cos B_0 \sin \theta,$$
$$\sin l = \frac{\cos \theta \sin \rho}{\cos B}.$$
(13.25)

The longitude L of the spot in the Carrington rotation system is found from

$$L = L_0 + l,$$
(13.26)

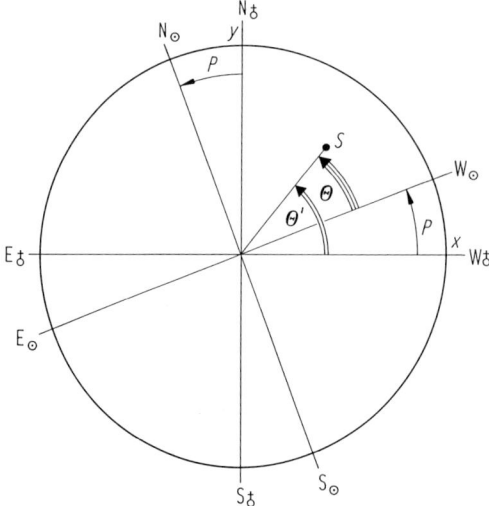

Fig. 13.49. Explanation of the relation $\theta = \theta' - P$. $N_{\oplus}$, $W_{\oplus}$ ...: directions on the Earth; $N_{\odot}$–$S_{\odot}$: projections of the solar rotation axis onto the sky; S: position of a sunspot.

where P, B_0, and L_0 are obtained from tabulated values by interpolating to the exact moment of observation. L_0 is the heliographic longitude of the central meridian referred to the Carrington rotation. The pair (B, l) gives the position of the spot relative to the instantaneous central meridian, whereas (B, L) refer to the Carrington system.

Formulae for calculating heliographic positions which differ in detail from those presented here can sometimes be found in the literature, because the trigonometric functions may change with the definitions from which axis and angle is counted. Therefore, formulae from various publications (which are self-consistent) should not be mixed (Junker [13.85]); the definitions should be studied from illustrations.

If a precision higher than about $\pm 0°\!.2$ is desired, the approximate formula of Eq. (13.23) for the angular distance ρ of the spot from the apparent center of the disk (as seen on the Sun) is replaced by the formula

$$\sin(\rho + \rho') = \frac{r}{R}, \qquad (13.27)$$

where

$$\rho' = \frac{r\Gamma'}{R}$$

and where Γ' is the apparent angular radius of the Sun, which averages $0°\!.27$ (tabulated for each day in various almanacs), and ρ' is the angular distance of the spot from the center of the disk as seen from the Earth. Figure 13.50 explains the small error of the approximation in Eq. (13.23): since the Earth is not infinitely distant from the Sun, we see at the solar limb not the meridian $90°$ from the central meridian, but one with the smaller distance $90° - \Gamma' \approx 89°\!.7$. This foreshortening effect is irrelevant for most observations, as the measuring errors are mostly larger.

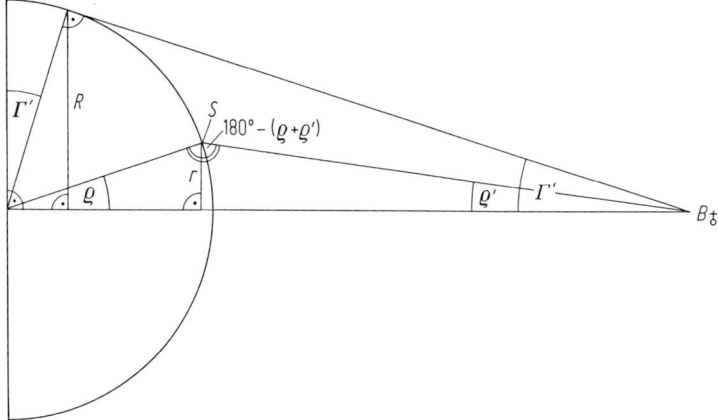

Fig. 13.50. Illustration of the correction for foreshortening for the angular distance of the sunspot S from the disk center for a terrestrial observer B_δ (see text).

Corrections for differential refraction when the Sun is at low altitude, and for distortion by lenses (eyepieces, Barlow lenses) can become important when the highest precision is demanded (Bendel [13.1], pp 345).

13.10 Solar Photography *(C.H. Jahn)*

13.10.1 Introduction

After visual observations of the Sun, solar photography provides a wide range of activities for amateur observers. Even a small telescope and moderate equipment can give pictures of interest. Particularly fine examples can be found in Fig. 13.51 a,b, where the photographers have obtained some remarkable photographs using only small refractors, a fact which can be taken as encouragement to others.

The solar photographer nowadays has at his/her disposal high-quality photographic materials which, with suitable processing, can be adapted to fit all needs. It is inevitable that the photographer must do the darkroom work him/herself.

The first unsuccessful attempts should not discourage but rather serve to point out where improvements in technique can be made. The beginner should be warned that solar photography is not an easy subject and requires quite a bit of experience. Focusing the image is just as difficult as assessing the atmospheric condition. Also, the choice of camera and film determine the quality of the negatives.

The beginning photographer will find in this section and additional literature ([13.1, 86–93]) many important hints on how to prevent mistakes before they happen. These ideas, however, often come from the experience of other amateurs and should not be considered the final word.

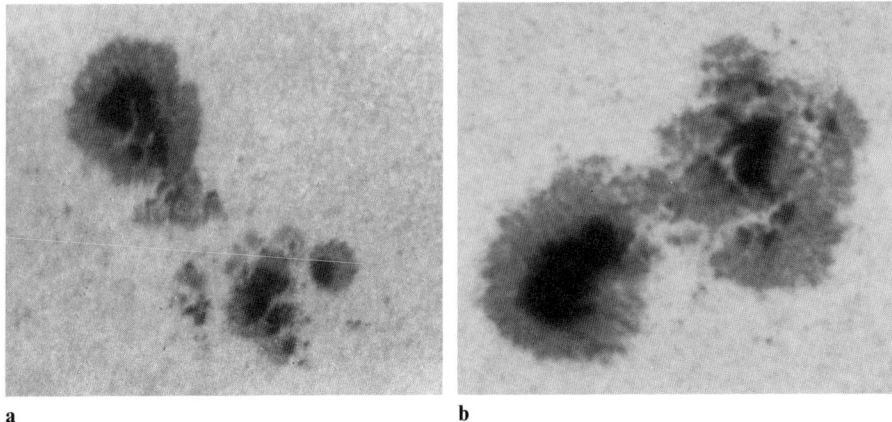

Fig. 13.51a, b. Sunspot groups: **a** photographed 1981 July 22 by U. Bendel with a 60/900 mm refractor; **b** photographed on 1979 May 13 by E. Remmert with a 80/1000 mm refractor. Both pictures were taken using eyepiece projection with a solar prism.

13.10.2 Suitability of Observing Instruments

As in many other areas of observation, the question of what is the most suitable instrument to use arises also in solar photography. The traditional answer that a reflecting telescope is unsuitable and only a refractor deserves recommendation is false. Certainly, the refractor has substantial advantages over the traditional reflector, but excellent solar photographs have in fact been obtained by amateurs using reflectors. These reflectors employ *uncoated* mirrors, which reflect only 3–4% of the light. These mirrors require high thermal stability in order to avoid geometric distortions caused by heat storage (distortion of the paraboloid causes an irregular change in optical properties). Such mirrors should be made from a heat-resistant material such as *Zerodur* (Hornung and Hückel [13.89]).

Suitable precautions (e.g., a filter placed in the entrance pupil) can render any reflector conditionally useful for solar photography. The temperature inside the instrument should be kept as low as possible, as internal turbulence distorts the image. A glass filter (Sect. 13.10.3) of moderate density is similarly suited for reflectors.

Unquestionably the best and most versatile telescope for solar photography is the refractor. It can be used with an objective filter, or with filtering methods at the eyepiece end. Thus it can be adapted to the desired purpose (focal or projection method) and the corresponding exposure times.

Fortunately, solar photography is not subject to the standard rule: "The larger the instrument, the better." Sunlight is so overwhelmingly abundant that the choice of instrument depends instead on other criteria. To image the Sun completely into the standard 24×36-mm film format, its diameter cannot be over 23 mm on the negative. As is well known, the diameter of the image varies during the year. At the primary focus, the diameter d of the image is given by

$$d(\text{mm}) = (0.009\,17\ldots 0.009\,48)f(\text{mm}), \qquad (13.28)$$

where f is the focal length of the instrument. The telescope with $f \geq 2.5$ m cannot without additional optics (e.g., a Shapley lens) image the entire solar disk. Only those instruments with focal lengths below above 2.5 m can be used for focal photography without loss of image.

When choosing the most suitable *objective diameter D*, one should consider that many small features on the Sun have sizes on the order of $1''$. Also, fine details (granules) can still be photographed in the range $2''-3''$. Calculating D according to the Rayleigh criterion for the resolving power (see Sect. 4.6.1), it is found that a 3-inch aperture suffices to photograph all interesting phenomena. The calculated resolution, however, is a theoretical one; in practice, the resolution is impaired by atmospheric and photographic influences. Thus, for solar photography, a 4- to 6-inch $f/15$ instrument is a better recommendation.

For larger instruments, increased atmospheric influences must be expected as a major part of scintillation is caused by turbulence cells with sizes on the order of 10 cm. A large aperture therefore imposes stringent demands on the observing conditions (see Sect. 13.2.1), which often negates the advantage of the increased resolution. It is only at well chosen sites and under tranquil atmospheric conditions that large refractors display their full potential (Fig. 13.56 a).

An equatorial mount is recommended for solar photography so that one can concentrate fully on the photography and not be bothered with the annoyance of continual guiding. Other mounts can be used, in which case the Sun is made to trail in the field of view and exposed at a suitable moment.

13.10.3 Options for Light Reduction

The numerous possibilities for reducing the intense light from the Sun include objective filters, foil filters, and solar prisms. *Never* use your telescope without such equipment. The use of eyepiece filters is dangerous, as the entire solar radiation is concentrated upon the filter, which becomes extremely hot. Should the filter suddenly crack under the thermal stress, the unprotected camera, film and eye would then be immediately exposed to the damaging solar rays. Aluminum-coated eyepiece filters provide a makeshift solution to the problem. Reducing the objective aperture or using standard grey (neutral-density) filters to diminish the light degrades the resolving power of the instrument, and is therefore not advisable.

Objective filters are plane-parallel glasses coated with a chromium or aluminum layer. They require a very good surface quality: the parallelism of the side of the filter should be at least 1/10 of the wavelength. Objective filters are excellently suited for focal photography. Their optical density should filter the sunlight in the ratio 1000:1 (0.1% transmission or "Density 3"), thus yielding exposure times of around 1/1000 second. When purchasing an objective filter, the buyer should be aware that many filters are designed specifically for visual observation, and often have a higher density (Density 5 or 0.001% transmission). Their use for photography is limited as the strong

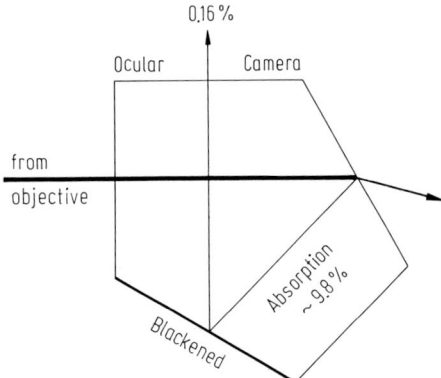

Fig. 13.52. The optical path in the Pentaprism.

filtering requires long exposure times. The best general advice is to buy a filter of Density 3, and to use additional grey filters for visual observation.

Objective Filters
Advantage:
– No heat in the instrument (important for reflectors).

Disadvantages:
– Eyepiece projection increases exposure times. The limit of 1/125 second should not be exceeded.
– Objective filters are expensive.

Foil filters are thin polyester foils vapored with an aluminum or gold layer. They are often mentioned as alternatives to the objective filter, and can be attached in front of the objective or the entrance pupil of reflectors in one or two positions without problems. Most users employ the *salvage foil* (used in first-aid kits) or the *Solar Skreen foil* which is marketed in the U.S. However, foil filters cannot match the quality of objective filters, a fact which should be considered particularly in photography.

Foil Filters
Advantages:
– Much cheaper than objective filters.
– Suitable for first acquaintance in test photographs.

Disadvantages:
– Foils are often inhomogeneous and thus do not filter smoothly.
– Foils contain microscopically small holes causing diffraction and light scatter which impairs the image quality.

The *solar prism* (Fig. 13.54 a) is also no alternative to the objective filter, but rather provides a sensible accessory in order to obtain short exposure times using the eyepiece projection method. However, the solar prism can be used just as well for solar photography. Problems are partly caused by air turbulence owing to heating inside the tube. During the observing session, the instrument must be covered at regular time intervals, (e.g., every 15 minutes).

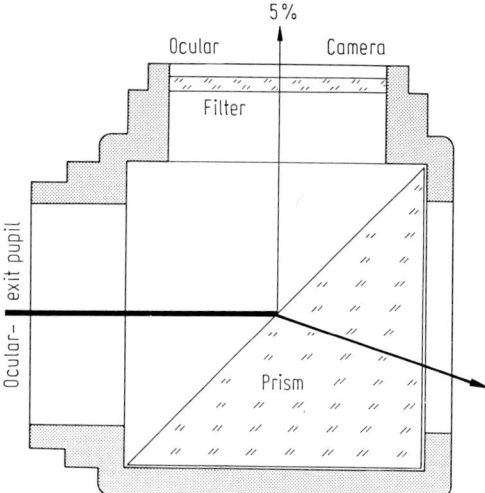

Fig. 13.53. The zenith prism as a solar prism.

The optimal idea would be then to pump all air out of the telescope and thus obtaining a *vacuum refractor*.

Solar prism
Advantages:
– Very well suited for projection photography.
– Cheaper than objective filter, especially for large instruments.
Disadvantages:
– Air turbulence sometimes present in the tube.
– Long optical path; thus often problems with the construction length of the tube.

A *zenith prism*, which undoubtedly many amateurs have bought along with their first telescope, is a special form of solar prism that can be used. The zenith prism is turned around and used according to Schröder [13.91] as a surface reflecting prism, so that only its housing must be constructed. That part of the housing toward the camera should allow for the insertion of color filters.

As a final comment, the optimal filtering consists of the combination of objective filter (Density 2–3) and a secondary filtering which can adjust the brightness of the solar image in a measurable way, for instance through polarization, in order to adapt it to the desired exposure time. This system requires polarization filters of high quality.

13.10.4 The Most Suitable Camera

In solar photography, the image must be viewed directly (with a finder), and thus a reflex camera with interchangeable lenses is needed. Usually 35-mm cameras are used, but cameras of larger format are not excluded. The camera must be firmly attached to the eyepiece tube in a mechanically simple fashion. Cameras with a bayonette fitting use a T2 adapter (other manufacturers use the term T- or P-adapter), which generally has an M42 thread and is screwed onto the telescope tube. Other methods of attachment should not be used, as they will only result in mechanical troubles and poor photographs.

Any commercial mirror reflex camera can be used for solar photography, but the more expensive models (with features such as interchangeable screens, multiple-exposure feature, etc.) are certainly more advantageous. Less-expensive cameras often have rather coarse viewing screens which may cause problems when focusing (Sect. 13.10.6). Owners of high-quality cameras are advised to purchase a screen with a clear spot containing a reticle (crosswire). Also, a clear-glass screen (together with an etched reticle) can be used, although the eye may have some difficulty in finding the correct focus. Exposure times shorter than 1/1000 second are sometimes needed, and can be obtained with special camera makes; alternatively, suitable neutral-density filters must be employed to effectively increase the exposure times.

Some older reflex cameras tend to expose unevenly at very short exposure times; the shutter does not move smoothly, which thereby results in vertical stripes on the negatives. Repairs being usually to no avail, the owner should then consider purchasing a new camera.

An angle finder with additional magnification connected with the eye-side lens of the finderscope can also be useful and practical (Fig. 13.54 c). Camera systems with interchangable screens are available from such well-known companies as Olympus (OM2SP), Nikon (F2,F3), Canon (F1, AE1-Program), Pentax LX or Praktika (VLC2, obtainable only second-hand). Some systems have interchangable screens only, while in others the entire finder system can be interchanged. The requirement of a manual camera is not as important for solar photography as it is in nighttime photography, where the batteries of the automatic camera discharge quickly.

With automatic exposures (i.e., automatic variation of the ASA/DIN setting) it is possible to be more consistent with exposure times of photographs, and thus keep the density of the negative fairly constant. Influences such as changing atmospheric transparency or solar altitude can thus be better compensated and allowed for in the exposure. Manual settings, on the other hand, often lead to negatives which are slightly under or overexposed, which must be corrected in the darkroom processing. Automatic exposures, however, become problematic when photographing the limb of the Sun. The offset location of the window which measures exposure time may then lead to improperly exposed images.

13.10.5 Comments on Photographic Materials

The smallest details to be photographed on the solar surface have apparent angular diameters on the order of $1''-2''$ (granulation), the largest about $10''-300''$ (sunspots; Paech in [13.1], pp 39). In order to photograph the smallest details the film must be exceptionally fine-grained, and this will mandate using films of low and middle speed. Amateurs have had very good experiences with the following:

- Agfa Ortho 25 (ASA 25/15 DIN)
- Kodak Technical Pan 2415 (ASA 50 – 80/18 – 20 DIN)

Fig. 13.54 a–c. Solar prism, eyepiece projection equipment, and angle finder. These accessories are indispensable for solar photography.

The average speeds or ranges thereof are given in parentheses. These quantities can be further varied by suitable development. Both of the films named are high-resolution material. When developed at high contrast (contrast range 1000:1), resolutions of 280–320 lines pairs/mm are obtained.

Agfa Ortho is an *orthochromatic film*, which means it is red-*insensitive*. For photography of sunspots, this makes little difference, but for prominences and for Hα photography, such film cannot be used. Instead, the Kodak TP 2415 is needed. Both films can be purchased in 35-mm format or in 6 × 6-cm rolls, separately or as "meterware." For high consumption, the purchase of larger quantities is of course advantageous.

Depending on the desired contrast, various developers can be used. Agfa Ortho 25 should be developed with Rodinal (see data in Hornung/Hückel [13.89] and Beck et al. [13.87]). Kodak TP 2415 has several options for developing: sunspots are usually better in harder contrast so that hard developers like Kodak D19 are employed. Prominence photography, on the other hand, must reveal tenuous detail, so that "soft" developers are preferred (e.g., Neophin Blue or Neophin doku from Tetenal; see also Jahn [13.90]).

For solar photography, as in all of astrophotography, the general rule is that all photo-processing should be do-it-yourself to obtain the most satisfactory results.

13.10.6 White-Light Photography

The term *white-light photography* refers to the photography of all photospheric phenomena on the Sun—spots, plages, light bridges, etc.—in visible light. Two techniques are distinguished: *focal photography* and *projection photography* (see [13.92]). The use of color filters and various films often limits the spectral range. The filters should act to enhance the contrast and suppress the influence of secondary spectra. They are placed near the focus so that the quality of even coarsely polished color glass will suffice. (They should not, however, be placed directly *at* the focus, as there is the risk of imaging dust particles.)

Focal photography records the focal image in the instrument, which gives good survey photographs of the Sun. In small instruments, the use of a 2× converter is advised to entirely fill the area of the negative (see Fig. 13.55).

Projection photography uses eyepieces to increase the imaging scale. The eyepiece is mounted in the optical path between solar prism and camera (see Fig. 13.54 b; the use of an objective filter, of course, dispenses with the need for a solar prism). The focal length of an eyepiece should not be chosen too short; otherwise focusing is considerably impeded by overmagnification. Also, when high magnifications are employed, the exposure times can easily exceed 1/125 seconds. For large sunspot groups, a longer focus eyepiece is advised.

The necessary color filters are inserted into the eyepieces, which should always be of good optical quality (e.g., orthoscopic). The basic set of yellow, orange, and green filters is recommended (e.g., Schott filters GG495, OG550, VG6). Occasionally,

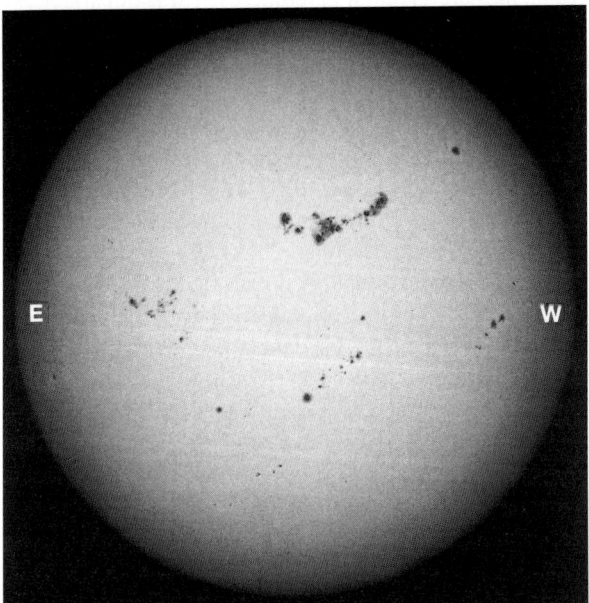

Fig. 13.55. Focal photograph of the Sun on 1984 June 26 by C.H. Jahn with a 100/1000 mm refractor and objective filter.

solar photographers also use wide-band interference filters (e.g., 550 ± 50 nm), which employ the optimal imaging qualities of the objective; such filters are not inexpensive, however.

The optimal exposure times must be determined for the various techniques and scales. They can be either computed as prescribed by Schröder [13.91] or determined by a series of trial exposures. In general, photographs of the Sun should not be exposed over 1/125 s, as then atmospheric effects begin to deteriorate the quality of the negative to an unacceptable level. Normally, 1/1000 s is a workable exposure time.

A problem frequently encountered by beginners in solar photography is difficulty in focusing the image. The image in the finder is very bright (sunglasses may help) and the untrained eye does not adjust well to these conditions. Many observers thus tend to focus the image once and use this setting for all subsequent exposures. This method invariably leads to poorly focused pictures, as the focus constantly changes with temperature and other influences.

In any event, when focusing with a clear glass screen or screen with a clear spot, the focusing is performed on the reticle, so that both sunspot and reticle must appear simultaneously sharp. Although there is some tolerance, this creates severe problems and produces blurred pictures. As a comparison, consider the photographs in Fig. 13.56 a and b. In both cases, air conditions were optimal, and the correctly focused picture of Fig. 13.56 a shows a high degree of sharpness while 13.56 b is rather fuzzy. In practice, it is thus necessary to refocus every picture. The motion of the object should also be watched at the finder, so that the exposure can be made as soon as a moment of steady air occurs. It takes supreme patience to watch through the finder and wait for just the proper moment to expose.

It is more difficult to focus a camera with a fixed screen. The low brightness in the finder does not permit the use of the intersection image indicator. As a makeshift technique, one-half of the intersect image can be focused. Tiny dust particles on the screen can serve as a reticle so that the eye can fix upon them.

All applications of solar photography should naturally contain a cable trigger to avoid any possibility of the image "jumping" if the camera is jolted.

13.10.7 Photography in Narrow Spectral Ranges

The amateur solar photographer uses a part of white-light photography and also special ranges such as prominence (Hα) photography, and violet (calcium-line) photography. These applications necessitate special instruments and filters which, depending on their application, cover a wide range of prices. Prominence photography uses prominence telescopes, attachments, or eyepieces as described by Nemec [13.46] or Lille and Dobrzewsky in [13.1], pp 89. It is characteristic of this area of solar photography that only the limb is imaged, while the disk is covered by the conic stop. As prominences are photographed with Hα-filters (0.2–10 nm at $\lambda = 656$ nm), only red-sensitive films can be used (e.g., Kodak TP2415). Special requirements when developing are found in Sect. 13.10.5, and it should be kept in mind that prominences sometimes display very low constrast. Figure 13.57 a shows a very impressive prominence photograph

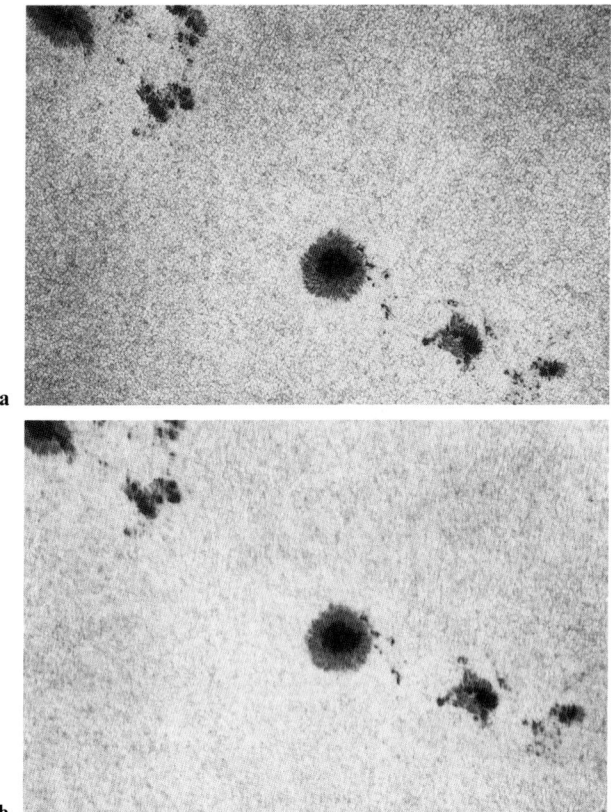

Fig. 13.56 a, b. Sunspot group 1983 June 24 photographed by C.H. Jahn with a refractor of the Astronomical Station of the University of Hannover with a solar prism. **a** Taken under very good conditions, correctly focused; **b** taken under the same conditions but poorly focused.

taken with a special prominence attachment. It may also be worthwhile to photograph the ascents of prominences (see hints by Völker in [13.1]).

In the practice of photography in narrow spectral ranges, the focusing of the image becomes particularly problematic. Owing to the strong brightness gradient and to ensure proper darkening, it is advisable to use a black cloth (placed over observer and camera) or other suitable method. Prominence photography need not be done with a conic diaphragm attachment. Observers using interference filters of width ≤ 0.1 nm may wish photograph the surface (chromosphere) of the Sun as well as prominences. Figure 13.57 b shows a composite image of the Sun in which surface and limb have been mounted together. These filters (e.g., DayStar filter) are rather expensive, and also need much care; for instance, UV and infrared protection filters are needed for DayStar filters.

For photography in monochromatic light (here referring in particular to photography of the chromosphere), Kodak TP2415 is used exclusively and developed to middle or high contrast. Exposure times vary for different instruments (guidelines: solar limb

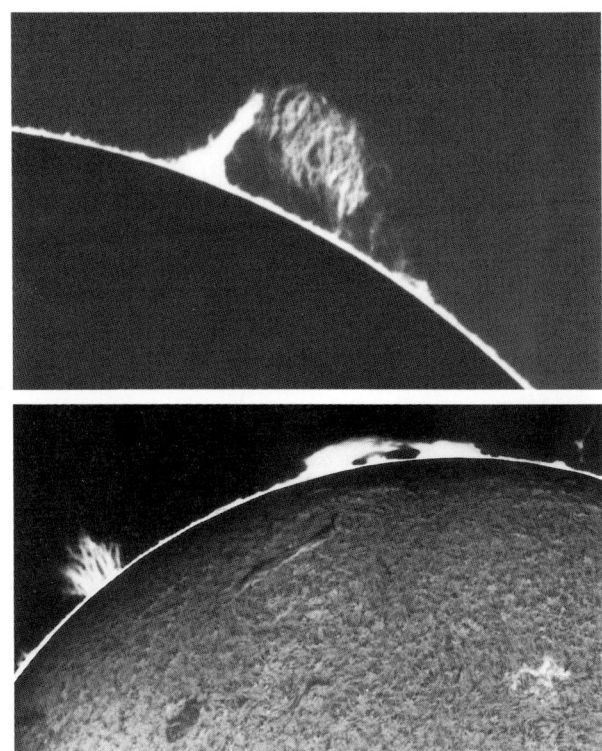

Fig. 13.57. a A photograph taken with a prominence attachment by W. Lille (Hα-filter with half-width 0.2 nm). **b** A composite image of the chromosphere by W. Paech, made with a DayStar filter.

1 s, surface 1/60–1/125 s) and atmospheric transparency. Also, the refractor has a demonstrated superiority in this area of solar photography.

The photographer working in Hα light usually prefers to use focal photography, employing a 2× converter. Eyepiece projection causes longer exposure times and perhaps fuzzy images. This conflict may be resolved by using another refractor which has, for instance, a longer focal length for photographing the details.

Photography in the calcium line at a wavelength $\lambda = 393$ nm is more difficult than that in Hα. Visual observations are virtually impossible at this wavelength as the human eye is insensitive to violet light. Focusing of the image must therefore rely exclusively upon trial photographs. Figure 13.58 a shows the Sun in calcium light, which reveals the active regions distinctly.

A special area of solar photography is the study of photospheric faculae (or plages) on the disk (see Fig. 13.58 b). While invisible on the disk itself, the faculae appear near the limb in white-light photography owing to limb darkening. Faculae can also be made visible on the disk by using, according to Lille in [13.1], pp 150, an uncoated reflector with polished and dull-black rear side together with the filter combination UG1/BG38

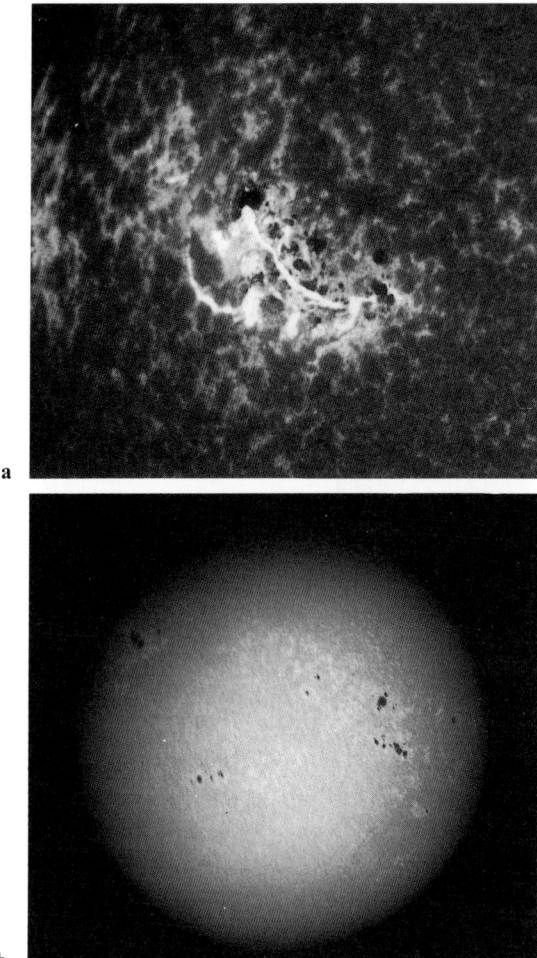

Fig. 13.58. a Photograph taken of the Sun in the light of calcium by G. Appelt. **b** Photograph taken by W. Lille using a UG1/BG38 filter and showing photospheric plages on the disk.

(Schott). Here, mirror systems are preferred as the glass lenses of refractors partly absorb ultraviolet light upon transmission. The observer should also be aware that refractor optics are corrected for the visual spectral range and have strong chromatic errors in the violet. Figure 13.58 b shows a photograph of the solar disk with plages. The focusing of the image in this method is, as with other studies, to be found by taking a series of test exposures.

13.10.8 Double Exposures as a Method of Measuring Positions

The photographic method of determining sunspot positions makes use of *double exposures* (Fritz et al. [13.88]), in which two focal images of the Sun are exposed into one negative (Fig. 13.44). The camera must permit true double exposures without shifting the film in between exposures as the shutter is cocked. Exposure times for both images are the same as those used for a focal photograph.

The procedure of observations can be outlined as follows:

1. Place the solar image in the finder, so that two images fit onto one negative.
2. Trigger the first exposure, and record the time for later processing.
3. Stop tracking and let the Sun trail through the field. Resume tracking when the Sun is in the other half of the field.
4. Operate the multiple exposure switch and cock the trigger.
5. Make a second exposure and record the time.

After photo-processing, the negatives will be measured, either by a measuring machine or, to limited precision, using the enlarging equipment. For a discussion of the evaluation and accuracy of the method, the reader is directed to Sect. 13.9.

13.10.9 Observing Programs

The programs listed in this section may serve as suggestions to interested observers and photographers regarding the processing of one's observations. In order to succeed with that intention, a certain degree of commitment and continuity is needed, as is the case in other areas of observational astronomy.

This section compiles the various possibilities which are available to the amateur and small college astronomer, and adds some "key words" characteristic of the subject. The reader should also compare these with suggestions from [13.1,86,90,93].

- *Photographic Wolf numbers*: determined from focal photographs, but using always the same instrument. A k-factor must be determined; see Sect. 13.3.4.
- *Measuring areas of sunspots and sunspot groups*: with grids in suitable scales, areas can easily be measured. Results must be corrected for distortions; see Sect. 13.3.1.
- *Morphological evolution*: photos and sketches describe changes in the group. There should be two or three observations made per day; see Sect. 13.3.1.
- *Measuring positions*: see Sect. 13.9.
- *Study of motions*: global displacements of sunspots and sunspot regions. Motions within sunspot groups. Coordinates to be found from position measures at the telescope or from photographs at intervals of from 2 to 3 hours.
- *Short-term variations*: rapid motions of separate pores within groups. Position photographs at intervals of a few minutes.
- *Study of faculae*: see Sect. 13.4.
- *Study of polar faculae*: see Sect. 13.4.
- *Wilson phenomenon*: Sect. 13.3.3, daily photographs.
- *Light bridges*: Sect. 13.3.2, visual and photographic observations hourly or daily.

- *Structures of umbrae and penumbrae*: long-term observations needed. Use of photo archives. See Sect. 13.3.1.
- *Granulation*: requires large instruments and very good atmospheric conditions. Photographs at every 10–15 minutes with motor camera.
- *Umbral dots*: requires large instruments. Expose one or two steps longer than for sunspots. See Sect. 13.3.1.
- *Equidensitometry*: brightness distribution within sunspots and other phenomena. Areas of equal brightness to be determined.
- *Photospheric plages on the disk*: see Sect. 13.10.7.
- *Measure sizes of filaments*: only in Hα-light. Sect. 13.6.
- *Prominence number*: in Hα-light and with prominence attachments.
- *Development of flares*: see Sect. 13.7.

13.11 Conclusion

The preceding sections have shown that there is a wealth of possibilities for observing the Sun, even though some items could be mentioned only briefly. Evaluating one's observations, whether individually or within working groups, cannot be described in general recipes because it would challenge the creativity of the individual. Solar physicists and other active observers are usually glad to help in this regard.

Further information can be obtained from one of following organizations:

- *The VdS Working Group on the Sun*. Contact: Peter Völker, Wilhelm-Foerster-Sternwarte, Munsterdamm 90, D-12169 Berlin, Germany.
- *The American Association of Variable Star Observers–Solar Division*. Contact: Peter O. Taylor, P.O. Box 5685, Athens, GA 30604-5685, USA.
- *The British Astronomical Association–Solar Section*. Contact: Bruce Hardie, 13 Glencree Park, Jordanstown, Co. Antrim, BT37 0QS, N. Ireland.
- *The Association of Lunar and Planetary Observers–Solar Section*. Contact: Paul Maxson, 8839 N. 30th Avenue, Phoenix, AZ 85051 USA.

References

13.1 Beck, R., Hilbrecht, H., Reinsch, K., Völker, P. (eds.): *Handbuch für Sonnenbeobachter*, Vereinigung der Sternfreunde e.V., Berlin 1982.
13.2 Kiepenheuer, K.O.: Solar Site Testing. In J. Rösch: *Site Testing* (IAU Symposium No. 19) Paris 1962, p. 193.
13.3 Müller, R.: Sichtbedingungen bei Sonnenbeobachtungen. *Sterne und Weltraum* **1**, 170 (1962).
13.4 Brandt, P.N.: Frequency Spectra of Solar Image Motion. *Solar Physics* **7**, 187 (1969).
13.5 Newton, H.W.: *The Face of the Sun*, Penguin Book No. A422, Harmondsworth 1958.
13.6 Bray, R.J., Loughhead, R.E.: *Sunspots*, Chapman and Hall, London 1964.
13.7 Wilson, P.R.: The Structure of a Sunspot. *Solar Physics* **3**, 243 (1968).
13.8 McIntosh, P.S.: The Birth and Evolution of Sunspots. In L.E. Cram and J.H. Thomas (eds.): *The Physics of Sunspots*, Sacramento Peak Observatory, Sunspot 1981.
13.9 Schüssler, M.: Neues zur Theorie der Sonnenaktivität. *Sterne und Weltraum* **19**, 331 (1980).

13.10 Giovanelli, R.G.: *Secrets of the Sun*, Cambridge University Press, Cambridge 1984.
13.11 Fritz, U., Treutner, H., Vogt, O.: Portrait einer langlebigen Sonnenfleckengruppe. *Sterne und Weltraum* **19**, 142 (1980).
13.12 Bendel, U.: Die Achsenneigung von bipolaren Sonnenfleckengruppen. *Sonne* **4**, 7 (1980).
13.13 Künzel, H.: Statistische Untersuchungen über Häufigkeit, Zonenwanderung und Lebensdauer von Sonnenflecken im 18. Aktivitätszyklus. *Astronomische Nachrichten* **285**, 169 (1960).
13.14 McIntosh, P.S.: The Classification of Sunspot Groups. *Solar Physics* **125**, 251 (1990).
13.15 Waldmeier, M.: Sunspot Numbers and Sunspot Areas. *Astronomische Mitteilungen der Eidgenössischen Sternwarte Zürich* No. 358 (1978).
13.16 Wilson, P.R., McIntosh, P.S.: What is the Wilson Effect? *Solar Physics* **10**, 370 (1969).
13.17 Künzel, H.: Hinweise für die heute übliche Zählweise von Sonnenflecken zur Bestimmung der Relativzahl. *Astronomie und Raumfahrt* **14**, 121 (1976).
13.18 Beck, R.: Eine neue Definition der Sonnenfleckenrelativzahl. *Sonne* **1**, 56 (1977).
13.19 Wiechoczek, R.: Beobachtungsprogramm Sonne. *Sonne* **1**, 30 (1977).
13.20 Pettis, H.S.: Eine systematische Studie von Sonnenflecken. *Saturn* (Astron. Arbeitsgem. Paderborn) **11** (1978).
13.21 Malde, K.I.: "Klassifikationswerte"—eine neue Messung der Sonnenaktivität? *Sonne* **9**, 159 (1985).
13.22 Keller, H.U.: "A"-Sonnenfleckenbeobachtungen von bloßem Auge. *Orion* **38**, 180 (1980).
13.23 Keller, H.U.: Der Sonnenfleckenzyklus No. 21 von bloßem Auge registriert. *Orion* **44**, 154 (1986).
13.24 Waldmeier, M.: Neue Eigenschaften der Sonnenfleckenkurve. *Astr. Mitt. Eidgen. Sternw. Zürich* **133**, 105 (1935).
13.25 Bendel, U., Staps, D.: Kurz- und mittelfristige Sonnenfleckenprognose mit der P17-Mittelung. *Sterne und Weltraum* **19**, 180 (1980).
13.26 Gleissberg, W.: *Die Häufigkeit der Sonnenflecken*. Akademie-Verlag, Berlin 1952.
13.27 Karkoschka, E.: Neue Relativzahl-Mittelung. *Sonne* **3**, 33 (1979).
13.28 Schatten, K.H.: A Solar Timing Predictor. *Solar Physics* **125**, 185 (1990).
13.29 Bracewell, R.N.: Simulating the Sunspot Cycle. *Nature* **323**, 516 (1986).
13.30 Wilson, R.M., Reichmann, E.J., Teuber, D.L.: An Empirical Method for Estimating Sunspot Number. In: *Solar-Terrestrial Predictions*, NOAA, Boulder 1986, p. 26.
13.31 Malde, K.I.: Der 22. Sonnenfleckenzyklus. *Sonne* **13**, 114 (1989).
13.32 De Jager, C.: Structure and Dynamics of the Solar Atmosphere. In: Flügge, S. (ed.): *Handbuch der Physik*, Springer, Berlin 1959, Band 52, p. 173.
13.33 Gericke, V.: Das SOLOS-Fackelprogramm. *Sonne* **2**, 59 (1978).
13.34 Gericke, V.: Fackelklassifikation 1978–1982. *Sonne* **7**, 82 (1983).
13.35 Reble, M.: Beobachtung der Sonne. *Sterne und Weltraum* **10**, 307 (1971).
13.36 Reble, M.: Zur Beobachtung photosphärischer Fackeln. *Sterne und Weltraum* **21**, 43 (1982).
13.37 Wadsworth, D.J.: Some Observations on the White Light Solar Faculae. *J. British Astron. Assoc.* **88**, 444 (1978).
13.38 Brown, G.M., Evans, D.R.: The Use of Solar Faculae in Studies of the Sunspot Cycle. *Solar Physics* **66**, 233 (1980).
13.39 Waldmeier, M.: Polare Fackeln. *Zeitschrift für Astrophys.* **38**, 37 (1955).
13.40 Brauckhoff, D., Delfs, M., Stetter, H.: Polfackeln, Neuland für den Amateursonnenbeobachter. *Sonne* **10**, 114 (1986).
13.41 Bruzek, A., Durrant, C.J.: *Illustrated Glossary for Solar And Solar-Terrestrial Physics*, D. Reidel Publ. Co., Dordrecht 1977.
13.42 Rothe, R.: Beziehungen zwischen chromosphärischen und photosphärischen Fackeln. *Sonne* **3**, 154 (1979).
13.43 Mangis, S.J.: *Introduction to Solar Terrestrial Phenomena and the Space Environment Services Center*, NOAA Technical Report ERL 315-SEL 32, Boulder 1975.
13.44 Nögel, O.: Ein Fernrohr zur Beobachtung der Protuberanzen für den Amateur. *Die Sterne* **28**, 135 (1952).

13.45 Nögel, O.: Das Protuberanzenfernrohr des Sternfreundes. *Die Sterne* **31**, 1 (1955).
13.46 Nemec, G.: Das Protuberanzenfernrohr als Hochleistungsinstrument. *Sterne und Weltraum* **10**, 171 (1971) and **11**, 17 (1972).
13.47 Hanisch, H.D.: Protuberanzenansatz für kleine Refraktoren. *Sterne und Weltraum* **14**, 370 (1975).
13.48 Richter, G.: Ein vereinfachtes Protuberanzenfernrohr. *Die Sterne* **50**, 105 (1974).
13.49 Hale, G.E., Ellerman, F.: Spectroheliograph. *Astrophysical Journal* **19**, 41 (1904).
13.50 Hale, G.E.: The Spectrohelioscope and its Work. *Astrophysical Journal* **70**, 285 (1929).
13.51 Veio, F.N.: An Inexpensive Spectrohelioscope by a Californian Amateur. *Sky and Telescope* **37**, 45 (1969).
13.52 Veio, F.N.: *The Sun in Hα-Light with a Spectrohelioscope*, Adams Press 1972.
13.53 Veio, F.N.: A Miniaturized Spectrohelioscope. *J. British Astron. Assoc.* **86**, 66 (1975).
13.54 Lyot, B.: Un monochromateur à grands champs utilisants les interférences en lumière polarisée. *Comptes Rendus* **197**, 1593 (1933).
13.55 Pettit, E., Slocum, F.: Observations of Solar Prominences with a Lyot-Telescope. *Publications of the Astronomical Society of the Pacific* **45**, 187 (1933).
13.56 Thiele, St.: Einige Fragen zu DayStar-Filtern. *Sonne* **6**, 138 (1982).
13.57 Völker, P.: Sonnenbeobachtung im Hα-Licht. Article series in *Sonne* **1–6** (1977–1982).
13.58 Tandberg-Hanssen, E.: *Solar Prominences*, D. Reidel Publ. Co., Dordrecht 1974.
13.59 Völker, P.: Die Protuberanzenbeobachtung des Amateurs. *VdS-Nachrichten* **19**, 14 (1970) in *Sterne und Weltraum* **9** (2/1970).
13.60 Wattenberg, D.: Die Statistik und Periodizität der Protuberanzen. *Die Sterne* **16**, 95 (1936).
13.61 Osservatorio Astrofisico di Catania: *Catania Solar Observations* (yearly).
13.62 Waldmeier, M.: *Astronomische Mitteilungen der Eidgenössischen Sternwarte Zürich* (yearly until 1980).
13.63 Stetter, H.: Darstellung der Verteilung der Protuberanzenaktivität durch die Protuberanzenrelativzahl R_P. *Sonne* **10**, 76 (1986).
13.64 Glitsch, I.: Schauspiel eines Riesenfilamentes. *Sonne* **10**, 8 (1986).
13.65 Grossmann-Doerth, U., Uexküll, M.V.: Spectral Investigation of Chromospheric Fine Structure. *Solar Physics* **20**, 31 (1971).
13.66 Svestka, Z.: *Solar Flares*, D. Reidel Publ. Co., Dordrecht 1976.
13.67 Neidig, D.F.: The Importance of Solar White-Light Flares. *Solar Physics* **121**, 261 (1989).
13.68 Kiepenheuer, K.O.: Erfahrungen mit einem Lyotschen Hα Filter sowie vorläufige Ergebnisse über die Struktur der Chromosphäre. *Zeitschrift für Astrophysik* **42**, 209 (1957).
13.69 Athay, R.G.: IAU Symposium No. 56: *Chromospheric Fine Structure*, D. Reidel Publ. Co., Dordrecht 1974.
13.70 Athay, R.G.: *The Solar Chromosphere and Corona*. D. Reidel Publ. Co., Dordrecht 1976.
13.71 Bray, R.J.: High-Resolution Photography of the Solar Chromosphere. *Solar Physics* **4**, 318 (1968).
13.72 Bray, R.J., Loughhead, R.E.: *The Solar Chromosphere*, Chapman and Hall, London 1974.
13.73 Zerm, R.: Bestimmung der differentiellen Rotation der Sonne. *Astronomie und Raumfahrt* **22**, 127 (1984).
13.74 Balthasar, H., Vásquez, M., Wöhl, H.: Differential Rotation of Sunspot Groups in the Period 1874–1976. *Astronomy and Astrophysics* **155**, 87 (1986).
13.75 Hammerschmidt, S.: Schmetterlingsdiagramm. *Sonne* **11**, 11 (1987).
13.76 Yallop, B.-D., Hohenkerk, C.Y.: Solar Butterfly Diagramm 1874–1976. *Solar Physics* **68**, 304 (1980) and *Sonne* **5**, 96 (1981).
13.77 Jahn, C.H.: Eigenbewegungen einer Sonnenfleckengruppe. *Sterne u. Weltraum* **25**, 340 (1986).
13.78 Pfister, H.: Spezielle Eigenbewegungen in Sonnenfleckengruppen. *Astron. Mitt. Zürich* No. 342 (1975); and Pfister, H.: Klassifikationsschema für Eigenbewegungen. *Sterne u. Weltraum* **28**, 598 (1989).
13.79 Mehltretter, J.P.: On the Proper Motion of Small Pores in Sunspot Groups. *Solar Physics* **63**, 61 (1979).

13.80 Brauckhoff, D., Delfs, M., Stetter, H.: Polfackeln—Neuland für den Amateursonnenbeobachter. *Sonne* **10**, 114 (1986).
13.81 Balthasar, H., Stark, D., Wöhl, H.: The Solar Rotation Elements i and Ω Derived from Recurrent Single Sunspots. *Astronomy and Astrophysics* **174**, 359 (1987).
13.82 Vogt, O.: Positionsbestimmung von Sonnenflecken. *Sterne und Weltraum* **16**, 58 (1977).
13.83 Treutner, H.: Sonnenpositionsfotografie. *Sonne* **1**, 141 (1977).
13.84 Joppich, H.: Graphische Positionsbestimmung nach K. Silber. *Sonne* **10**, 120 (1986).
13.85 Junker, E.: Drei Wege zur richtigen Sonnenfleckenposition. *Sonne* **12**, 87 (1988).
13.86 Beck, R.: Sonnenfotografie mit kleinen Refraktoren. *Sterne und Weltraum* **15**, 252 (1976).
13.87 Beck, R., Paech, W., Remmert, E.: Seminar über Sonnenfotografie. *Sonne* **3**, 138 (1979) and **4**, 93 (1980).
13.88 Fritz, U., Treutner, H., Vogt, O.: Positionsbestimmung von Sonnenflecken. *Orion* **33**, 38 (1975).
13.89 Hornung, H., Hückel, P.: Sonnen-, Mond- und Planetenfotografie mit Amateurteleskopen. *Sterne und Weltraum* **23**, 96 and 393 (1984).
13.90 Jahn, C.H.: Einführung in die Sonnenfotografie. In: *Einführung in die Himmelsfotografie*, VdS-Fachgruppe Astrofotografie 1985.
13.91 Schröder, K.P.: *Einführung in die Astrofotografie*, Gesellschaft für volkstümliche Astronomie e.V., Hamburg.
13.92 Sherrod, P.C.: *A Complete Manual of Amateur Astronomy*, Prentice-Hall Inc., Englewood Cliffs 1981.
13.93 Remmert, E.: Sonnenfotografie und ihre Probleme. *Sterne und Weltraum* **24**, 158 and 606 (1985).
13.94 Taylor, P.O.: *Observing the Sun*, Cambridge University Press, Cambridge 1991.

14 Observations of Total Solar Eclipses

W. Petri

14.1 Photography of the Solar Corona

14.1.1 The Coronal Continuum

The finest offering of a total solar eclipse is the view it affords of the solar corona. From the scientific point of view, too, the corona is still the most important object of study during an eclipse. When the Moon shields the observer and the Earth's atmosphere surrounding him against the profusion of photospheric light, the outer wreath of solar rays appears in all its vast extent and subtlety, much richer than the best Earth-based coronagraph can ever show in the absence of an actual eclipse. The powerful spectrographs employed by large expeditions record multitudes of spectral lines whose analysis requires many months but provides clues as to the composition and characteristics of the outermost solar atmosphere.

The occasional observer with his much simpler equipment can record only the "white" corona, which is the continuous part of the coronal spectrum produced by light scattered by the photosphere. The scattering takes place on the free electrons in the coronal *plasma*, which is a highly ionized gas. Another part of the continuum spectrum is caused by diffraction of photospheric light by interplanetary particles. These are in reality the same kinds of particles which at larger angular distances from the Sun give rise (by reflection) to the *zodiacal light*, but the transition between diffraction and reflection is largely unexplored. Here, long-exposure photographs taken at large angular distances from the Sun during total eclipses and made possible by wide-angle optics and a very transparent sky will be most helpful.

14.1.2 The Structure of the Corona

The shape of the corona as seen on photographs depends essentially on the particular phase within the cycle of solar activity at the time of the eclipse. It varies between an extended, moustache-like corona that stretches along the equatorial plane at minimum, and a roundish one with pronounced rays extending all around and into space at activity maximum. On a smaller scale, the corona has a tenuous structure which is determined by the prevailing local active regions (sunspots, prominences, faculae, etc.) near the limb and by the electromagnetic fields associated with these phenomena. Thus, at each eclipse the corona displays individual traits which can be preserved in the photograph. The coronal continuum is essentially radially polarized, so photographs taken with polarization filters show the raylike structures most clearly if the filter has been oriented correctly. Exposures are therefore made with various filters at different position angles. Additional information can be obtained by color filters with suitably corresponding emulsions. Since the degree of polarization of the corona varies somewhat in the different spectral ranges, taking photographs with a combination of polarization *and* color filters is the best recommendation.

14.1.3 Processing

In order to obtain illustrative photographs of the brightness distribution in the corona, a series of enlargements with different exposure times is made on extra-hard copying paper. The coronal rays, which are so conspicuous to the eye, do not show clearly in an isophotic representation. A thorough photometric evaluation of coronal photographs requires not only much experience and skill, but also the facilities afforded by a large observatory. Those who intend to obtain scientifically useful material should at the start contact an experienced professional astronomer. Clouds and inclement weather can (and in fact do) sometimes deprive large eclipse expeditions of the reward for their efforts; subsequently, they will often turn to a less-well-equipped observer who has successfully obtained useful photographs while working at a different site.

In order to evaluate eclipse photographs scientifically, it is important that the negatives be calibrated. There are several options, but the simplest one is to record a step wedge onto the original negative in a location where it does not interfere with the image of the eclipse itself; the uneclipsed Sun at known zenith distance (to correct for extinction) serves as the calibration light source.

14.1.4 Photography with the Telescope

Any telescope can serve for sky photography. Since here long focal length is more important than light-gathering power, an astrographic refractor is generally preferred over reflectors. In order to have as large a field of view as possible, the observer should forego the use of eypiece magnification (which is by far the preference for planetary pictures) and use the primary focus only. Single-lens reflex cameras (SLRs) are very convenient; after removal of the objective, they are attached with a flange or gasket at the eyepiece collar. Caution is advised, however, because the solar image generates intense heat, and a strong neutral-density filter must be used when focusing!

Tracking can be controlled outside totality by means of a simple projection finder. It consists of a complex lens of about 1 m focal length (a spectacle lens) which is attached to the outside of the tube but near the objective. It casts next to the eyepiece the solar image, which must coincide with a previously drawn circle on the projection screen.

Large instruments are usually not equatorially mounted, but rather are set up in a fixed horizontal mount and illuminated with the Sun's light by a heliostat. For eclipse observations, the latter function can be achieved using a single plane mirror which is driven by clockwork around the polar axis in 48 hours. When positioning it, the observer must consider well ahead of time in which azimuth the camera is to be pointed, so that during eclipse the solar image will appear at the center of the field. The heliostat must be adjusted to the geographic latitude of the observing site. Since the relatively heavy mirror at the bottom of the tube will affect the clockwork unequally owing to its changing position, the rate should be regulated for the "eclipse position."

14.1.5 Exposure Times

The exposure times may vary widely—as they should! There is a tendency to overexpose photographs of the solar corona. Because of the very steep radial brightness gradient, a complete set of exposures, whose exposure times differ successively by a factor of 3 to 5, should be made. For example, even with a refractor of focal ratio $f/10$, one can start at 1/5 s. This yields the series 0.2–1–5–25–125 seconds. The long exposures are measured according to the beats of a metronome or with a stopwatch. The use of filters requires that factors to increase the exposure times be determined in advance and then allowed for.

14.1.6 Amateur Photographs

Ordinary cameras as well as special astrographic equipment can be used to photograph eclipses. Because of their high light-gathering power and the wide field of view, they can record the outermost corona. Tracking can be dispensed with unless extreme red photographs are intended. When working with several short-focus cameras, it is best to attach them to a common optical bench which may be screwed tightly at the proper tilt onto a heavy box. This will be more comfortable—and certainly much safer—during the haste and semidarkness than an array of several individual photographic tripods. It is important that transport of film be smooth and rapid. These photographs should also be calibrated by recording density marks onto a part of the unexposed and uncut film.

Color film has been used for many years in eclipse photography. In principle, slides are preferred over negative print film because they give truer colors. To reach high brightness contrasts, special emulsions are now available. As progress in the field of photographic chemistry has been rapid, the latest information on that subject should be consulted.

Photographs taken with movie cameras, on the other hand, are of little value except for demonstrations and/or entertainment purposes. It is imperative that the aperture be opened at the second contact (i.e., the beginning of totality). The use of time-lapse techniques will help to conserve film. Telephoto optics are needed but should not narrow the field of view so much that during totality the Sun moves out of the field.

14.2 Special Astronomical Programs

14.2.1 The Chromosphere

During totality, prominences at the solar limb extend out beyond the lunar disk and can be observed with the naked eye or, better still, with binoculars, as pinkish (owing to $H\alpha$ emission) features. They also appear on coronal photographs exposed short enough that the density is not excessive even at the solar limb. It is quite difficult, on the other hand, to photograph the chromosphere, as there are just a few available moments around the second and third contacts. Spectrographic equipment which can be used to obtain a useful spectrogram of the "flash spectrum" is usually beyond the means of the amateur astronomer. He/she may, however, try to see the flash phase with its dramatic line-reversal with a spectro*scope* (i.e. visually, using a prism or replica grating).

14.2.2 The Times of Contact

If the geographic position of the observing site is precisely known, then the recording of the times of contact is as crucial as it is when registering stellar occultations (Chap. 17). Of course, the first and fourth contacts—when the disks of the Moon and Sun just touch from outside—cannot be sharply determined even if their position angles are known from an almanac. At the second and third contacts the profile of the lunar limb becomes conspicuous in the shape of the "string of pearls" phenomenon, also known as "Bailey's beads." This occurs when the mountains at the lunar limb protrude beyond the solar disk; the thin solar crescent then breaks into various individual dots of light. The appearance is that of a diamond ring, the ring being formed by the distinct inner corona, and the diamonds consisting of photospheric light streaming through lunar valleys.

14.2.3 The Partial Phase

Instead of observing the partial eclipse phases directly in binoculars, which would be feasible only with very strong and optically flawless filters, it is usually preferable to project the image of the solar crescent through binoculars onto a screen. This helps to keep the observers abreast of the progress of the eclipse, and has the additional advantage of distracting the attention of the unavoidable onlookers away from the eclipse telescopes.

Photographic series of exposures of the partial phase are certainly attractive, but they have little or no scientific value. Only with a very large imaging scale could they be used to examine the *limb-darkening law* of the Sun. The facts are, first, that the scattered light in the Earth's atmosphere is already strongly diminished owing to a partial covering of the solar disk, and second, it can be photometrically determined most favorably at an eclipse because the lunar disk actually should appear completely dark (apart from the brightening by Earthlight, which is here entirely negligible). The scattered light from the lunar disk can thus be immediately determined. Since limb darkening strongly depends on the wavelengths of light for such photographs, one should use various filters.

14.2.4 The Star Field

If the total eclipse takes place in a sufficiently clear sky, the brightest stars will suddenly become visible. One should therefore find out in advance which stars may be seen and where; this also holds for the inner planets Mercury and Venus. With some luck, a comet may even be spotted. Such instances are not at all rare, and are favored by the fact that comets develop substantial tails when they near the Sun. (In the past some astronomers even used eclipses to search the sky near the Sun for a planet "Vulcan" which was suspected to have an orbit within that of Mercury.) It may also be possible to locate an artificial Earth satellite traveling outside the central path of the Moon's shadow.

A very special aspect of eclipse astronomy is the determination of the relativistic deflection of the path of light by the Sun's gravitational field, as predicted by Einstein's General Relativity. The instrumental requirements for testing this effect are exceedingly high, quite out of reach of the typical amateur observer. Nevertheless, the stars appearing on photographs of the corona can serve to determine the position angles of prominences and coronal rays.

14.3 Special Terrestrial Programs

14.3.1 Brightness and Color of the Sky

During a total solar eclipse, the eerie character of the twilight darkness that rapidly develops shortly before the second contact can often make a very strong impression on the observer. The Sun "sets," so to speak, in the middle of the sky, while the surrounding horizon is brightened by partial photospheric light. When prominence activity is strong, a reddish tint may be added. Color photographs—both of the sky and the surrounding landscape—can then be most impressive. The changing brightness of the sky can be measured with a photoelectric cell (any good exposure meter will do) attached to a rod placed about 6 ft or more above ground level and directed toward the zenith. If the meter has several sensitivity ranges, the observer should not forget to adjust it as time progresses. Soon after the third contact, the observer will feel subjectively that it is "day" again. When making exposures, the observer should not rely solely on this feeling, but should consult the exposure meter.

14.3.2 Flying Shadows

During the minutes near totality, when the Sun's disk has been reduced to a narrow crescent, there is a good chance of observing so-called "flying shadows" as narrow, dark streaks moving across the ground with the typical speed of a moving automobile. These are created when light from a narrow, slitlike light source (the crescent) passes through inhomogeneities in the Earth's atmosphere, and are best viewed at some distance from a plane surface of uniform, bright-gray tone.

Not to be confused with the flying shadows is the actual lunar shadow cone, which moves with the order of 1 km/s across the Earth's surface. Its approach is an exciting and awesome spectacle, especially when seen through thin clouds and from an elevated position.

14.3.3 Meteorological Observations

Nearly every eclipse expedition also collects some meteorological data, which, when combined with records of the preceding and following days, provides the basis for

Table 14.1. Solar Eclipses from 1988 to 2000 A.D. From Mucke and Meeus, *Canon of Solar Eclipses* −2003 *to* +2526 (1983).

Year	Date	Type[a]	Location[b]
1988	Mar. 18	t	in the Pacific
	Sep. 11	a	
1989	Mar. 07	p	
	Aug. 31	p	
1990	Jan. 26	a	
	Jul. 22	t	northern Russia and the Arctic
1991	Jan. 15	a	
	Jul. 11	t	Central America
1992	Jan. 04	a	
	Jun. 30	t	the South Atlantic, to Cape Town
	Dec. 24	p	
1993	May 21	p	
	Nov. 13	p	
1994	May 10	a	
	Nov. 03	t	South America
1995	Apr. 29	a	
	Oct. 24	t	southern Asia
1996	Apr. 17	p	
	Oct. 12	p	
1997	Mar. 09	t	the North Pacific
	Sep. 02	p	
1998	Feb. 26	t	Central America
	Aug. 22	a	
1999	Feb. 16	a	
	Aug. 11	t	middle Europe
2000	Feb. 05	p	

[a]Types are t = total, a = annular, p = partial.
[b]Location given only for total eclipses.

judging weather prospects of future expeditions to regions with similar climates. The rapid and dramatic reduction of solar radiation strongly affects not only the air temperature, but also wind, cloudiness, and humidity. A well-known phenomenon is the "eclipse wind," which sometimes at the very last moment literally blows away the veil of clouds blocking the Sun. It also increases the cooling as well as the demands on the stability of instruments. Any loose papers should be gathered up prior to the event, or they may be blown away! In order to follow both the air and radiation temperatures, an ordinary thermometer in the shade is compared with another one whose bulb has been blackened and which is exposed to direct solar radiation.

14.3.4 Biological Observations

Plants react to the "twilight" and "night" of the eclipse as they would during the normal decrease of daylight, but animals often react by either going to sleep or becoming frightened. More details on biological observations will not be given here, but it should be mentioned that the behavior of domesticated pets, particularly of dogs, may reflect the excitement of the observer more than that of the animal.

14.3.5 The Ionosphere

Finally, the close dependence of the state of the ionosphere on the activity on the Sun is revealed during eclipses by various effects. Amateur radio astronomers will find a rich field of activity here, even if they are situated outside the zone of totality. The occultation of certain active regions on the Sun by the Moon alone suffices to cause effects which may substantially affect worldwide shortwave radio transmission.

14.4 The Observing Station

14.4.1 Devising the Program

It goes without saying that a proper solar eclipse expedition can be quite expensive, if only because of the long journey that is usually required. To use the seconds of totality to best advantage, the leaders of the expedition will usually have a prearranged program, which fixes the functions of every participant in their time sequence precisely. With this program, everything is rehearsed until all operations work correctly. A stopwatch can be used to optimize the procedure. At least one "dress rehearsal" should be performed in dark twilight. A prepared cassette tape may be helpful, because then a recorder can tell the observers the proper moment for the beginning of the operation and continually instruct them as to the time and action. Experience shows that the end of the eclipse (the third contact) often comes as a surprise. The spatial arrangement of the instruments is also important, particularly if the same observer has to operate different instruments, for instance taking short exposures with a miniature camera while keeping the refractor busy with long exposures.

If the intent is to insert an entire stack of plateholders, it is advisable to place, as a memory aid, a cardboard marker at those positions where another operation is due (e.g., a change of filter). One must be careful in the use of double plateholders to avoid getting "out of synch." This "foolproof" arrangement cannot be overdone!

Exposures that require a particularly dark sky should be arranged for the middle of totality. If no time reserve in the schedule is planned for unforeseen events (e.g., shutter jams; trigger not cocked!), then there will be time at the end of totality to add a few less important or repeated exposures whose loss can more easily be borne.

Of course, the expedition will keep a complete log book, which may be supplemented by photographs.

14.4.2 Site Selection

The choice of site is essentially prescribed by the path of the line of totality, as can be found in the astronomical almanacs. There are also special publications that appear usually several years before each eclipse and that can be subscribed to or consulted in observatory libraries. Accessibility and general weather prospects further narrow the choice of sites. However, it would be unfortunate if everyone were to converge on the theoretically best location. Only by uniformly distributing the expeditions will the risk of being "wiped out" by inclement weather be diminished.

Many considerations will go into making the final choice, such as protection against the wind, location on the leeward side of a mountain, ease of guarding the station, and avoidance of any unusual and undesirable characteristic of the region, such as rainy climates, mosquitoes, and so on. One should plan on being as independent as possible of the standard electrical network. A homemade darkroom is very convenient, as the joint use of professional photographic laboratories rarely works out well. However, there might be in the vicinity a refrigerator with a little space for infrared photographic material; the films and plates can be inserted into their holders on the night before the eclipse.

14.4.3 Equipment

The specific items of eclipse equipment are chosen according to the circumstances. Flashlights should always be on hand, although it will rarely be so dark that the instruments cannot be operated without artificial illumination. The observer should choose suitable clothing to prepare for the expected cooling. The instruments should be protected by waterproof covers that are secured against wind; windscreens and radiation shields are also a good investment. The importance of filters has already been discussed.

If one intends to make star counts during totality, or to follow visually the outermost coronal rays, then he/she will want to wear adaptation spectacles such as are used in X-ray laboratories. The dark adaptation of the eye can also be improved by special prescription drugs. A timer (clock with time signals) or a theodolite are optional, a portable radio is convenient, and a first-aid box is a must.

References

14.1. Billings, D.E.: *A Guide to the Solar Corona*, New York 1966.
14.2. Cicco, Dennis, di: Photographing the Moon's Shadow. *Sky & Telescope* **53**, 323 (1977).
14.3. Espenack, F.: Isophotes of the Sun's Corona. *Sky & Telescope* **58**, 96 (1979).
14.4. Film the Eclipse. *Astronomy* **11**, 44 (1978).
14.5. Leavens, P.A.: Hints on Photographing the Eclipse. *Sky & Telescope* **6**, 358 (1972).
14.6. *Solar Eclipse Photography for the Amateur*. Kodak Publication No. AM-10.
14.7. Young, A.T.: The Problem of Shadow Band Observation. *Sky & Telescope* **43**, 291 (1972).
14.8. Young, A.T.: Shadow Band and the March Solar Eclipse. *Sky & Telescope* **39**, 176 (1970).
14.9. Zirker, J.B.: Total Solar Eclipses of the Sun. *Science* **210**, 1313 (1980).

The above are general references on solar eclipses. A general survey is given by H. Mucke and J. Meeus in *Canon of Solar Eclipses* −2003 *to* +2526, Astronomisches Büro Wien 1983. Still useful is also *Oppolzers Kanon der Finsternisse* (1887), republished by Dover Publ. (ed. by O. Gingerich) 1962. Extensive up-to-date information is contained in magazines such as *Sky & Telescope* and *Astronomy Now*, which are accessible to everyone.

15 The Moon

G.D. Roth

15.1 Problems and Ideas for Lunar Observations

15.1.1 The Moon as a Test Object for Telescopic Work

As the nearest celestial body, the Moon is justifiably a favorite target for amateur observers. Even without a telescope, the surface distinctly shows bright and dark spots. Binoculars reveal to the observer the richness of observable features which are displayed in larger telescopes in even more detail. These features provide an almost inexhaustible test field for observers and their instruments. Concerning instrumentation, see also C.J. Watkis in [15.5], pp 11, and Chap. 19, "Observations of the Planets," in this volume. Köhler [15.1] describes how even a 1-inch telescope at 40× shows so many features that no one can grasp them on the first attempt. And even uninitiated observers using moderate-quality optics can find something to be seen on the surface of the Moon. So, even when testing, one should always aim for the small and fine detail. This is most advisable when the observer has a definite program in mind (Price [15.18]).

15.1.2 Previous Studies and Space Missions

The Moon is the first heavenly body upon which humans have landed (Neil Armstrong and Edwin Aldrin with *Apollo 11* on 1969 July 20). For over 30 years now, the Moon has been explored by space missions, the first close-up study having taken place in 1959 with the USSR's *Luna 3* probe. Apart from the direct inspection of the lunar surface by humans, there are also the rock samples, radiation measurements, and photographs at the site itself, which have not only supplied new materials for research, but gave the term "lunar exploration" new meaning. Schaifers and Traving [15.2] note that within a few years, the *Ranger*, *Surveyor*, *Luna*, and *Apollo* missions proceeded to explore the Moon so as to render it now a domain of geoscience (morphography, geology, petrography, mineralogy). Numerous books and magazine articles report on this fact, and the amateur observer interested in the Moon should consult them [15.3,4].

The observer is encouraged to study the lunar surface, not because of the hope of immediate scientific reward, but rather to gain personal experience during and after work on an observational project. Systematic series of observations, such as monitoring possible changes over parts of the lunar surface, can also bring about scientifically useful results [15.5,6,7]. To better one's own understanding, the serious observer should try to know and understand methods of lunar geology. According to Guest and Greeley [15.8], "To the year 1960, the bulk of work on the Moon executed according to purely geological principles, was not large. But since then, geologists have shown that such principles, established during a century of research on Earth, can readily be applied to the Moon. "

The status of scientific knowledge about the Moon as achieved in the frame of the *Apollo* program is important today for the better understanding of solar–terrestrial relations, and is indispensible for the serious observer [15.9]. The lunar surface is exposed to a constant bombardment by high-energy ultraviolet and X-rays, and to the solar corpuscular radiation (chiefly electrons and protons). According to Z. Kopal [15.50], "The Sun represents the hot cathode and the Moon the anode in an ion tube of cosmical dimensions and whose glass walls are formed by Earth's atmosphere." A more detailed knowledge of the structure of the lunar surface was also of value for the early exploration of the solar system, particularly in the study of crater-shaped surface structures on the planets Mercury, Venus, and Mars. Finally, the origin of the Moon can be understood only in the context of the origin and structure of the entire planetary system.

In spite of this, lunar observations are still attractive to the amateur, even after the "*Apollo* shock." There are so many objects on its surface that "only amateurs will have time to monitor the Moon accordingly" [15.10] (cf. also [15.5]). One may even succeed in photographing the short flash when a meteorite strikes the surface [15.11]. Caution is advised as a blink of sunlight reflected from the surface of an artificial satellite may mimic a flash [15.54].

15.1.2.1 Representation in Maps. In topographic studies, the highest perfection is found in the preparation of special maps of certain lunar regions and in overall visual representations of the entire visible surface. The latter has reached, in the work of amateur observers Philipp Fauth and H.P. Wilkins, a height of cartographic artistry which will not easily be surpassed [15.12,13,14]. It will not be necessary to surpass them, however, because excellent-quality photographic maps of the entire lunar surface have meanwhile become available. A particularly important work is the *Photographic Lunar Atlas* by G.P. Kuiper [15.15]. A huge aggregate of lunar maps based on visual and photographic observations was compiled in the U.S. and in the U.S.S.R. in preparation for the space missions to the Moon [15.16].

For amateur use, various atlases have been published, partly from earlier observers, also based on visual and photographic observations. The survey and orientation maps in particular still have significance for the amateur observer. As an exemplary work, the *Fotographische Mondatlas* by W. Schwinger (1983) should be mentioned: "The diameter of the individual lunar pictures is on the average 19 cm. Enlarged partials of the north and south hemispheres when put together yield a diameter of around 40 cm. Instruments used were the refractor (100/1600 mm) and a Cassegrain (200/1000/3000 mm). The photographs were made at the focus and by the eyepiece

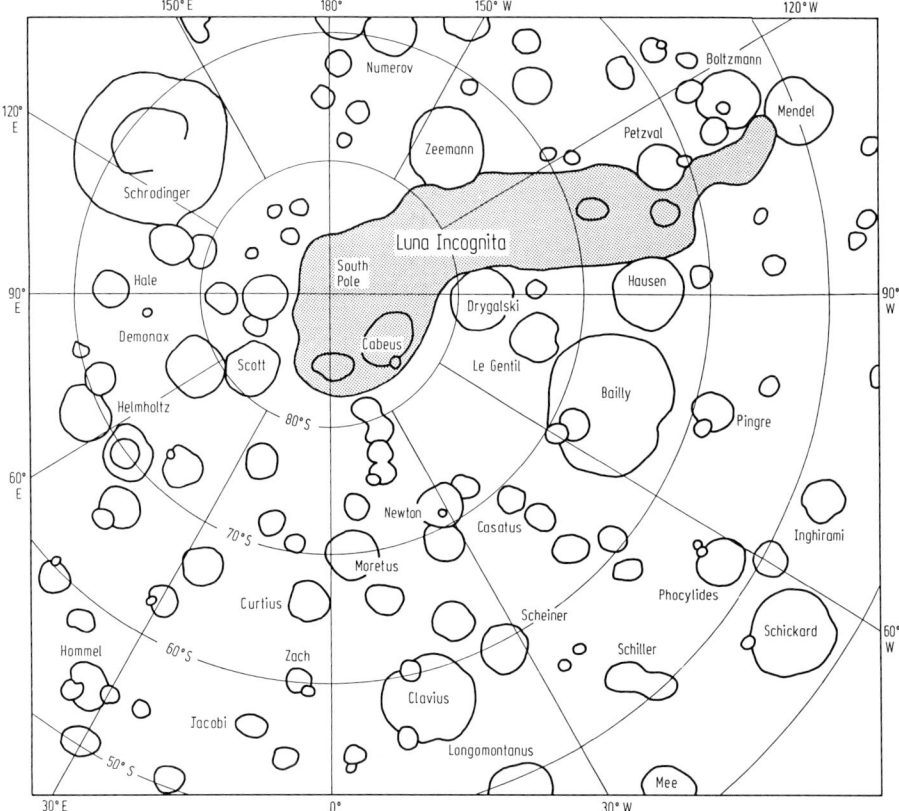

Fig. 15.1. Survey map of the region "Luna Incognita" (*shaded*). After J.E. Westfall [15.32].

projection method. The individual exposure times ranged from 2 to 1/250 s." [15.17]

For every charting, good photographs should supplement the visual results. Each chart, even when showing only a very tiny lunar region, should have a scale and should be ordered into a degree grid. Every observer who systematically monitors a certain lunar region should attempt to compile his sketches into a chart. This is needed to help understand more deeply the two primary events on the Moon: impacts and vulcanism [15.8].

Observational tasks whose results suggest the preparation of a special chart are:

- observing fine structures in the maria;
- observing a ray system;
- observing a region at the lunar limb, and discussing libration;
- observing a lunar rille and its accurate shape;
- observing the distribution of small craters in the vicinity of a walled plane or mountain ring;
- observing yet-unmapped regions (luna incognita) [15.32].

Such work will be based primarily on visual observations, but these may be usefully supplemented by control photographs if suitable equipment is available [15.56,57].

15.1.2.2 Critical Lunar Topography. The Moon shows numerous formations whose real shape and extension are inadequately known or which are suspected to show apparent or real changes. It is now known that the Moon's surface is everywhere covered with a fine layer of dust and debris 1 to 20 m thick, and every fresh meteorite impact smashes rock and reshuffles the lunar soil. Thus changes of some formations are conceivable.

Critical lunar topography therefore includes the comparison of recent with older observations in order to detect possible changes. This may include the use of photographs. In this context the craters Plato, Aristarchus, and Alphonsus are especially interesting. The occurrence of dark spots and luminescence phenomena should also be considered; see Sect. 15.1.2.3.

Yet critical lunar topography consists not merely in demonstrating changes, but rather in most cases simply in describing each object more accurately; it requires intense observing and solid knowledge of previously published series of observations. The effects of the light and shadow that are prevalent especially on the Moon may play tricks with the observer's eyes, causing some surface formations to appear, when viewed under certain angles of illumination, different from how they actually are. Of course, the quality of the optics may contribute their own problems when one works near the limit of their resolution. An observer should not be discouraged from working with 2- or 3-inch apertures, although these instruments are in the final analysis quite limited when it comes to perceiving fine and hyperfine detail. In fact, an instrument of 6-inch aperture is the lower limit for critical topographic observations.

New detailed information from such topographic observations is also the foundation by which special maps are made; see Sect. 15.1.2.1.

15.1.2.3 Observations Relating to Physical Conditions. These include work on the following surface properties [15.5,18,19]:

– Brightness changes;
– Color changes;
– Polarization of the light scattered by the surface;
– Occurrence of luminescent phenomena (transient lunar phenomena, abbreviated TLP or LTP, also called moonblinks).

Luminescence features are caused by gases emanating from the interior and triggered by solar radiation (photoluminescence). The observer will have various perceptions, as J. Classen [15.20] says: "The emanating gases veil the view of the surface by absorption or scattering, and give the observer the impression of a darkening or a 'gray cloud.' On the other hand, the gases may display luminescence. Colored spots will be seen on the illuminated part of the surface. When observed in the 'ashen' moonlight, whitish patches are perceived as the foveal color vision is then replaced by the quite sensitive, but uncolored, extrafoveal vision."

TLPs have been observed in the following craters: Alphonsus, Aristarchus, Atlas, Censorinus, Copernicus, Eratosthenes, Eudoxus, Gassendi, Grimaldi, Herodotus, Ke-

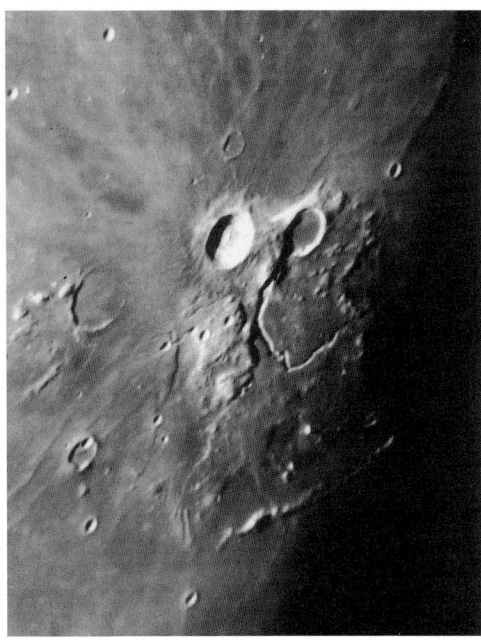

Fig. 15.2. Aristarchus and Schröter Valley—one of the suspected TLP regions. Photograph taken on 1980 November 19 20^h55^m UT, Cotonou (Benin) by J.Dragesco with a Celestron C14, $f/d = 50$, exposed 2 s on TP2415, developed in Ilford Microphen.

pler, Lichtenberg, Linné, Menelaus, Peirce, Picard, Pickering–Messier A, Piton, Plato, Posidonius, Procius, Ptolemaeus, Theophilus, Tycho.

In some craters (e.g., Alphonsus and Atlas), dark spots can be observed which show changes in intensity. Small craters are observed in their centers. Dark spots can be monitored using even a small telescope. Like TLPs, they provide the opportunity to collect data on changes on the lunar surface, which is helpful when interpreting the origins of formations.

J. Classen reported dark colorations of the soil to the southwest and west of the crater Copernicus in Sinus Aestum, and north of the Hyginus rille in Mare Vaporum [15.21]. These color changes have diameters of 1 arcminute each and are best observed with a small binocular telescope (magnification $\sim 10\times$). The centers of these colorations are given by Classen in the following table.

Table 15.1. Centers of dark coloration on the lunar surface. From Classen [15.21].

Longitude	Latitude
$-14°$ W	$+ 6°$ N
$- 8°$ W	$+ 6°$ N
$- 5°$ W	$+12°$ N
$+ 6°$ E	$+11°$ N

Overall, the objects dealt with here are of interest because of their photometric behavior. In addition to visual observations, photographic studies offer many opportunities, which are increased when color and polarization filters are used to take the photographs. H.J. Gramatzki pointed out 50 years ago that it would be worthwhile photographing the topographically less interesting regions in four spectral ranges (infrared, yellow–green, blue, ultraviolet) at different phases. This project would be within the reach of a 4-inch reflector [15.22]. Much has been published in recent years on TLPs, dark spots, colorations, and similar phenomena [15.23,24]. The British Astronomical Association in their *Guide to Observing the Moon* [15.5] recommends the use of a sequential color glass equipped with red, blue, and neutral filters, which is to be inserted between Barlow lens and eyepiece. This is known as a *crater extinction device* (CED).

Moonblinks also include the bright flashes originating at the site of an impact of a meteorite on the lunar surface and which are observable from Earth [15.11,54,55].

Persons interested in an observing project should contact an amateur observatory or working group in their area. To find the location of the nearest such group, readers in the U.K. should refer to the *Handbook for Astronomical Societies*, while those in North America should consult the *1989 Directory of Astronomy Clubs* which was published in the May 1989 issue of *Astronomy* magazine. See also Sect. 12.3.8 in Vol. 1.

15.1.2.4 Observations of the Dark Side of the Moon. Luminescence phenomenon are also observed on the night side of the Moon. A good 5-inch telescope and sufficient magnification (around 150×) will enable the experienced observer to do the job [15.26]. In this context, the monitoring of the "secondary light" (*ashen light*) is advised. It can be observed up to 6 days of lunar age (sometimes longer) and again after 22 days. W.M. Tschernow states that brightness fluctuations of an amplitude of about 0.6 mag can be found with maximum brightness in March to May and minimum in June to August. Also, the mean annual brightness varies by about 0.8 mag, which can be ascribed to, among other things, dimming and turbidity in the Earth's atmosphere. Color changes (green tints) have also been noted. Various intensity scales have been recommended in the literature [15.26,27].

15.2 Conditions of Visibility

The orbit of the Moon about the Earth is elliptical and is also subject to large perturbations. The average distance between the centers of Earth and Moon is 384,402 km, but the actual distance varies between 356 410 and 406 740 km, which causes the apparent diameter of the Moon to vary between $33'30''$ and $29'22''$. Thus, whereas the apparent diameter of the Moon is about the same as that of the Sun, the variations in the former are larger. Figure 15.3 illustrates the difference between the apparent diameters at *perigee* and *apogee* (respectively the points nearest to and farthest from the Earth).

The Moon moves quite rapidly against the backdrop of stars, describing an arc of about 13° per day from west to east. Observers should be aware that the apparent

Fig. 15.3. Change in apparent lunar diameter owing to the changing distance from the Earth. From W. Schwinge, *Fotografischer Mondatlas*.

motion does not coincide with the geocentric one given in almanacs, because it depends significantly upon the observer's specific location on Earth (i.e., on the position of the Moon relative to him/her). As astronomers who study the occultations of stars well know, an occultation event differs with respect to time, duration, and orientation for every place on Earth. Lunar observations are similarly affected: the instantaneous apparent semidiameter must be calculated whenever specific observations are made; for example, measuring heights of features at the lunar limb requires a measure of the lunar semidiameter in order to convert them to meters; cf. also [15.38] and [15.50].

The position of the Moon can be referred to the place of observation (the *topocenter*) by a formalism found, for instance, in Graff [15.28]: let a coordinate system be placed at the Earth's center with one axis directed along the Earth's rotation axis

($+z$), one in the direction to the vernal equinox ($+x$), and one in the equator at right ascension 90° ($+y$), and let a parallel coordinate system be centered upon the place of observation. In the latter, the so-called *topocentric lunar coordinates* can be found from the geocentric ones as follows:

$$r' \cos \delta' \cos \alpha' = r \cos \delta \cos \alpha - \rho \cos \varphi' \cos t,$$
$$r' \cos \delta' \sin \alpha' = r \cos \delta \sin \alpha - \rho \cos \varphi' \sin t, \qquad (15.1)$$
$$r' \sin \delta' = r \sin \delta - \rho \sin \varphi'.$$

Here, r is the geocentric distance of the Moon (in units of the Earth's equatorial radius) and α, δ are the geocentric celestial coordinates of the Moon, all found in the *Astronomical Almanac*. Also, r' and α', δ' are the corresponding topocentric coordinates of the Moon, ρ the distance of the observer from the Earth's center, φ' the geocentric latitude of the observer, and t the local sidereal time = hour angle of the vernal equinox.

The parallax π of the Moon, also listed in the *Almanac*, is related to r via the relation

$$r = \frac{1}{\sin \pi}. \qquad (15.2)$$

The local constants ρ and φ' can be found from Chap. 4. Thus, the right-hand sides of Eqs. (15.1) are known, and their division yields $\tan \alpha'$, $\tan \delta'$, and r'.

Thus, α', δ' give the apparent positon of the Moon at the observing site, and the increased topocentric semidiameter s' can easily be found from the geocentric s:

$$s' = s \frac{r}{r'}. \qquad (15.3)$$

15.2.1 The Phases of the Moon

The visibility of lunar formations depends on their illumination by the Sun. The lunar phases are well-known under the names *new moon, first quarter, full moon,* and *last quarter.* The phase refers to the illuminated arc of the visible lunar hemisphere. An approximate angle g for the phase is given by

$$g = \lambda - \lambda_\odot, \qquad (15.4)$$

where the ecliptic longitudes of the Moon and Sun are λ and $\lambda_\odot$, respectively. Thus, for instance, $g = 0°$ corresponds to new moon, and $g = 180°$ to full moon.

More often, the lunar phases are expressed by the time elapsed since new moon, called the *age of the Moon*. For instance, an age of about $7\frac{1}{2}$ days corresponds to first quarter.

In order to determine the lunar phases for given dates in past and future years, one can refer to special tables. The well known tables by J. Meeus use the fact that 251 *synodic* months equal approximately 269 *anomalistic* months: $7412\overset{d}{.}1776$ and $7412\overset{d}{.}1741$, respectively. If a full moon takes place at perigee, then it will be in that same position again after 251 synodic months. More precisely, if the difference between the full moon (opposition) and perigee is exactly $0°0'$ at the beginning, then it will be $0°2\overset{m}{.}76$ at the end of the 251-month cycle. The deviations between actual and mean lunar phases are attributable to the eccentricity of the lunar orbit, and repeat in each cycle at the same positions with only very small shifts. Simplified tables have been published by P. Ahnert [15.29] in his *Astronomisch-chronologische Tafeln für Sonne, Mond, und Planeten*.

15.2.2 The Terminator

Of particular interest to the observer is the position of the *terminator*, which forms the boundary of the illuminated portion of the Moon. Waxing and waning phases taken together, the terminator passes through every object on the surface 25 times per year, corresponding to at least 25 periods of changing illumination angles. Many formations can be observed only when at the terminator. The recording of the position of the terminator is an essential part of any observation of the Moon. It is defined as the selenographic longitude at which, during the instant of observation, the Sun rises (waxing moon) or sets (waning moon), and is tabulated in almanacs and calendars for each day of the year.

The western or eastern longitude L of the terminator can be graphed in the rectangular (ξ, η) coordinates of Fig. 15.9. At the western (+) or eastern (−) terminator longitudes L,

$$\xi = \cos b \cos L, \tag{15.5}$$

where b is the selenographic latitude of the parallel observed.

The terminator can also be defined by the colongitude of the Sun at the center of the Moon, which is the complement of the selenographic longitude of the terminator at the lunar equator. It is found in various almanacs (e.g., in the *Astronomical Almanac* or *Handbook of the B.A.A.*) under the heading "Sun's Selenographic Colongitude." It equals approximately 270° at new moon, 0° at first quarter, 90° at full moon, and 180° at last quarter.

15.2.3 Libration

Although the Moon always keeps the same face toward the Earth, up to 60% of the entire lunar surface can be observed. The cause is the *libration* phenomenon, which is composed of three effects: (1) the *libration in longitude* (maximally ±7°.9), which results from the ellipticity of the lunar orbit; (2) *libration in latitude* (maximally ±6°.9), caused by the tilt of the Moon's rotation axis with respect to the perpendicular to the plane of the orbit (one can sometimes see past the north pole and sometimes past the south pole of the Moon); and (3) the *parallactic libration* (maximally ±1°), caused by the fact that an observer on the surface of the rotating Earth views the Moon from changing directions and always from a point which is spatially separated from the geocenter. These three effects may add up to a shift of the lunar limb by as much as 10° and which is particularly pronounced at position angles 45°, 135°, 225°, and 315° of the limb; see Fig. 15.5. However, the foreshortening of the perspective impedes observations at the limb and in the libration zones.

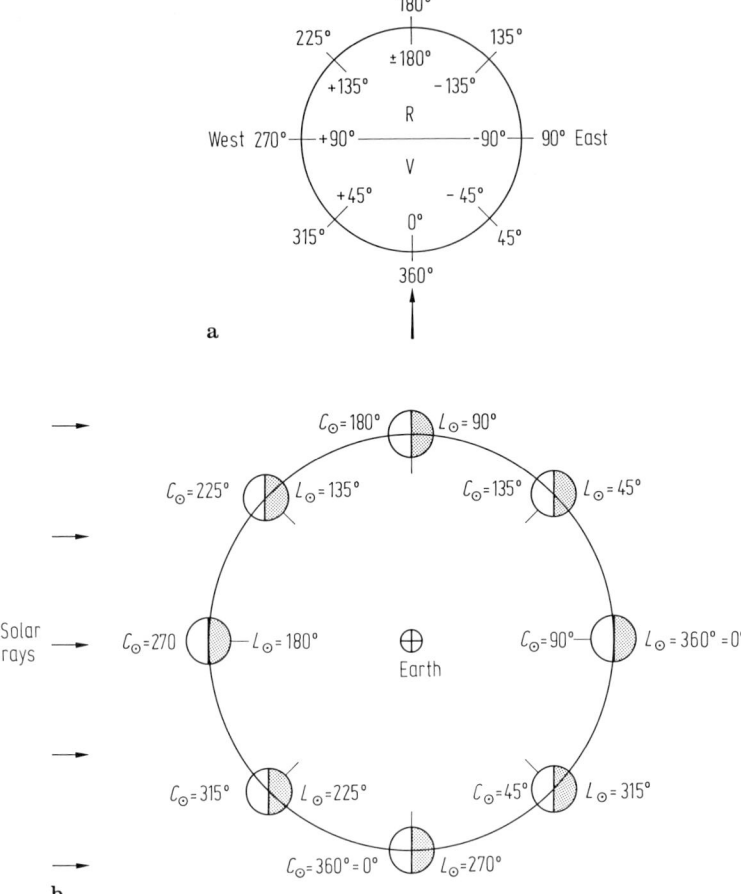

Fig. 15.4a, b. The colongitude. **a** View of the Moon in an astronomical telescope; V = frontside, R = rear side, inner circle = conventional count of longitudes, outer circle = 0°–360° count. **b** Viewing the lunar orbit from North, 8 positions of the Moon are shown with selenographic longitude 0° marked with a bar toward Earth. The practically parallel solar rays are represented by the arrows on the left. The Sun is almost 400 times further away than the Moon, its diameter nearly twice that of the lunar orbit. The graphed direction to the Sun ($L_\odot$) coincides at full moon with the longitude 0° of the selenographic coordinates. The selenographic longitude of the Sun $L_\odot$ and corresponding colongitude $C_\odot$ are shown for each position. From P. Ahnert, *Kalender für Sternfreunde*.

Astronomical almanacs and some amateur yearbooks (e.g., Ahnert's *Kalender für Sternfreunde*) list for every day the selenocentric longitude and latitude of the Earth, which are also the longitude and latitude of the center of the lunar disk as seen from the center of the Earth. If the latitude of the center point is positive, the observer sees more from the northern limb, and if negative, more from the southern limb; similarly, when its longitude is positive, he sees more of the western limb, and if negative, more

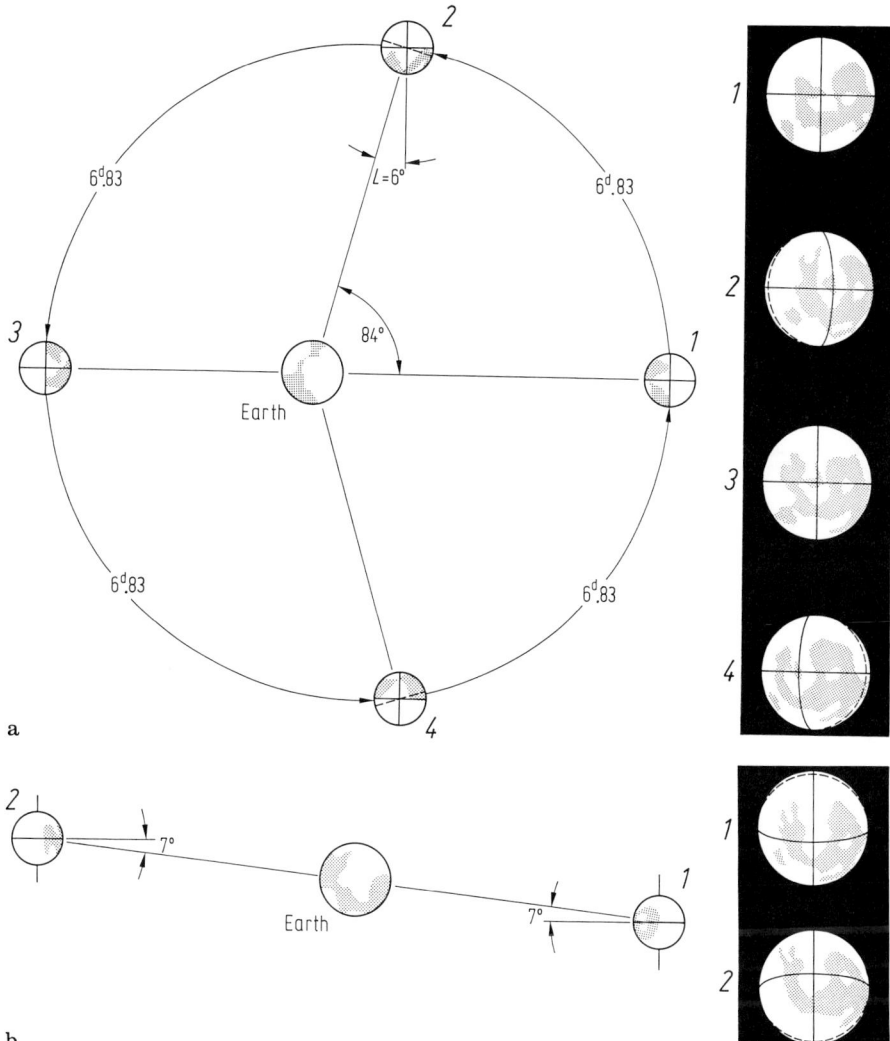

Fig. 15.5. a The rotation of the Moon about its axis is very regular, performing in $27.32 \div 4 = 6.83$ days a rotation of $90°$. The motion of the Moon around the Earth is nonuniform, owing to the elliptical nature of the orbit. At the orbital point nearest to the Earth (*perigee*), the Moon moves fastest, and at the most distant point (*apogee*) slowest. Position 1 is the apogee. 6.83 days later (position 2), the Moon has rotated $90°$ but moved $84°$ with respect to the Earth, thus making visible a small portion of the far side of the Moon. This is called *libration in longitude*. As the eccentricity of the lunar orbit is subject to perturbations, the maximum longitudinal libration varies. **b** The rotation axis of the Moon is not perpendicular to the orbit, which results in a libration in latitude. Sometimes more of the north polar region (position 1), sometimes more of the south polar region (position 2), can be seen. From B. Koch, *Sternenführer*, Treugesell-Verlag, Düsseldorf 1987.

Table 15.2. Relations between selenographic longitude of the Sun, colongitude, and terminator. The + and − signs distinguish East and West longitude (sunrise and sunset terminator), respectively.

Lunar Phase	$L_\odot$	$C_\odot = L_\odot + 90°$	Terminator
New Moon to First Quarter	180° to 270° e.g. 200°	270° to 360° then $C_\odot = 290°$	$360° - 290° = +70°$
First Quarter to Full Moon	270° to 360° e.g. 295°	0° to 90° then $C_\odot = 25°$	$0° - 25° = -25°$
Full Moon to Last Quarter	0° to 90° e.g. 60°	90° to 180° then $C_\odot = 150°$	$180° - 150° = +30°$
Last Quarter to New Moon	90° to 180° e.g. 120°	180° to 270° then $C_\odot = 210°$	$180° - 210° = -30°$

of the eastern limb. C. Albrecht has published photographs of libration regions and has shown that 4- to 6-inch refractors are well-suited for such tasks; see Figs. 15.15–17.

One region which is much affected by libration is the "Luna Incognita" (Fig. 15.1), whose cartographic processing was begun in 1972 by the Association of Lunar and Planetary Observers (ALPO) in the U.S. [15.32]. This area was not photographed during the space missions to the Moon, and thus it provides amateur observers with the opportunity to contribute to the completion of the lunar map.

15.2.4 The Lunar Coordinate Grid

An observer's view of the Moon closely matches the orthographic projection, which is a parallel projection centered on the intersection of the planes of lunar equator and zero meridian. Latitude parallels are projected as line segments parallel to the equator, dividing the circular limb meridian into equal parts; the zero meridian bisects the disk at right angles to the equator, and the meridians of other longitudes become ellipses with the zero meridian as the common major axis. The selenographic latitude b or β is the angular distance of, say, a crater from the lunar equator; north is *up* for a naked-eye observer in the northern hemisphere, *down* in an inverting telescope. The selenographic longitude l or λ is the angle between the planes of the crater and the zero meridian. As for planetographic coordinates, its direction is now defined (by IAU decision in 1961) in the "astronautical" sense—a person on the Moon facing north has west to the left; for the terrestrial observer west is the side toward which the terminator moves but it is seen as east in the sky; cf. Fig. 15.7. (Older maps are often labeled differently.) When ± signs are used for longitudes, "+" denotes east.

Strict orthographic projection maps the surface details of the Moon (or of a planet) as if projected by a parallel beam onto a plane.

Draw a circle with a suitably chosen radius for the Moon, say 10 cm. Divide the circumference from the west point toward north and south in 10° intervals (Fig. 15.8), and the same from the

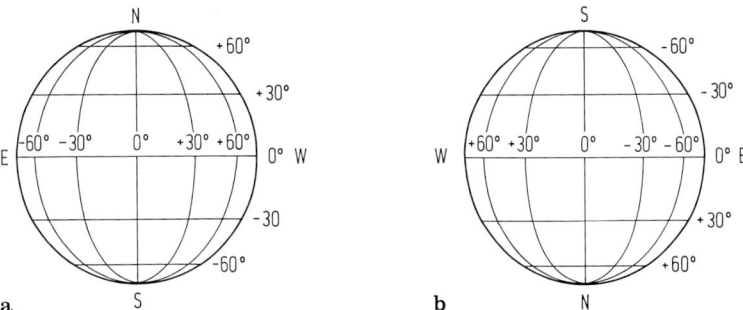

Fig. 15.6a, b. Lunar grid showing the four points of the compass: **a** with the naked eye; **b** with a telescope.

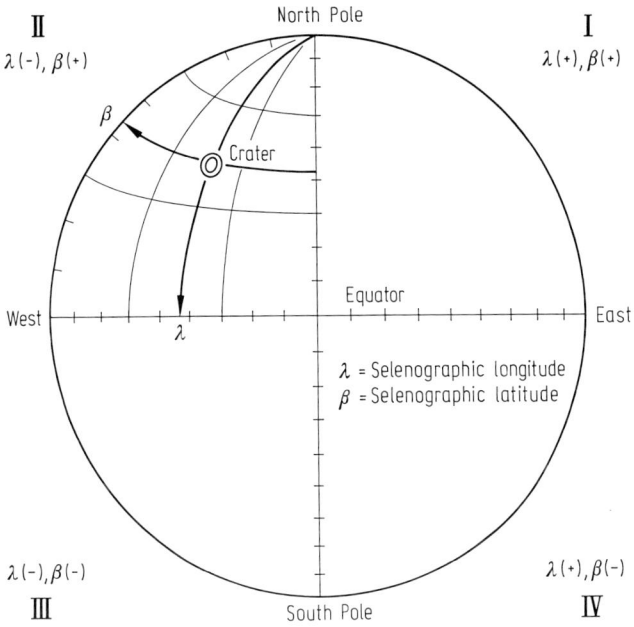

Fig. 15.7. Selenographic coordinates (east-west in the astronautical sense).

east point. P_1, P_2, P_3, etc. give the lines which represent the latitude circles. The meridians are circles projected into ellipses; they are graphed from the dashed lines at right angles off P_1, P_2, ..., also at 10° intervals. They have the semimajor axis $MS = MN$, and the semiminor axes MM_1, MM_2, etc. Graph them to latitude 80°, not all the way to the poles as their convergence would clutter the map.

The areas are unequal and not to scale. Note the closer spacing of meridians and latitude lines toward the limb.

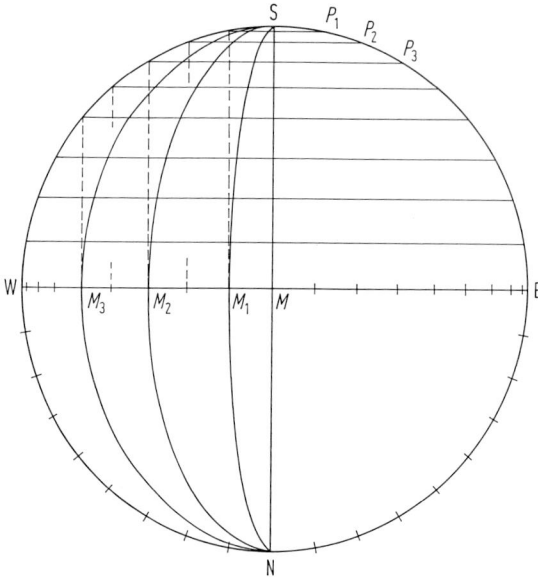

Fig. 15.8. Construction of a 10° grid.

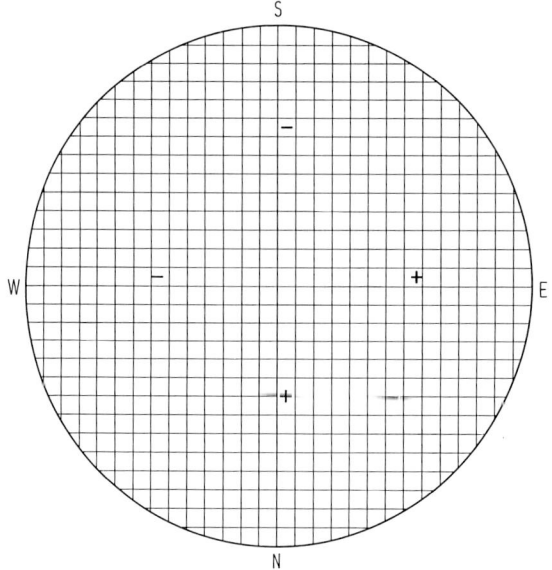

Fig. 15.9. Right-angled map grid for a lunar map.

While spherical coordinates are useful for general mappings of the entire surface, observers of smaller regions may prefer rectangular coordinates ξ, η which can be thought of as dividing the surface into numerous small squares (Fig. 15.9), oriented for zero libration and scaled for strict orthographic projection. The third spatial coordinate, the radial distance ζ of a point from the plane of projection, will also be needed, for instance, to determine positions of lunar surface formations.

The selenographic and the rectangular coordinates are related thus (lunar radius = 1):

$$\xi = \sin l \cos b, \qquad \eta = \sin b. \tag{15.6}$$

and

$$\zeta = \cos l \cos b, \tag{15.7}$$

Reliable coordinates are now known for many surface points, without significant errors from foreshortening even in near-limb regions. A list of coordinates for 15 fundamental crater positions, given in Appendix Table B.13 in Vol. 3, can serve as a basis for further position measuring and mapping; see also H.R. Hatfield "Directions and Coordinates" and K. Koziel "Libration of the Moon" in [15.5].

15.3 Lunar Formations

15.3.1 Maria

A *mare* (latin for "sea," plural *maria*) is a large, dark surface region which can be seen even without a telescope. About 40% of the side of the Moon that faces the Earth is occupied by maria. They are the essentially flat areas formed in the belt of maria in one contiguous system: Mare Imbrium, Mare Serenitatis, Mare Crisium.

The surfaces of maria are not without structure. Three kinds of objects will interest observers: (1) the *wrinkle ridges*, which are slightly wavy, damlike mounds or elevated regions on the mare surface; (2) the cone- or dish-shaped *craterlets* found in maria, which are interesting objects with respect to the evolution of the Moon; and (3) the *domes*, which are observed only at low illumination and which possess holes perhaps similar to volcanic calderas. The rock samples from maria returned during the Apollo space missions have proven to be of the *basaltic* type; apparently maria consist of large lava flows which have inundated low-lying regions primarily on the Earth-facing side of the Moon. The view of solidified lava streams is particularly evident in the Mare Imbrium. The streams show slight differences of tint, the older being more reddish, the younger bluish [15.33].

15.3.2 Terrae

Terrae, or highlands, refer to the topography of the brighter areas of the lunar surface which are, as a rule, more elevated than the maria. The terrae occupy more than one-half of the front and almost all of the far side of the Moon's surface. By contrast

Fig. 15.10. Some typical formations on the lunar surface. A: young crater Aristarchus, diameter 40 km, with rays; B: small, dish-shaped crater; C: crater Prinz, partially inundated by mare lava; D: mare surface; E: bent rille; F: straight rille; G: mare–ridge ; H: pre-mare material standing out from mare lavas. Some bent rilles follow older ditches and graben structures. Photo: NASA/ULO Planetary Image Center (J.E. Guest).

with the maria, they are mountainous regions which lie over a crust many kilometers thick.

Most *craters* are found in highlands. Craters are considered the outcome of huge bombardments by meteorites in whose course older craters have been hit and damaged or destroyed by more recent impacts. Thus, recent craters are often found within or near old craters.

15.3.2.1 Ring Structures

(a) *Walled plains* are craters whose ringed wall enclose a plane of diameter 50–200 km. Examples include Abulfeda, Archimedes, Clavius, Fra Mauro, Grimaldi, Maurolycus, Plato, and many others.
(b) *Ring mountains* are similar to the walled plains, but with higher walls and smaller extent of the lower enclosed plain. They are often found to contain central peaks. The best-known examples are Eratosthenes, Copernicus, Petavius, Theophilus, and Tycho.
(c) *Craters* in the narrow sense include all circular forms on the Moon which do not distinctly show any elevated features around them. Their diameters are smaller than those of walled plains and ring mountains. There are also the tiny minicraters and craterlets, which, according to lunar orbiter photographs, have diameters of 50 cm and even less (crater holes). Some groups of craters are arranged in string-of-pearls fashion, and in a small telescope can often be mistaken for rilles. Terraced inner slopes, the existence of one or several mountains, strata of ejecta, secondary craters, and bright rays in the vicinity make many craters interesting targets of observation. The visibility of the different structures, however, depends strongly on the state of illumination.

15.3.2.2 Mountain Chains. These are large mountainous ranges with substantial elevations, the best-known examples of which include the Alps (heights up to 4000 m) and the Apennines (up to 6000 m). Concerning the history of formation, however, they are not to be compared with mountain ranges on Earth. There are convincing reasons to suggest that the lunar structures have been "piled up" as a result of powerful impact events.

15.3.2.3 Linear Formations

(a) *Rilles* are indentations up to 1000 m wide and up to several hundred km long. Their floors often show many craterlets. Examples are the Ariadeus and Hyginus rilles.
(b) *Crevaces* are cracks in the lunar crust which cut into the surface up to several hundred meters deep.
(c) *Valleys* are narrow notches in mountain ranges. The most striking example is the Alpine Valley.
(d) Bright *rays* are particularly distinct at full moon when no shadows interfere with their visibility. They are luminous stripes, usually directed outward from a crater. The most conspicuous crater with rays is Tycho; other notable rayed craters include Kepler, Copernicus, and Olbers.

15.3.2.4 Unusual Formations

(a) *Ghost craters* appear only shadowy in the dark soil of a mare. These are craters which have been covered by lava flows. A particularly remarkable object is the half-inundated crater Guericke in the Mare Cognitum.

(b) *Central peaks* are conic mountains or systems thereof lying inside walled plains and mountain rings, often appearing similar in structure to shield volcanoes on Earth.

15.3.3 On the History of Lunar Nomenclature

The names of many lunar regions and formations originate largely from the early days of visual observations of the Moon. A detailed historical survey is given by Z. Kopal and R.W. Carder in [15.16]. Langrenus in the 17th century was the first to assign names to lunar formations, taking names from the Bible or religious stories about saints. His younger contemporary Johann Hevelius (1611–1687) used names of terrestrial landscapes, of which "Alps" and "Apennines" have remained. The most popular naming scheme in current use was introduced by Riccioli, who assigned names of known scientists and philosophers to craters and mountain rings. The names given to lunar maria show some astrological influence. The German observers Schröter (1745–1816), Beer (1797–1850), and Mädler (1794–1874) further refined the nomenclature of formations. Beer and Mädler were the first to use letter codes for craters. Under the auspices of the International Astronomical Union (IAU), an internationally accepted nomenclature was prepared in 1935 by Mary A. Black and Karl Müller [15.34]. The naming after personalities and after geological formations on Earth has been retained to the present, for instance when naming the recently discovered features of the far side of the Moon (Soviet Mountains, Lomonossow, Joliot-Curie, etc.).

The cartographic recording of the far side and, in recent years, very detailed mappings of the near side of the Moon (scales of 1:250 000 and larger) have created new demands on nomenclature. The large-size maps are divided into 144 regions, each of 16 subregions, each one of these being referred to by the name of a prominent local crater. This organization, however, is not needed for the maps used by amateur observers.

It is still customary to assign names to lunar formations. In the past decade, a number of small craters, formerly coded by letters, have been given names; for instance, Messier G became "Lindbergh." The old letter codes, however, are still in use.

Since 1976, additional names for structures other than craters have been used. For example: Mare Insularum = Sea of Islands ($b = 7°$ north, $l = 22°$ west) or Sinus Amoris = Bay of Love ($b = 7°$ north, $l = 38°$ east). Updates on the nomenclature are published in the triennial transactions of the IAU, which is—by endorsement of the UN and the International Council of Scientific Unions—in sole charge of naming astronomical objects or of changing names. The appearance of many small and very small lunar formations required also new generic names which are taken from the Latin; for instance: anguis = curved rille, catena = crater chain, dorsum = ridge, fossa = straight rille, mons = mountain, montes = mountain range, promontorium = cape, rima = rille, rupes = ditch, vallis = valley.

15.4 Observational Projects

15.4.1 Visual Observations

For visual work, long-focus refractors and reflectors ($f/10$ to $f/20$) have been used to advantage for optimal image definition. The brightness of the Moon sometimes dazzles the observer. Sensitive eyes will find help in neutral-density (gray) filters or by observing during twilight. Other observational tools include color and polarization filters and also—for instruments with 6-inch or larger apertures and clock drive—a filar micrometer.

The situation concerning the proper magnification to use is similar to that in planetary observations. The noted lunar observer Philipp Fauth wrote on this subject, "I work best on the perception of finest surface features at my 163-mm objective with magnification 160×, at the 176-mm apochromat with 176×, at the Medial at 300-mm aperture with 300× and 350×, and at 385-mm aperture with 350× and 430×. Younger observers with more sensitive eyes may wish to increase magnifications by 50% over these, image steadiness permitting. To go still further has little purpose; although the image grows larger, it also becomes paler and more diffuse, and on the Moon clarity is more important than image size" [15.35] (cf. also [15.5]).

Clarity is aided by binocular observations, which are particularly well suited for lunar observation and provide not only a lasting impression but also relaxed viewing. According to the author's experience with Baader binoculars (Munich), details on the Moon are recorded not only faster but also more safely and less ambiguously when using both eyes. G. Miller writes in a report: "It is confirmed that fine details were seen better and apparently larger when using both eyes in binoculars than by using only one of the two eyepieces." [15.37]

15.4.4.1 Drawings at the Telescope. According to Philipp Fauth, "A patient acquaintance with a limited region, taking every available opportunity, makes one so engrossed in its structure that ultimately one perceives every detail and records it on paper."

A prerequisite for any topographic work is the knowledge of the basic surface structure. Skeleton maps for orientation and drawings of the outlines (silhouette maps) can serve as a starting point. Both are obtained by tracing reliable lunar maps or, better still, photographs. Low-contrast copied and enlarged photographs can be used as templates for sketching. Every silhouette map should show exact markings of position so that all details can be entered with their true orientation. The scale to be chosen should not be too small. P. Fauth drew his map at a scale of 1:1 million, which corresponds to a lunar diameter of 3477 mm. This is an appropriate order of magnitude, which should, in the interests of clear presentation, be used as a guide. K.W. Abineri, who observed for many years with a 200-mm reflector, adds in instructions for observers [15.5] the advice: a relaxed posture (seated, if possible), a stable support for the drawing pad, an adjustable red lamp for illumination, and a soft pencil. See also [15.25, 33].

The technique of sketching depends on the ability of the observer. One distinguishes between simple line sketches and the more expressive full-tone drawings, in which

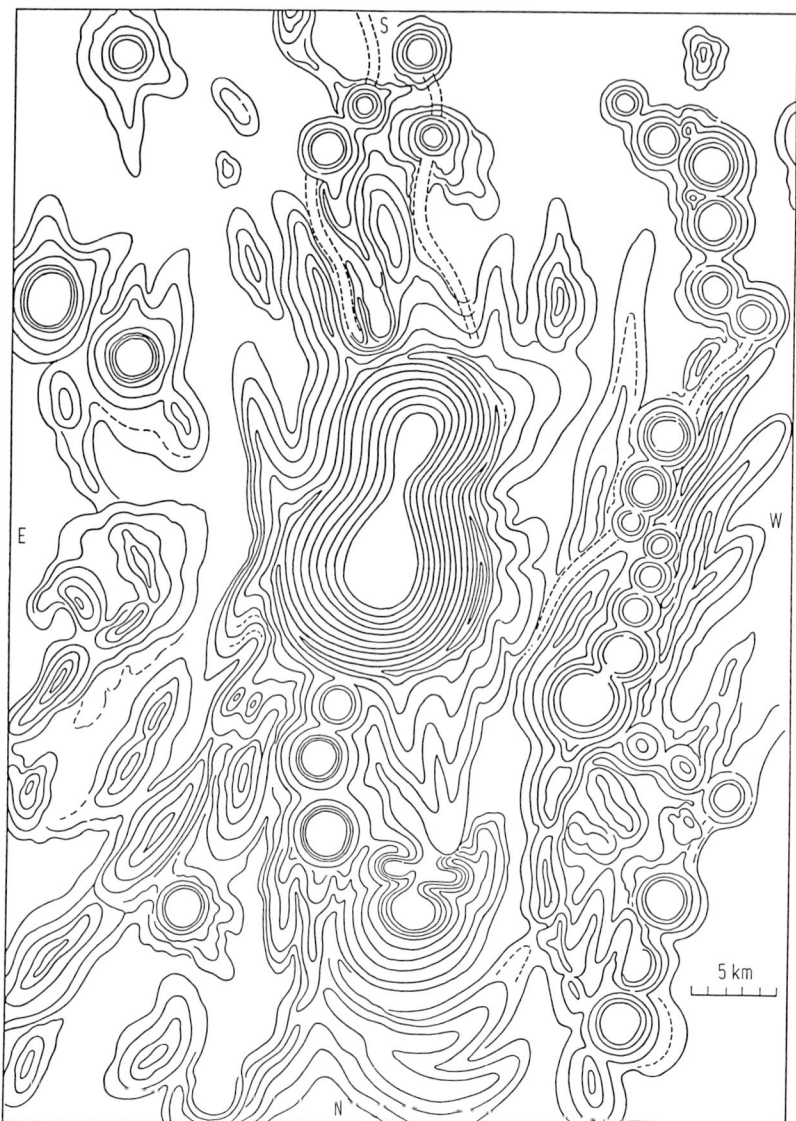

Fig. 15.11. Example of a topographic representation of the lunar landscape. The picture is from P. Fauth and represents the walled mountain named after him. From *Die Sterne* **33**, 158 (1957).

one tries to incorporate a range of light and dark densities into the picture. Fauth drew in contour lines, or *isohypses* (Fig. 15.11), and expressive hachure.

Figure 15.12 illustrates the transition of a simple line drawing into a tone drawing which clearly shows shadows and light intensities.

As an example of a plan for systematic observations, several points recommended by F. Billerbeck-Gentz [15.39] for the observation of lunar ray systems will be mentioned here. The observer should try to answer the following questions:

(a) Where does the ray begin? (b) What is its course? (c) Where does it end? (d) How wide is it? (e) How bright is it? (f) Does the brightness of the ray vary along its length and width? If so, at which points and when? (g) What is the general surface structure of the ray? Could the structure of the ground perhaps be simulated by a series of bright points, "white" craters, or small surface warps? (h) What is the shape of the ray? Is it straight or does it bend, and if so, where? Is it interrupted, and if so, by how much? (i) Do any parts of the ray show a change of color or tint at times when the ray itself is not visible? (k) At which times do the rays appear and disappear?

15.4.1.2 Measurements at the Telescope. Direct telescopic measurements make use of the micrometer to determine relative lunar dimensions (crater diameters, mountain heights) and positions. The method of deriving elevations on the Moon was already well known in the 18th century. When the length of the shadow is measured and the altitude of the Sun known, the height of a lunar formation relative to the surrounding terrain can be found. At low solar altitudes (i.e., near the terminator), differences in elevation are enlarged by a factor of 100. The accuracy reached is only 200 to 300 meters. These heights are called "relative" heights because they refer to the terrain onto which the end of the shadow falls. These are to be distinguished from the *absolute* heights referred to the radius of a chosen reference sphere. J. Hopmann chose the mean radius of the limb of the lunar hemisphere turned toward Earth, as it is free of irregularities of the limb profile and libration [15.44].

Most altitude data on lunar maps before 1960 are derived from shadow measures, and are relative. Better accuracy of relative heights is possible with photographs taken on space missions and evaluated according to photogrammetric principles. For instance, the *Lunar Orbiter* maps, which appeared after 1967, have relative heights with precisions of a few meters [15.16].

The measurement of the shadow lengths of certain lunar features in order to derive heights is a useful exercise for the observer. Shadows are always measured at right angles to the line joining the horns (or cusps). The fixed horizontal wires are oriented along the shadow, which also defines the mountain peak exactly. The precision obtained will depend on the length of the shadow; the measurement of shadows which are too short leads to unreliable results. The altitude of the Sun during such measurements should be in the range $5°-15°$. If it is less than $5°$, there is the danger that the very long shadow will merge with the shadows of other formations or run into the unilluminated part of the Moon.

J. Hopmann has determined with a 21-cm refractor (2.92 m focal length) 175 heights of 163 points on the lunar surface [15.42]. He reported as follows on the measuring technique: "To measure the shadow length, the fixed 'horizontal' wire was oriented along the position angle of the terminator from the ephemerides for physical observations of the Moon in the *Almanac*. Then it can be always clearly identified which part of a crater, ridge, etc. casts the shadow. Because of overexposure, these places appear so bright in photographic atlases (Paris, Kuiper) that identification of the shadow-casting point is not possible. For objects a few arcseconds wide, the eyepiece still shows the shadow peak, whereas on the photographic copy, and probably

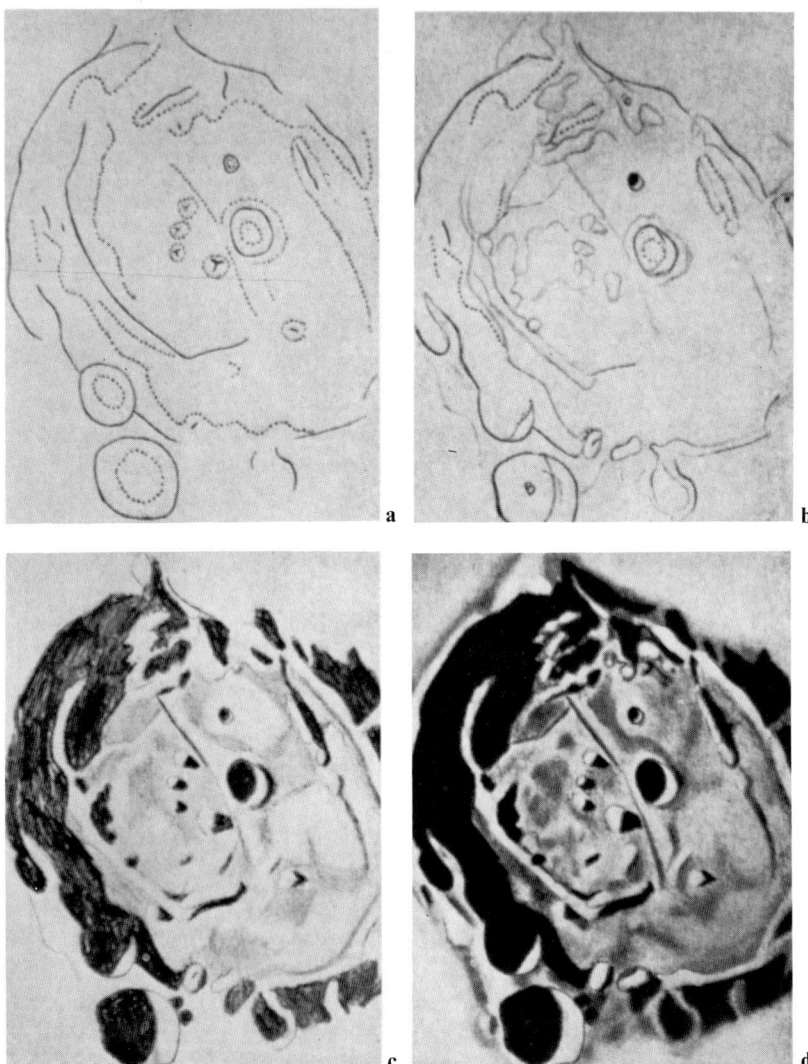

Fig. 15.12a–d. Drawing of the walled plain Posidonius by C.R. Chapman (ALPO): **a** shows the outline taken from a photograph, **b** suggests illumination and libration, **c** adds shadows, and **d** completes the contours and intensities. From: *The Strolling Astronomer* **17**, 47 (1963).

also on the original negative, the entire region over 2″ to 3″ appears white or dark. When observed at the same phase, the peak may be equally well marked visually and photographically, but then the micrometer measurement gives a more precise length of the shadow. The measuring procedure follows that of double-star observations: double distances are determined with 3 settings on either side. The time of observation at the midpoint is recorded to the nearest minute."

For the theory and numerical reduction, the papers by K. Graff [15.43], J. Hopmann [15.44], and G. Schrutka-Rechtenstamm [15.45] should be referred to. The latter author also reports on the determination of the absolute heights of lunar mountains in another publication [15.46]. See also "Determination of Altitudes on the Moon" in [15.5].

A practical method for amateurs is recommended by P. Ahnert [15.47] (see also Zimmerman [15.48]). This method makes use of some permissible simplifications. The least accurate of the measurements is that of the position of the terminator. Moreover, the shadows usually fall onto uneven terrain; consequently the observer obtains different values for the same height when, at different solar altitudes, the shadow peaks are projected onto different ground levels. Therefore, recording the solar altitude together with the measurement is quite important. The observer must measure both the length S of the shadow and the distance A of the particular mountain from the terminator. This can easily be accomplished using an eyepiece with a reticle. One then observes the time between the transits of the mountain peak, the end point of the shadow (shadow peak), and the terminator. This is best done with a stopwatch. The calculation then requires the following quantities, listed here with their symbols:

ρ: Radius of the Moon in arcseconds;
β: Selenographic latitude of the mountain;
L: Ecliptic longitude of the Sun;
l: Ecliptic longitude of the Moon;
b: Ecliptic latitude of the Moon;
P: Position angle of the lunar axis;
δ: Declination of the Moon;
E: Angle at the Earth in the plane triangle Sun–Earth–Moon;
η: Angle at the Sun in the plane triangle Sun–Earth–Moon;
M: Angle at the Moon in the plane triangle Sun–Earth–Moon;
ϑ: Angle between terminator and line joining the peaks of the horns;
ε: Angle between terminator and the meridian of the mountain;
φ: Altitude of the Sun above the mountain;
ψ: Selenocentric angle between the direction to the peak of the mountain and the shadow peak;
s: Shadow length as fraction of the Moon's radius;
h: Height of the mountain in meters.

The first seven quantities are obtained from a good lunar map and from the ephemerides in astronomical almanacs and calendars; S and A are measured at the telescope. The measured transit time is converted into seconds of arc via

$$S \text{ or } A = 14\rlap{.}''46 \frac{D \cos \delta}{\cos P}, \tag{15.8}$$

where D is the transit time in seconds of time. The next step is the determination of the angles E, η, M, and ϑ:

$$\begin{aligned}\cos E &= \cos b \cos(l - L), \\ M &= 180° - E - \eta, \\ \vartheta &= M - 90° \quad \text{for Moon less than half-illuminated,} \\ \vartheta &= 90° - M \quad \text{for Moon more than half-illuminated.}\end{aligned} \tag{15.9}$$

The angle η amounts to at most $9'$ (Earth–Moon distance as seen from the Sun). Assuming the orbits of Moon and Earth to be circular, the angle η can be tabulated as a function of E, as is shown in Table 15.3.

The unevenness of the lunar surface prevents an exact determination of the terminator, but the following approximations are permitted:

$$\begin{aligned}\vartheta &= 90° - (l - L) \quad \text{for Moon less than half-illuminated,} \\ \vartheta &= (l - L) - 90° \quad \text{for Moon more than half-illuminated.}\end{aligned} \tag{15.10}$$

Table 15.3. Angle η as a function of angle E, where η and E are, respectively, the angles at the Sun and Earth in the plane triangle Sun–Earth–Moon.

E	η	E
9°.7 to 16°.3	2′	163°.7 to 170°.3
16.4 to 23.2	3	156.8 to 163.6
23.3 to 30.5	4	149.5 to 156.7
30.6 to 38.3	5	141.7 to 149.4
38.4 to 47.2	6	132.8 to 141.6
47.3 to 57.9	7	122.1 to 132.7
58.0 to 73.9	8	106.1 to 122.0
74.0 to 90.0	9	90.0 to 106.0

Neglecting η and $\cos b$ in ϑ results in an error of at most 0°.2; this is usually smaller than the uncertainty in defining the terminator.

The position of the measured mountain is now graphed onto a map, and the approximate selenographic latitude determined. The calculation proceeds as follows:

$$\sin(\vartheta \pm \varepsilon) = \sin \vartheta \pm \frac{A}{\rho \cos \beta}, \tag{15.11}$$

where the positive sign holds when the Moon is less than half-illuminated, the negative sign when it is more than half-illuminated. The altitude of the Sun as seen from the top of the mountain is then determined thus:

$$\sin \vartheta = \sin \varepsilon \cos \beta. \tag{15.12}$$

The length of the mountain's shadow expressed as a fraction of the lunar radius becomes:

$$s = \frac{S}{\rho \cos \vartheta}. \tag{15.13}$$

All of these calculations are to be carried out to 4-digit accuracy. Only the last two formulae should be calculated to 6 digits:

$$\sin \psi = s \cos \varphi, \tag{15.14}$$

$$h = \left[\frac{\cos(\varphi - \psi)}{\cos \varphi} - 1 \right] \times 1.738 \times 10^6 \text{ m}. \tag{15.15}$$

15.4.1.3 Visual Photometry. Changes in the relative brightness of the Moon are distinctly seen in the natural course of the phases. However, the reflectivity varies over the different regions of the lunar surface. Therefore the change of brightness of the Moon does not occur quite symmetrically with the phases. Some recommendations for observers concerning the "integrated" photometry of the whole Moon have been published and are referred to in [15.48]; see also E.A. Whitaker "Lunar Photometric and Colorimetric Properties" in [15.5]. For advice regarding measurements to be taken during a lunar eclipse, see Chap. 16.

One task for visual estimates is to find the change in brightness of a certain formation, for instance the surface inside a crater, as a function of the position of the terminator. All estimates are carried out on a relative scale which is in principle similar to that of the step-estimation used for variable stars (see Chap. 8, Vol. 1). Of course, the objects with which the scale is calibrated must be absolutely invariable during the entire course of a lunation.

On a scale recommended by H.J. Klein and H.K. Kaiser, the darkness of the shadow is expressed by 0°, while 10° corresponds to the brightest point on the visible lunar surface. This gives the following other degrees of calibration:

0°.0 = shadow on the Moon;
1.0 = the darkest part of the inside of the walled mountains Grimaldi and Riccioli;
1.5 = inner surface of De Billy;
2.0 = inner surface of Endymion and J. Caesar;
2.5 = inner surface of Pitatus and Vetruvius;
3.0 = Sinus Iridum;
3.5 = inner surface of Archimedes and Mersinus;
4.0 = inner surface of Ptolemy and Guericke;
4.5 = Sinus Medii, surface around Aristyllus;
5.0 = surface around Archimedes, walls of Landsberg and Bullialdus;
5.5 = walls of Timocharus, rays of Copernicus;
6.0 = walls of Macrobius and Kant;
6.5 = walls of Langrenus and Theaetetus;
7.0 = Kepler;
7.5 = Ukert and Euclid;
8.0 = walls of Copernicus;
8.5 = walls of Proclus;
9.0 = Censorinus;
9.5 = interior of Aristarchus;
10.0 = central peak of Aristarchus.

The very similar "Elger's Albedo Scale" is also cited in the literature, and was refined some time ago by P. Hedervari [15.49]. Observers may find it worthwhile to verify the scale and amplify it by using other objects and intermediate steps.

The scale for visual estimates can be extended by linking the observed brightnesses to photographically determined photometric values of certain objects [15.50]. Aside from visual estimates, visual intensities can be measured by photometric means with the aid of neutral wedges and polarization filters which dim an artificial light source to match the brightness of a lunar formation (see Chap. 8, Vol. 1).

15.4.2 Photographic Observations

Reflectors as well as refractors are suitable for lunar photography. A telescope of 1 m focal length produces an image of the Moon of about 10 mm in diameter; the image is correspondingly reduced or enlarged for shorter or longer focal lengths.

The essentials of technique and of the selection of emulsions is discussed in Chap. 6, Vol. 1. As in the procedure for planetary photography (see Chap. 19), the eyepiece collar carries either a special Moon or planet camera or a single-lens reflex camera (Nikon, Olympus, Pentax, etc.), but minus the optics. For focal photography,

Fig. 15.13. Mare Humorum and the walled plain Gassendi. Photograph taken on 1981 October 09 18^h24^m UT, Cotonou (Benin), by J. Dragesco with a C14, $f/d = 60$, exposure 2 s on XP 400 film.

the eyepiece of the telescope is not used. If, however, it is desired to obtain an enlarged image of the Moon, then the eyepiece is used to project the primary image of the Moon onto the emulsion. The essential differences between the two methods can be summed up nicely as follows:

– *Focal photography*—gives a bright, but relatively small, image requiring a short exposure time and subsequent enlargement;
– *Projection photography*—yields a dimmer, but larger, image (again see Chap. 6), and hence requires longer exposure and less subsequent enlargement.

Excellent photographs have been published of late in various sources, for example J. Dragesco (see Fig. 15.13), G. Viscardy [15.52], B. Flach-Wilken [15.51], and W. Schwinge [15.17]. The photographs by G. Nemec [15.53], which demonstrate the outstanding performance of the refractor as well as the achievements of the projection method, are still eminently serviceable.

The changes in the apparent brightness of the Moon over the phase cycle pose substantial problems to the photographer. The brightness does *not* increase or decrease

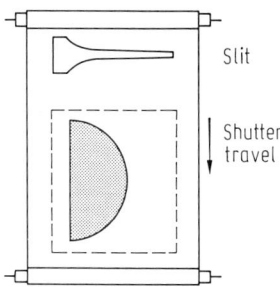

Fig. 15.14. Shutter with logarithmically shaped slit. After E.A. Whitaker.

proportionally to the extent of the illuminated surface. In comparison with pictures taken of the full moon, the exposure time at first and last quarter should be increased by a factor of 4, and for the 3- or 24-day-old crescent by as much as a factor of 12. This is done with the understanding that the "correct" exposure succeeds only near the terminator, while the regions with steeper light incidence are considerably overexposed.

While intensity ratios of the order 50:1 are recorded for the full moon, these ratios can range up to 1000:1 for other phases. To successfully bridge such ranges, one must apply quite different exposure times for the zones near the terminator than for the well-illuminated regions. At first and last quarter, this may be achieved with a specially shaped slit (Fig. 15.14). The slit may be shaped, for example, such as to give 1/4 second near the limb, and a logarithmic increase up to 2 seconds toward the terminator. Of course, this method requires the highest precision in setting and tracking.

There is no fixed rule regarding the exposure times, as much depends upon the instrument used, the lunar phase, the altitude of the Moon, and the type and speed of emulsion. The following may be used as a crude approximation: a picture of the full moon using the focal method (without filter etc.) can be taken at $f/15$ and ISO 50/18° emulsion with 1/50 to 1/10 s, a moderately large (10–15-cm) projected image with 1/2 to 3 s. Use of color filters requires application of the corresponding factors of increase. Any photograph whose exposure exceeds 1 s demands faultless tracking of the telescope. See Appendix B, Table B.14.

In two examples in his book *The Moon* ([15.50], p 223), Z. Kopal illustrates how the structure revealed very close to the terminator increases with exposure time. The definition seen at the terminator gains still more when the other surface is already very much overexposed. Depending on the phase, such terminator photographs should be made using exposure times of 1/2 s to 1/2 minute, again considering the speed of the film chosen. See the section "Lunar Photography" by J.F. Pedler in [15.5].

The key to successful lunar photographs is careful focusing. Using a magnifying glass of from 10× to 15× for focusing on the screen is a much-used aid. The screen itself should be nearly free of grain. H.J. Gramatzki recommends making the screen by distributing a drop of diluted seccotine on a clear glass plate. What remains is a very fine opaque layer which makes it possible to focus the lunar image reliably, when viewed from the side.

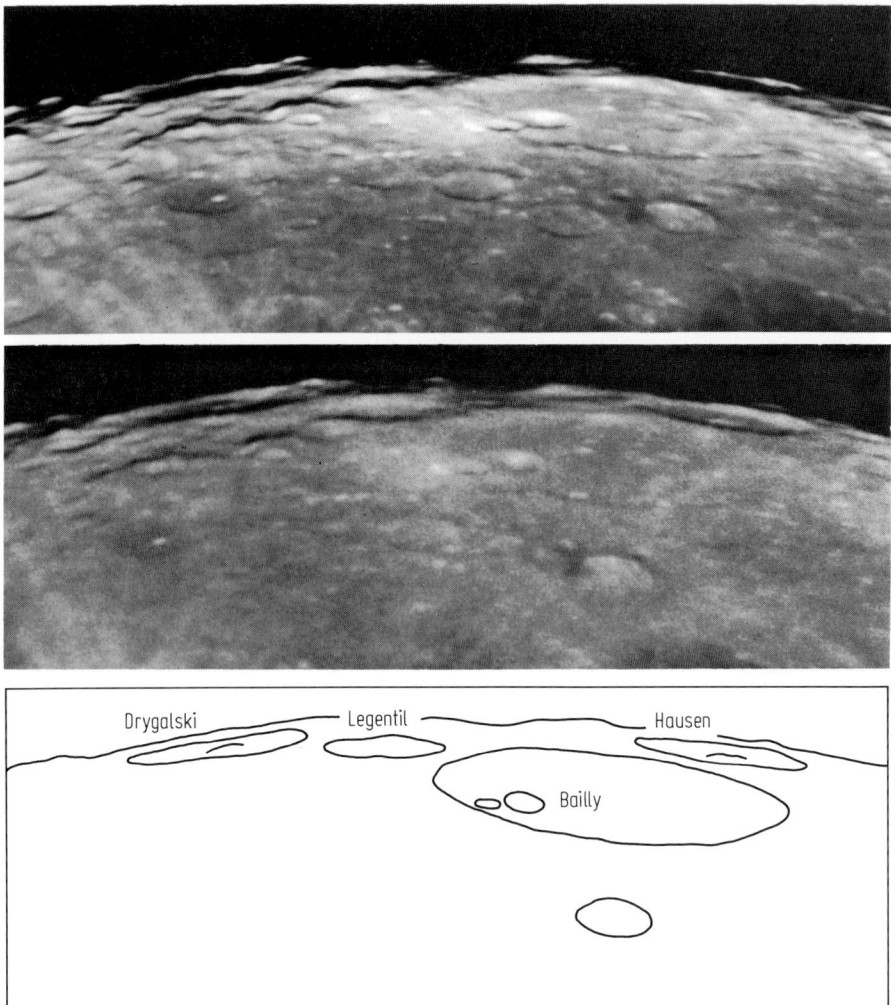

Fig. 15.15. Craters Drygalski, Bailly, Hausen. Photos 71 11 02-23.00, 73 09 12-22.00 by C. Albrecht; libration in latitude: $+2°\!.2$, $-4°\!.8$; libration in longitude: $-6°\!.5$, $-6°\!.5$.

Any vibration or disturbance at the telescope must be avoided during the exposure. It is helpful to operate the shutter indirectly (e.g., with a cable release) or to perform the exposure with a rhythmically moved cardboard in front of the objective lens (guided by the tick of a wristwatch).

The emulsions chosen should combine fine grain, steep contrast, and the highest possible sensitivity (speed). The development can influence sharpness and tint by the choice of developer (e.g., special fine-grain developing). Special developers (Neophin) act favorably toward a subsequent sensitivity increase of fine-grained but slower emulsions. Well-exposed negatives on which the lunar disk has a diameter of 20 mm (focal images) can be enlarged to a full moon diameter of up to 50 cm. Several trials will be needed to obtain the best results [15.50].

The successful photographer of the Moon and the planets, G. Nemec, writes, "It is absolutely necessary to work very precisely, and time pressure is always detrimental to the results. In the case of the Moon, this may be the major reason why photographs of the waning Moon are often more successful than those of the waxing Moon, which culminates in the early evening when the astronomer is still occupied with his instrumental preparations and the telescope has not yet cooled down sufficiently. Atmospheric conditions often change irregularly during the night, so that one can never be sure that the second half of the night will guarantee the best seeing. Besides scintillation, it is the altitude of the object on which the definition essentially depends. Photographs taken more than one or two hours before or after transit yield less satisfactory results." [15.53].

15.4.2.1 Tasks for Lunar Photography. The lunar photographer with a skilled hand and good optics may want to supplement or revise visual observations with photographs. Naturally, he/she will also compare his/her plates with those of other observers. Good negatives can be investigated by photometric methods. The *Fotografische Mondatlas* [15.17] by W. Schwinge has demonstrated that amateur instruments (apertures 50–200 mm) are capable of high-resolution photographs and that a good-quality atlas can be prepared based on photographs taken over almost an entire lunation. Newly developed objectives such as the high-efficiency optics APQ 100/1000 (Carl Zeiss, Jena) remove the difficulty that visual and photographic foci differ for the observer photographing with a refractor.

Pictures taken of a particular region, say one suspected of luminescence activity, can be taken with color filters, for instance green (545.0 nm) and red (672.5 nm) interference filters. When performed systematically, they supply useful material for comparative processing. This may be achieved with a good 5-inch refractor.

The photographic determination of heights of lunar formations is of special interest, but affected by the drawback that shadow measurements on a photograph are rather inaccurate owing to the influence of bad seeing, diffuseness of the images, and so on. For such research, visual observations are preferable (see Sect. 15.4.1.2).

15.4.3 Photoelectric Observations

Recent technological advances have rapidly made photoelectric observations accessible to the amateur (see Chap. 8, Vol. 1, for information on equipment and technique); such methods can be applied to the Moon. Measurements described in the following sections are also suited for the search for transient lunar phenomena (TLPs) as mentioned in Sect. 15.1.2.3; see also [15.5, 49].

15.4.3.1 Photometric Measurements. The scattering of sunlight off the lunar surface does not follow Lambert's law of reflection ([15.5] and Fig. 15.18). Photometry of comparable surface formations on various parts of the disk and at various solar altitudes permits the determination of the backscattering behavior of the lunar surface as

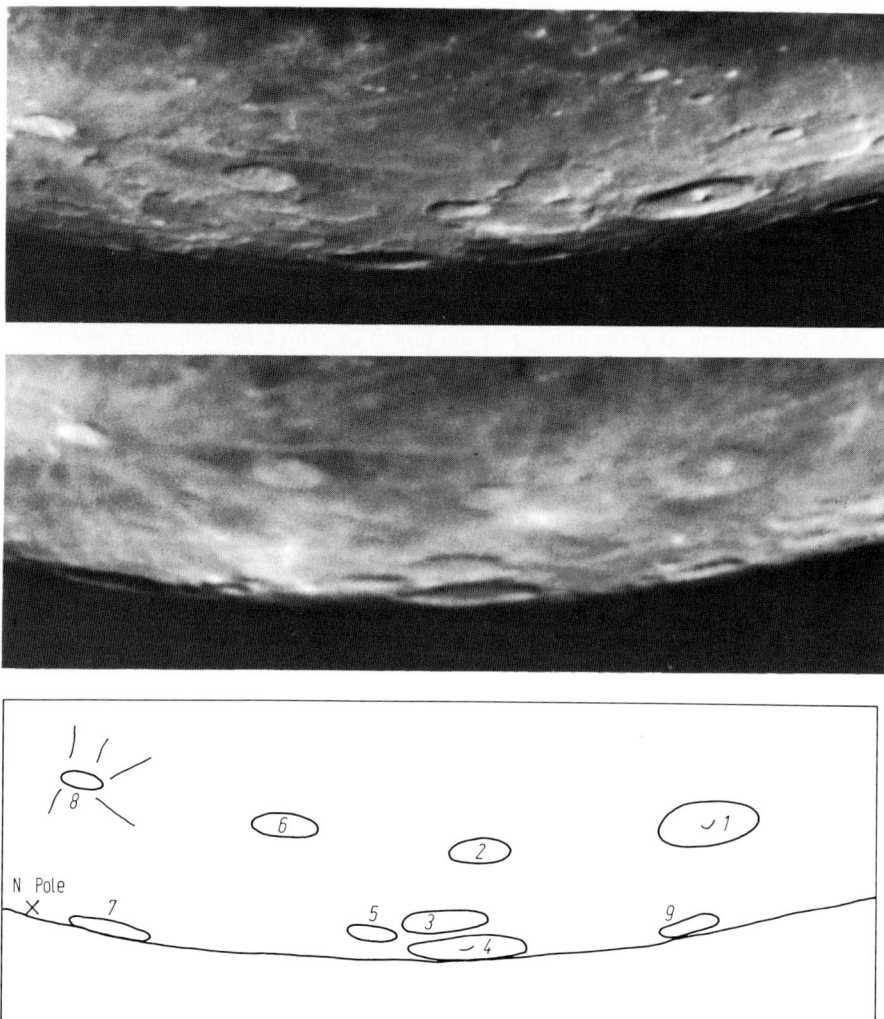

Fig. 15.16. Phythagoras, Pascal, Brianchon. Photos 73 02 16-22.00, 71 05 11 by C. Albrecht; latitude $+3°9$, $+6°3$; longitude $+4$, $-3°7$. *1* Phythagoras; *2* Carpenter; *3* Pascal; *4* Brianchon; *5* Mouchez; *6* Anaxagoras; *7* Hermite; *8* Cremona.

a function of the angles of incidence and reflection (phase curve). The characteristics of scattering are determined primarily by three surface properties:

1. Distribution of scattering, and unevenness and cavities on the ground. Sunlight may reach the surface features from very different angles.
2. The albedo of the surface material; maria have a lower albedo (about 0.05) than the highlands (about 0.10 to 0.15).

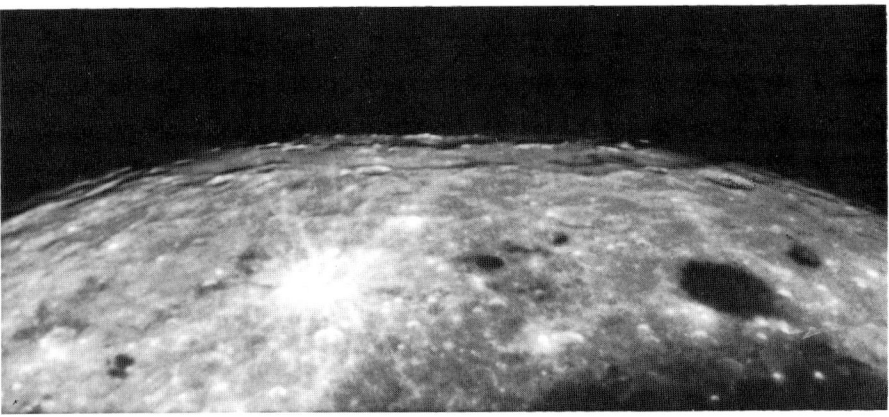

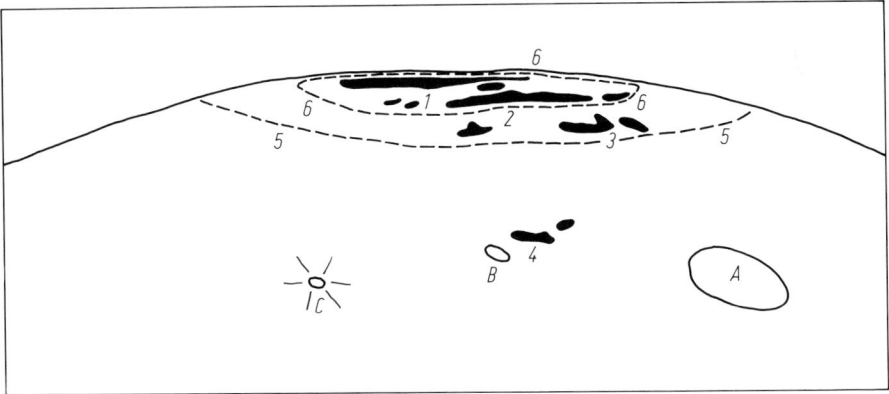

Fig. 15.17. Mare Orientale. Photo 72 09 22-21.30 by C. Albrecht; libration in latitude $-4°2$; libration in longitude $-5°8$. *1* Mare orientale; *2* Lacus veris; *3* Lacus autumnala; *4* Lacus aestalis; *5* Cordillera Mountains; *6* Rook Mountains; *A* Grimaldi; *B* Crüger; *C* La Paz.

3. Scattering functions of various components of this material such as rocks, grains, and dust particles.

Observers may wish, for instance, to monitor the phase curves of selected objects such as craters, valleys, rilles, bright rays, and dark spots. Regions of interest include, for instance, the Long Wall, the Alpine Valley, Vallis Rheita, and the "Cobra Head" of the Schröter Valley. See also Hedervari [15.49].

15.4.3.2 Color Measurements. Differences of color and tint within the Mare Imbrium have already been mentioned (Sect. 15.3.1). The view of the full moon makes it difficult to gain an impression other than that of a silvery white disk with gray spots. But spectrophotometric studies indicate that the lunar surface reflects about 8% of incident blue light and 12% of red light. The backscattered light thus has a brownish tint, and various regions of the Moon show small, but distinct changes of tint. By comparison, for example, Mare Tranquilitatis has a more steel-grey coloring, Mare

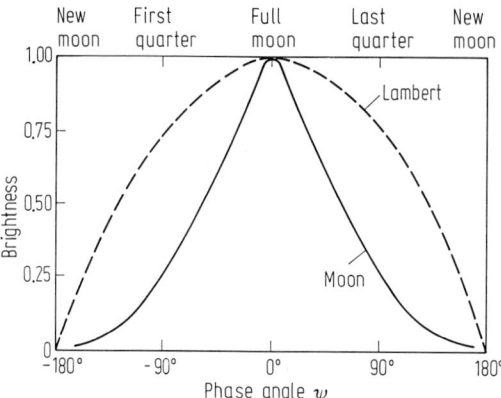

Fig. 15.18. Photometric phase curve of the Moon. *Dashed curve*: predicted total brightness from Lambert's law of reflection. *Solid curve*: actual brightness. From R.H. Giese, *Erde, Mond und benachbarte Planeten*, Mannheim, Wien, Zürich 1969.

Serenitatis more yellow–brown. Spectrophotometry uncovers the fact that both Maria have about the same blue backscatter (8%) but differences in red: Mare Serenitatis reflects red slightly better (about 12%) than Mare Tranquilitatis (about 11%).

Gray formations in brown surroundings include, for instance, the floor of the crater Boscovich and the floor and central peak of Aristarchus. Examples of brown formations with gray surroundings are the craters Moltke and Plinius, and the floor of Bullialdus. Other good examples can be found in [15.5].

15.4.3.3 Polarization Measurements. The available measurements indicate that the dark maria polarize light more strongly than the bright terrae. Polarized radiation can be found with the aid of a polarization filter rotatable by 135° and placed directly in the photometer beam. The individual measurements should be accurate to 0.01 mag, which is achievable only with a photoelectric photometer.

Various albedo patterns of the waxing and waning moon are manifested in the polarization behavior. Some formations show distinct differences in polarization over a lunation [15.16]; see also Fig. 15.19.

A polarimeter connected to a photoelectric indicator permits polarization studies of individual regions on the lunar surface, with the intention of detecting changes of the degree of polarization P, defined as

$$P = \frac{I_1 - I_2}{I_1 + I_2} \times 100\%. \tag{15.16}$$

Here, I_1 is the light intensity polarized in the plane Sun–Moon–Earth and I_2 the intensity in the plane perpendicular to it.

15.4.3.4 Observations of Luminescence. Certain areas on the Moon fluoresce when subjected to solar irradiation. This luminescence can be identified visually, photoelectrically, and by filter photography. The photoelectric measures are preferably carried

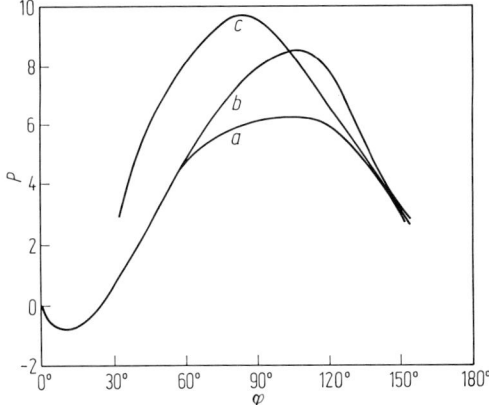

Fig. 15.19. Change in the degree of polarization depending on the lunar phase (after Lyot). *a* waxing moon; *b* waning moon; *c* ashen light. From D. Böhme [15.40].

out in three colors—green, red, infrared; cf. Sect. 15.1.2.3 on transient lunar phenomena. For further reading, the section "Luminescence of the Lunar Surface" in [15.50] is recommended.

15.4.3.5 Studies of Surface Structures. A detailed knowledge of the shape of the lunar surface can be achieved only with a variety of techniques. A combination of the various measurements mentioned under Sect. 15.4.2.4, if carried out under as many different illuminations of the Moon as possible, can lead to the determination of a kind of "microrelief" of the lunar surface. Unfortunately, the requisite accuracy admits only of photoelectric measures.

Also suggested are albedo measures of the Moon taken in the infrared, and the search for formations which, when located in the unilluminated part of the Moon, radiate infrared radiation [15.8, 58]. The peak spectral sensitivity of the measurements should be in the range 700–900 nm. The heat absorbed by the lunar surface in the daytime is reradiated and can be observed in the infrared most distinctly during the lunar night. Lunar eclipses (Chap. 16) are especially useful in this regard. At the onset of the eclipse, the visible surface is more or less evenly heated, but as the eclipse progresses, the incident solar energy is gradually blocked off. The degree of cooling can be determined for the entire lunar surface, and different rates of cooling have indicated a range of physical properties of the surface material.

References

15.1 Köhler, U.: Mondspaziergang mit Liebhaberfernrohren. *Sterne und Weltraum* **26**, 346 (1987).
15.2 Schaifers, K., Traving, G.: *Meyers Handbuch Weltall* (6th ed.), Bibliographisches Institut, Mannheim, Wien, Zürich 1984, p. 93.
15.3 Royal Society (ed.): The Moon—A New Appraisal from Space Missions and Laboratory Analyses. London 1977.

15.4 French, B.M.: What's New on the Moon? *Sky & Telescope* **53**, 164 and 257 (1977); and Classen, J.: Fortschritte der Mondforschung von 1976 bis 1978. *Die Sterne* **56**, 111 (1980).
15.5 British Astronomical Association (ed.): *Guide to Observing the Moon*, Enslow Publishers, Hillside, Aldershot 1986.
15.6 Cameron, W.S.: Report on the ALPO Lunar Transient Phenomena Observing Program. *The Journal of the Association of Lunar and Planetary Observers* **25**, 1 (1974).
15.7 Stadler, L.: Neues über die Dunkelflecken auf dem Mond. *Sterne und Weltraum* **15**, 28 (1976).
15.8 Guest, J.E., Greeley, R.: *Geology of the Moon*, Wykeham Publ., London 1977.
15.9 Taylor, St. R.: *Lunar Science—A Post-Apollo View*, Permagon Press, Oxford, New York 1975.
15.10 Hilbrecht, H.: Mondbeobachtung nach dem "Apollo-Schock." *Sterne und Weltraum* **24**, 93 (1985).
15.11 Starzynski, H.: Moonblink photographiert. *Sterne und Weltraum* **26**, 719 (1987).
15.12 Fauth, H.: Ph. Fauth and the Moon. *Sky & Telescope* **XIX**, 20 (1959).
15.13 Mondkarten aus Anlaß der Herausgabe der Großen Mondkarte von Philipp Fauth. *Nachrichten der Olbers-Gesellschaft Bremen* 60 (1964) and 63 (1965).
15.14 Roth, G.D.: Die Mondkarte von H.P. Wilkins. *Die Sterne* **31**, 180 (1955).
15.15 Kuiper, G.P., *et al.*: *Photographic Lunar Atlas*, Chicago 1960.
15.16 Kopal, Z., Carder, R.W.: *Mapping of the Moon*, D. Reidel, Dordrecht 1974.
15.17 Schwinge, W.: *Fotografischer Mondatlas*, Johann Ambrosius Barth, Leipzig 1983, p. 7.
15.18 Price, F.W.: The Moon Observer's Handbook, Cambridge University Press, Cambridge 1988, p. 309.
15.19 Roth, G.D.: *Taschenbuch für Planetenbeobachter* (3rd ed.), Verlag Sterne und Weltraum, Munich 1987.
15.20 Classen, J.: Das Innere des Mondes (I). *Die Sterne* **50**, 157 (1974).
15.21 Classen, J.: Das Innere des Mondes (II). *Die Sterne* **50**, 213 (1974).
15.22 Gramatzki, H.J.: *Das Weltall* **39**, 145 (1939).
15.23 Cameron, W.: *Lunar Transient Phenomena Catalog*, NASA-Goddard Spaceflight Center, Greenbelt 1978.
15.24 Küveler, G.: Internationales Mondprogramm der Sternwarte Gummersbach. Lunar Transient Phenomena 1972. *Sterne und Weltraum* **12**, 231 (1973).
15.25 O'Meara, S.O.: The Lunar Straight Wall. *Sky & Telescope* **73**, 639ff (1987).
15.26 Günther, O.: Zur Sichtbarkeit von Einzelheiten aud der Nachtseite des Mondes. *Die Sterne* **42**, 1 (1966).
15.27 Küveler, G., Klemm, R.: Neue Intensitätsskalen für das sekundäre Mondlicht. *Sterne und Weltraum* **11**, 239 (1972).
15.28 Graff, K.: Der Mond. In: *Astronomisches Handbuch*, Franckh'sche Verlagshandlung, Stuttgart 1924, p. 131.
15.29 Ahnert, P.: *Astronomisch-chronologische Tafeln für Sonne, Mond, und Planeten* (5th ed.), J.A. Barth, Leipzig 1971.
15.30 Ahnert, P.: Was bedeutet der Begriff "Colongitude"? In *Kalender für Sternfreunde 1988*, Johann Ambrosius Barth, Leipzig 1987, p. 164.
15.31 Albrecht, C.: Selten sichtbare Mondlandschaften IV. *Sterne und Weltraum* **13**, 238 (1974).
15.32 Westfall, J.E.: Lunar Incognita: Completing the Map of the Moon. *Sky & Telescope* **67**, 284 (1984).
15.33 MacRobert, A.M.: Three Lunar Challenges. *Sky & Telescope* **77**, 520f (1989).
15.34 Rükl, A.: Selenodäsie und Selenographie. *Der Sternenbote* **24**, 68 (1981).
15.35 Fauth, Ph.: *Unser Mond*, Breslau 1936.
15.36 Handweiser für Beobachter am Fernrohr (from Philipp Fauth's *Unser Mond*). Zweites Sonderheft Mondkarten. *Nachrichten der Olbers-Gesellschaft Bremen* **63** (1965).
15.37 Miller, G.: Erfahrungen mit dem Baader-Binokular. *Sterne und Weltraum* **26**, 716 (1987).
15.38 Kopal, Z.: *Physics and Astronomy on the Moon*, Academic Press, New York/ London 1962.

15.39 Billerbeck-Gentz, F.: *Die Sterne* **27**, 156 (1951).
15.40 Böhme, D.: Polarimetrie für Amateurastronomen. *Sterne und Weltraum* **25**, 544 (1986).
15.41 Ahnert, P.: Messung von Berghöhen auf dem Mond. In: *Kalender für Sternfreunde 1973*, J.A. Barth, Leipzig 1973, p. 194.
15.42 Hopmann, J.: Die Genauigkeit der Angaben von relativen Höhen auf dem Monde. Neue Werte Für 163 Punkte. *Mitteilungen der Sternwarte Wien* **12**, no. 19 (1965).
15.43 Graff, K.: Formeln zur Reduktion von Mondbeobachtungen und Mondphotographien. *Veröffentlichungen des Astronomischen Recheninstituts zu Berlin* Nr. 14, Berlin 1901.
15.44 Hopmann, J.: Ermittlung von Höhen auf dem Monde. *Mitteilungen der Sternwarte Wien* **12**, no. 8 (1965).
15.45 Schrutka-Rechtenstamm, G.: Relative Höhenbestimmungen auf dem Mond mittels des Pariser Mondatlasses und visuelle Messungen am Fernrohr. *Mitteilungen der Univ.-Sternwarte Wien*, Bd. 7, Nr. 11, Wien 1955.
15.46 Schrutka-Rechtenstamm, G.: *Mitteilungen der Univ.-Sternwarte Wien*, Bd. 8, Nr. 17, Wien 1958.
15.47 Ahnert, P.: In: *Kalender für Sternfreunde 1952*, J.A. Barth, Leipzig 1952, p. 133.
15.48 Zimmermann, O.: *Astronomisches Praktikum* (4th ed.), Verlag Sterne und Weltraum, München 1987.
15.49 Hedervari, P.: Lunar Photometry. In: Genet, R.M. (ed.): *Solar System Photometry Handbook*, Willmann-Bell, Richmond 1983, p. 4-8.
15.50 Kopal, Z.: *The Moon*, D. Reidel Publ. Co., Dordrecht 1969. Contains an extensive list of books and papers.
15.51 Flach-Wilken, B.: Guter Mond du stehst so stille.... In: *Festschrift 18. VdS-Tagung in Bochum*, Vereinigung der Sternfreunde e.V., Bochum 1987, p. 46.
15.52 Viscardy, G.: *Altas-Guide photographique de la Lune. Ouvrage de référence à haute résolution*, Masson, Paris 1985.
15.53 Nemec, G.: Planetenfotografie. *Mitteilungen der Volkssternwarte Köln*, **10**, Part 1 (1966).
15.54 Lunar Flash Revisited. In: *Sky & Telescope* **79**, 590 (1990).
15.55 Mysterious Flash Photographed on the Moon. In: *Sky & Telescope* **77**, 468 (1989).
15.56 Dunlop, S., Gerbaldi, M. (eds.): *Stargazers. The Contribution of Amateurs to Astronomy*, Proceedings of Colloquium 98 of the IAU, June 20–24, 1987, Springer, Berlin Heidelberg New York 1988.
15.57 MacRobert, A.M.: The Moon Shall Rise Again. *Sky & Telescope* **76**, 478f (1988).
15.58 Fischer, K.: Die elektrooptischen spektralen Bildwandler und ihre Anwendung in der Selenographie. *Die Sterne* **38**, 181 (1962).

Current observational tasks and possibilities of international cooperation are reported in *The Journal of the British Astronomical Association*, *The Journal of the Association of Lunar and Planetary Observers* (USA), *Astronomy*, and *Sky & Telescope*. Advanced observers will find important information in *Icarus* (an international journal of solar system studies).

16 Lunar Eclipses

H. Haupt

16.1 Introduction

Eclipses of the Moon are among those celestial phenomena which can be impressive even to the layman. During such an occurrence, the full moon slowly darkens and takes on a peculiar illumination that does not correspond to any normal phase. It also often assumes paler tints of brightness and a dusky, often reddish-brown color, so that even modern-day observers will feel that they are witnessing a rare and memorable event (Fig. 16.1).

In former ages when such phenomena occurred, for instance, in the climatically favored regions of the Middle East or in China, where records date back to the second millenium B.C., the sense of awe and wonderment must surely have been even greater. Tradition and historical experience soon revealed that, although lunar eclipses occur rarely, they repeat at regular intervals. The efforts of ancient astronomers were directed over many centuries to investigating this regularity and deviations therefrom. Important insights were gained while progressing from the purely kinematic description of the motions to the dynamic explanation, since the Moon as the "second hand of the celestial clock" is the most sensitive and reliable indicator for events involving

Fig. 16.1a, b. Color photographs of the lunar eclipse 1982 January 09. **a** 18.30^h UT (partial phase); **b** 19.35^h UT (totality). Photography by H. Sussmann, Graz.

motion and for the gravitational effects in the solar system; lunar eclipses provided opportunities for testing such effects.

It was not until the 20th century that lunar eclipses were observed photometrically. The degree of darkening and the accompanying color changes have been interpreted as indicating the state of the Earth's atmosphere, in particular the ozone content and the pollution of the upper layers. The enlargement of the Earth's shadow which falls on the Moon also gives information on the state of the upper atmosphere and the potential contamination by meteoric dust.

Amateur observers have cooperated on all of these questions, which usually lie outside the scope of the large professional institutes. Thus, the observation of a lunar eclipse can be not only a marvelous experience but also a contribution to scientific knowledge.

16.2 The Origin and Frequency of Lunar Eclipses

16.2.1 Principles

The Moon can be eclipsed only when accurately aligned with Sun and Earth. The plane of the Moon's orbit is inclined to that of the Earth (the ecliptic) by an average of $5°8\overset{m}{.}7$ (cf. Fig. 16.4). Thus, not every full moon leads to an eclipse, but instead the eclipses can occur only in the vicinity of one of the two intersection points, called *nodes*, of the two orbital planes. The Moon may become completely immersed in the Earth's shadow. In the strictest sense, a "fully" illuminated Moon at phase angle $0°$ (see Fig. 16.2 for a definition) can never be observed from the Earth because it must then be undergoing an eclipse. The commonly visible "full moons" have phase angles of at least $1°\!.5$, with the Moon exactly opposite the Sun in longitude but at a nonzero latitude.

On the night side of the Earth opposite to the direction of the Sun, there is a converging shadow cone (the *umbra*) of length 1.4 million km and bounded by the external tangents t_1, t_2 of Sun and Earth. Around it is the zone of a diverging *penumbra*, which is characterized by the inner tangents t'_1, t'_2 (Fig. 16.3).

The diameter of the umbra at the distance of the Moon is about 3 times the Moon's diameter, and therefore an eclipse can occur to an angular distance of up to $10°$ away from the node. A *total eclipse* occurs when the entire Moon passes through the umbra,

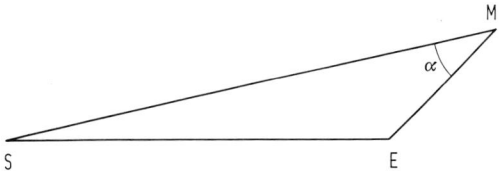

Fig. 16.2. Definition of the phase angle α in the triangle formed by Sun (S), Earth (E), and Moon (M). At full moon $\alpha = 0°$, and at new moon $\alpha = 180°$.

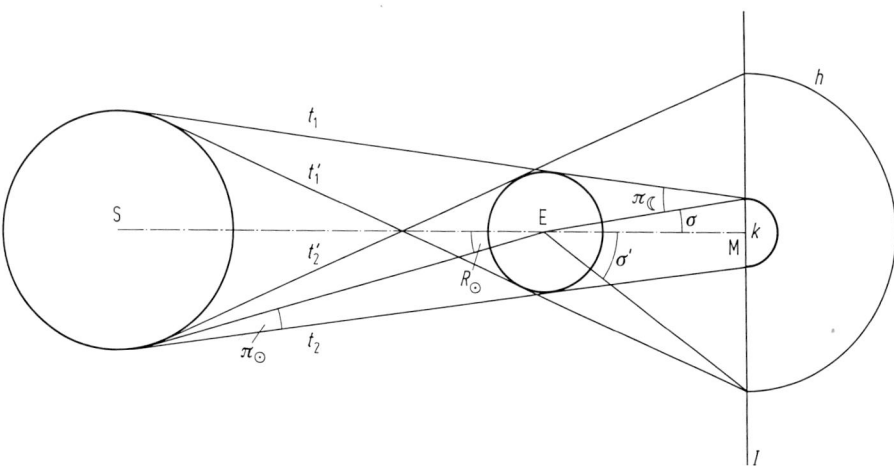

Fig. 16.3. Geometric parameters of a lunar eclipse (schematic, not to scale). See text for an explanation of symbols.

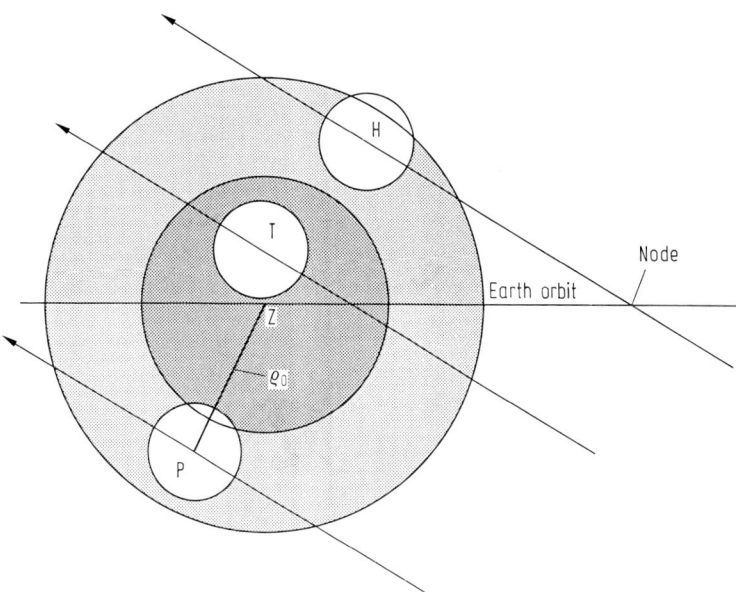

Fig. 16.4. Kinds of lunar eclipses: T = total, P = partial, H = penumbral. Z is the center of the shadow and ρ_0 the minimum distance between the center of the Moon and Z.

a *partial eclipse* when only a portion of the Moon passes through it, and a *penumbral eclipse* when the Moon misses the umbral cone entirely and passes through only the penumbral shadow (Fig. 16.4).

A total lunar eclipse is of longest duration when the Moon passes through the center of the Earth's shadow (a central eclipse). The duration also depends on the mutual distances of Sun, Earth, and Moon. Considering the fact that the Moon when near perigee moves faster through the shadow (Kepler's second law), the eclipse which occurs nearest to Earth is not necessarily the longest. The longest duration of totality is about 100 minutes. The purely penumbral eclipses are barely noticeable, but they nonetheless yield scientifically useful results (see Sect. 16.6.4).

16.2.2 Historical Studies and the Saros Cycle

A lunar eclipse can be observed at the same time and in the same way by the innumerable observers scattered over that fraction (slightly more than one-half) of the Earth's surface from which the Moon appears above the horizon.[1] A solar eclipse, on the other hand, is visible only within a narrow path on the Earth. Consequently solar eclipses—although actually more frequent over the entire Earth—occur much more seldom in specific locations than do lunar eclipses. While the passage of the Sun through a node produces at least one solar eclipse (at the preceding or following new moon, or both, to a maximum of 5 solar eclipses per year), not every full-moon season near a node yields a corresponding lunar eclipse, because the Earth's shadow is a smaller target than the Earth itself, and the fast-moving Moon may miss it. This consideration means that in some years there will not be an umbral eclipse of the Moon at all, while in others up to three eclipses may occur.

Based upon a series of observations spanning far longer than a human lifetime, the ancient Chaldeans discovered that the recurrence of eclipse conditions, i.e., the coincidence of *syzygies* (full moon or new moon) with the nodes, causes the return of eclipses in predictable cycles. As has long been known, the lunar nodes move in *retrograde* fashion along the ecliptic, that is, in the opposite sense to the motion of the Moon itself. The time between two consecutive passages of the Moon through the same node (the *draconitic month* = 27.212 22 days) is, on the average, shorter by 0.109 44 days or about 2.5 hours than the mean sidereal month of 27.321 66 days. The lunar phases, however, occur with the *synodic month* (=29.530 59 days), so that eclipse cycles are conditioned by the draconitic and synodic months. A search for common multiples of these two intervals reveals several periods of recurrence, one of which is quite precise. It is termed the *saros cycle*, and was known and used for predicting eclipses even in antiquity.

242 draconitic months = 6585.357 days ≈ 223 synodic months = 6585.322 days. After an interval of 6585.3 days or 18 years and 11 days, eclipses thus repeat themselves in the same sequence and at the same node under quite similar conditions. The latter is the case because, first, the recurrence in the same season eliminates the nonuniformity of the solar motion, and, second, because the cycle also matches— coincidentally—a multiple of the *anomalistic month* (the interval between successive

---

[1] In the rare case when the eclipse occurs just after moonrise or before moonset, the *Sun* may also still be above the horizon as a result of atmospheric refraction.

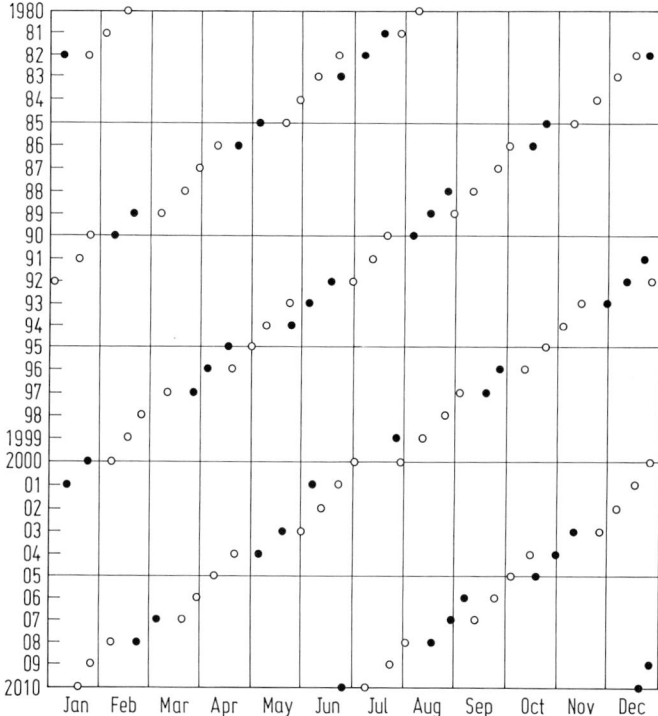

Fig. 16.5. Graph of the eclipses occurring in the period 1980–2010 showing the repetitive nature after 18 years, 11 days. *Filled dots*: lunar eclipses; *open circles*: solar eclipses.

passages through perigee, namely 239 × 27.554 55 = 6585.54 days), which also guarantees nearly the same position of the Moon in its elliptical orbit.

Figure 16.5 illustrates the occurrence of eclipses over $1\frac{1}{2}$ saros cycles. It shows how eclipses occur every year at earlier dates, owing to the retrogression of the nodes, and repeat after 18 years and 11 days in the same sequence. Table 16.1 gives a listing of the dates and times of lunar eclipses which are predicted for the interval 1988–2000.

16.2.3 Canons: Statistics of Eclipses

Eclipse data as given in the preceding section can be found in a compendium or "canon." For over a century, the most-cited source was the famous *Canon der Finsternisse*, computed by the Viennese astronomer von Oppolzer [16.1] and coworkers, all calculations having been performed by hand. This monumental work, published by the Vienna Academy, tabulates all the relevant data for approximately 8000 solar and 5200 lunar eclipses; for the former, the interval covered is 1208 B.C. to 2961 A.D., and for the latter, from 1207 B.C. to 2163 A.D. Included for solar eclipse work are maps showing the approximate central line for latitudes +90° and −30°. A bibliogra-

Table 16.1. Lunar eclipses occurring during the period 1988–2000: T = total, P = partial, TDT = dynamical time at mid-eclipse, g = magnitude of eclipse.

Year	Date	Type	TDT*	g
1988	Aug. 27	P	11^h04^m	0.92
1989	Feb. 20	T	15 36	1.28
	Aug. 17	T	03 08	1.60
1990	Feb. 09	T	19 12	1.07
	Aug. 06	P	14 11	0.68
1991	Dec. 21	P	10 33	0.09
1992	Jun. 15	P	04 57	0.68
	Dec. 09	T	23 44	1.27
1993	Jun. 04	T	13 01	1.56
	Nov. 29	T	06 25	1.09
1994	May 25	P	03 31	0.24
1995	Apr. 15	P	12 18	0.11
1996	Apr. 04	T	00 10	1.38
	Sep. 27	T	02 54	1.24
1997	Mar. 24	P	04 40	0.92
	Sep. 16	T	18 46	1.19
1999	Jul. 28	P	11 33	0.40
2000	Jan. 21	T	04 44	1.33
	Jul. 16	T	13 56	1.77

*Universal Time = TDT $-\Delta T$.
(for 1990, $\Delta T \approx 1\overset{m}{.}0$).

phy of canons for specific regions and time intervals is given by Mucke [16.2]; see also van den Bergh [16.3].

Today canons are prepared using computers and an improved lunar theory. Mention should be made of two volumes, also published in Vienna, one on solar eclipses (Mucke and Meeus [16.4]) and the other on lunar eclipses (Meeus and Mucke [16.5]); the latter is the source of the data used in this chapter. Additional data on the circumstances of individual eclipses are given, including region of visibility, tracks, and also a statistical survey.

It is found that the number total of lunar eclipses (including the barely noticeable penumbral eclipses) almost matches that of solar eclipses, notwithstanding what has been said on the local and long-term appearances. Some other statistical figures show, for instance, that the number total of lunar eclipses per century varies between 60 (18th century) and 84 (21st century). Even more differences are found in the occurrence of 4 consecutive eclipses all of which are total (*tetrades*): there are none from 1502 to 1908, but 16 in the interval 1909 to 2156, and thereafter again a "gap" for 300 years. Many other relations on the occurrence of eclipses could be read from the tables of the canons, but such considerations would lead us too far astray at this point.

16.3 Theory and Prediction of Lunar Eclipses

16.3.1 Geometric Theory

To repeat what has been said in connection with Fig. 16.3, the common outer tangents t_1, t_2 form a convergent conic umbra k, and the inner tangents t'_1, t'_2 the divergent penumbral cone h. This is how the solar rays would run in the absence of Earth's atmosphere, and so this case will be considered first. No light would penetrate into the umbra, and an observer there would find the Sun covered by the nontransparent Earth. An observer in the penumbra would receive some fraction of the sunlight, and see the Sun partially eclipsed by the Earth.

On a plane I placed at the distance of the Moon at right angles to the shadow axis SEM, the umbral and penumbral cones cut the corresponding circles with the apparent (angular) radii as seen from the Earth's center:

$$\sigma = \pi_{\mathrm{C}} + \pi_{\odot} - R_{\odot} \quad \text{for the umbra,} \tag{16.1}$$

$$\sigma' = \pi_{\mathrm{C}} + \pi_{\odot} + R_{\odot} \quad \text{for the penumbra.} \tag{16.1'}$$

These relations follow directly from the geometry of Fig. 16.3, with π_{C} being the horizontal parallax of the Moon, $\pi_{\odot}$ that of the Sun, and $R_{\odot}$ the apparent solar radius. These equations were known to Hipparchus as early as around 150 B.C.

The Moon projected onto the plane I will pass through various parts of umbra and penumbra (cf. Fig. 16.4). The instantaneous "depth" of an eclipse can be expressed by a quantity g, which is given by

$$g = \frac{\sigma - (\rho_0 - R_{\mathrm{C}})}{2 R_{\mathrm{C}}}, \tag{16.2}$$

where $g = 0$ at the beginning of the partial phase, $0 < g < 1$ during partiality, $g = 1$ at the beginning of totality, and $g > 1$ during totality. During the partial phase, g denotes the eclipsed fraction of the lunar diameter, but during totality it expresses the penetration into the umbra, again in units of the lunar diameter. The maximum of eclipse occurs at the instant of minimum distance ρ_0 between the centers of the Moon and the shadow. This moment is the middle of the eclipse, and the quantity g then is called the "magnitude" of the eclipse, and is listed in astronomical ephemerides.

As an illustration, assume the following representative values: $\pi_{\mathrm{C}} = 57'$ and $R_{\mathrm{C}} = 16'$, the penumbra has a radius $\sigma' = 63'$, and the umbra $\sigma = 41'$. For a central eclipse ($\rho_0 = 0$), the magnitude from Eq. (16.2) is then about $g = 1.8$.

16.3.2 Photometric Theory of Lunar Eclipses

The geometric approach neglects the presence of the Earth's atmosphere and thus fails to explain why the Moon is still visible in the umbra. Kepler pointed in the early 1600s that the explanation rests with the atmospheric refraction; but only after 20th-century exploration had obtained adequate data on the composition of various layers of the terrestrial atmosphere up to 100 km was Paetzold [16.6] able to prepare an

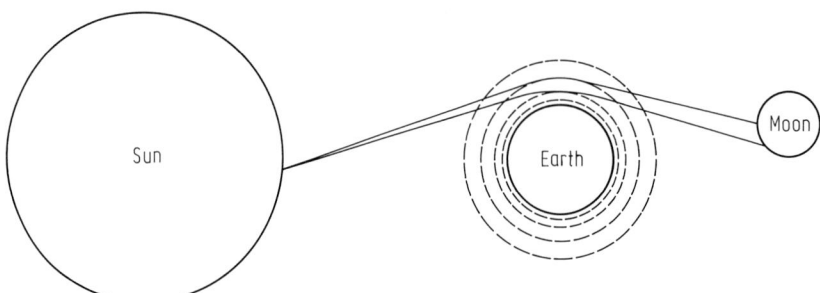

Fig. 16.6. Action of the Earth's atmosphere during a lunar eclipse.

exact photometric theory of lunar eclipses. These data primarily concern the change of density with height and permit one to calculate the refraction, i.e., how much a ray passing horizontally at height h above ground is deflected, and how strongly the bending depends on h.

Let a beam of sunlight traverse the Earth's atmosphere (Fig. 16.6): the deflection is seen to affect the upper ray more than the lower one, resulting in an increase of the (otherwise very small) divergence of the beam, and in a refractive dispersion diminishing the flux. This is a very sensitive effect as the very low air density in atmospheric layers around 50 km still causes some apparent dimming. Refraction weakens the light almost neutrally. Lower layers of the atmosphere contribute to a dimming mainly because of molecular scattering of the light. This extinction depends on the *air mass* penetrated, and is distinctly wavelength-dependent with λ^{-4} (Rayleigh's law). How these two components contribute to the total light reduction depends upon the minimum height of the ray, or on its position on the plane I. The accurate photometric theory also allows for the nonuniform brightness distribution across the solar disk.

Figure 16.7 shows some curves representing the illumination in the Earth's shadow, expressed in terms of the so-called depth D (Sect. 16.5.1), and illustrating the photometric situation under various conditions, assuming an atmosphere with ideal Rayleigh scattering:

(a) A nearly neutral coloring of shadow in the outer parts, where the extinction is low;
(b) Dominance of red light in the inner parts of the shadow;
(c) Dependence of brightness in the shadow on the lunar parallax.

Eclipses which occur at perigee ($\pi_{\mathrm{C}} = 61'$) are noticeably darker than those which take place at apogee ($\pi_{\mathrm{C}} = 54'$). This is because atmospheric refraction shortens the length of the darkest part of the umbral cone to only about 40 Earth radii from the Earth's center, and thus the Moon at perigee is more deeply immersed in this region than when at apogee.

A description of the appearance of the lunar landscape for an observer located on the Moon's surface will aid in explaining the photometric situation in the penumbra. An observer located within the umbra would see in the sky in place of the Earth a dark

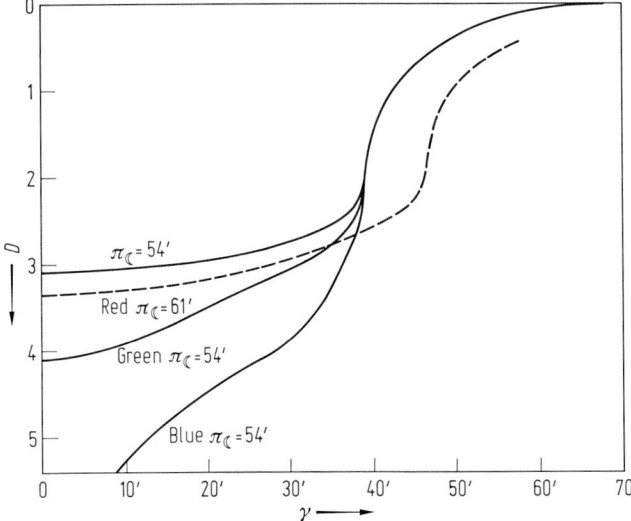

Fig. 16.7. Shadow depth D as a function of the angular distance γ from the shadow center.

circle with an apparent diameter of about $2°$ ($= 2\pi_{\mathbb{C}}$), surrounded by a bright ring, which is the sunlight deflected by refraction into the umbra. It is comparable with the *cusp phenomenon* often noted (and photographed) on Venus near inferior conjunction.

A partial eclipsing of the Sun by the dark Earth would be seen by an observer located in the penumbra. The illumination is then computed from the ratio of the uncovered to the total solar disk area, allowing for the effects of limb darkening on the Sun. Any modern photometric theory is guided by the need to further study the effects caused by Earth's upper atmosphere. A comparison with observations yields some interesting facts which, in conjunction with the results obtained from meteorological satellites, contribute to finding solutions to many still unanswered questions.

16.3.3 Prediction of Lunar Eclipses

Calculating the parameters of a particular lunar eclipse can be quite instructive, and may, in some respects, supplement the brief data found in astronomical almanacs. For the complete calculation of the event, the following data must first be obtained:

1. Times of ingress into and egress from the penumbra (external contacts); beginning and end of the partial eclipse, which are ingress into and egress from the umbra (again external contacts); beginning and end of totality (inner contacts with the umbra) and the middle of eclipse.
2. The graphic progress of the eclipse, i.e., the passage of the Moon through both shadows, in projection onto the plane I at the Moon's distance and lying perpendicular to the shadow axis.

3. The position of the terminator (edge of shadow on the Earth), to be determined in relevant phases or at arbitrary intervals during the eclipse.
4. The instant of passage of craters and other observable lunar formations through the edge of the umbra.
5. Topocentric data for the site of observation, such as azimuth and zenith distance, in relevant phases of the eclipse.

These data are useful for estimating the observing conditions as well as for later processing the results of measurements.

16.3.3.1 Numerical Calculation of the Course of an Eclipse. To this end, one begins with the following ephemeris data:

$$T = \text{time of opposition in right ascension;}$$
$$\alpha_\odot = \alpha_\mathleft(\pm 12^h = \text{right ascension of Sun and Moon at that instant;}$$
$$\delta_\odot, \delta_\mathleft(= \text{respective declinations at that instant;}$$
$$\Delta\alpha_\odot, \Delta\alpha_\mathleft(, \Delta\delta_\odot, \Delta\delta_\mathleft(= \text{hourly changes in these coordinates;}$$
$$\pi_\odot, \pi_\mathleft(= \text{parallaxes of Sun and Moon;}$$
$$R_\odot, R_\mathleft(= \text{apparent semidiameters of Sun and Moon.}$$

The center of shadow has the coordinates $\alpha_\odot \pm 12^h$, $-\delta_\odot$ and its hourly changes are

$$\Delta\alpha_\odot, \ -\Delta\delta_\odot.$$

Let a plane be placed through the center of shadow O and oriented at right angles to the shadow axis, and a cartesian coordinate system placed with origin at O, the positive abscissa $+x$ oriented toward west, and the ordinate $+y$ toward north. At opposition in right ascension, the position of the Moon's center M_0 is given by the coordinates (Fig. 16.8):

$$X_0 = 0,$$
$$Y_0 = \delta_\mathleft(+ \delta_\odot. \tag{16.3}$$

One hour after opposition, the Moon is located at position M, with the coordinates

$$x' = -15\cos\delta_\odot(\Delta\alpha_\mathleft(- \Delta\alpha_\odot),$$
$$y' = Y_0 + \Delta\delta_\mathleft(+ \Delta\delta_\odot = \delta_\mathleft(+ \delta_\odot + \Delta\delta_\mathleft(+ \Delta\delta_\odot. \tag{16.4}$$

The arc Δs described in one hour by the Moon in its orbit thus is

$$\Delta s = -\sqrt{x'^2 + (y' - Y_0)^2} = -\frac{\Delta\delta_\mathleft(+ \Delta\delta_\odot}{\sin i} = \frac{15\cos\delta_\odot(\Delta\alpha_\mathleft(- \Delta\alpha_\odot)}{\cos i}, \tag{16.5}$$

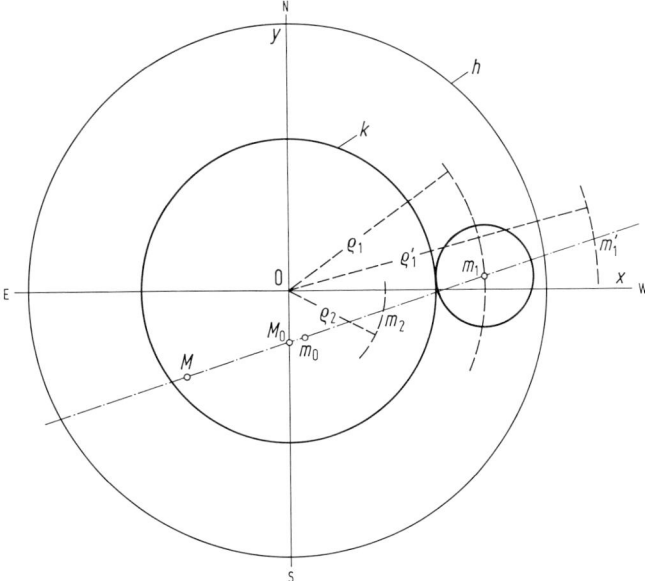

Fig. 16.8. Illustration of the predictive calculation of a lunar eclipse.

and its inclination to the $+x$-axis is

$$\tan i = -\frac{\Delta\delta_{\mathbb{C}} + \Delta\delta_{\odot}}{15\cos\delta_{\odot}(\Delta\alpha_{\mathbb{C}} - \Delta\alpha_{\odot})}. \tag{16.6}$$

At mid-eclipse, the distance between the centers of Moon and shadow becomes

$$\rho_0 = (\delta_{\mathbb{C}} + \delta_{\odot})\cos i, \tag{16.7}$$

and the length of the path which the Moon moves in its orbit between opposition and mid-eclipse is

$$\overline{m_0 M_0} = (\delta_{\mathbb{C}} + \delta_{\odot})\sin i,$$

and thus the instant of mid-eclipse occurs at

$$t_0 = T - \frac{\delta_{\mathbb{C}} + \delta_{\odot}}{\Delta\delta_{\mathbb{C}} + \Delta\delta_{\odot}}. \tag{16.8}$$

The beginning and end of the partial phase occurs when the distance of the Moon from the shadow center is

$$\rho_1 = \sigma + R_{\mathbb{C}}.$$

The segment of lunar orbit therefore is

$$\overline{m_0 m_1} = +\sqrt{\rho_1^2 - \rho_0^2}$$

and the corresponding instants are

$$t_0 \pm \frac{\sqrt{\rho_1^2 - \rho_0^2}}{\Delta\delta_{\mathbb{C}} + \Delta\delta_{\odot}} \sin i. \tag{16.9}$$

Correspondingly, the beginning and end of totality occur at

$$\rho_2 = \sigma - R_{\mathbb{C}},$$

and are calculated from the term

$$t_0 \pm \frac{\sqrt{\rho_2^2 - \rho_0^2}}{\Delta\delta_{\mathbb{C}} + \Delta\delta_{\odot}} \sin i. \tag{16.10}$$

Finally, the external contacts with the penumbra are obtained:

$$\rho_1' = \sigma' + R_{\mathbb{C}},$$

and hence

$$t_0 \pm \frac{\sqrt{\rho_1'^2 - \rho_0^2}}{\Delta\delta_{\mathbb{C}} + \Delta\delta_{\odot}} \sin i. \tag{16.11}$$

In these computations, the values σ and σ' are increased compared with Eqs. (16.1) and (16.1') owing to the increase of shadow size to be mentioned below. Link [16.7] and the *Astronomical Almanac* use the increase value of 1/50 (factor = 1.020), while the value 1/85 (factor = 1.0118) preferred by Danjon, Meeus, and Mucke is used in the present text. Finally, the magnitude of the eclipse follows from Eq. (16.2).

It is not difficult to write a program to calculate eclipses on a personal computer, but such programs can also be obtained commercially, in some cases even free of charge. Using commercial diskettes may render even the reading of coordinates, parallaxes, etc. from the *Almanac* redundant, as these data are also calculated. Various programs are offered in astronomical magazines, and at this point only the brief instruction by Duffet-Smith [16.8] will be mentioned as an example.

16.3.3.2 Graphical Representation of a Lunar Eclipse. The graphed picture of a lunar eclipse illustrates the basic numerical process. The same coordinate system as above is used, along with a suitable scale, such as $1' = 5$ mm. Circles centered on O are drawn with the radii 1.01σ (umbra) and $1.01\sigma'$ (penumbra). The positions X_0, Y_0, and x', y' of the Moon are graphed in this system. The segment between these two points is the hourly rate Δs of the Moon's motion. If opposition T occurs m minutes after a full hour, the Moon's position at that hour is found by extrapolating the segment back by $m\Delta s/60$ from M_0. Graphing Δs in both directions of the point thus found gives the Moon's position in full hours of UT, and by subdivision at intervals of 10 or 5 minutes.

Auxiliary circles are then drawn around the shadow center, with radii $1.01\sigma' + R_{\mathbb{C}}$ and $1.01\sigma \pm R_{\mathbb{C}}$. Their intersections with the Moon's path determine the instant of the contacts with umbra and penumbra. The position angles of the contacts are found directly using a protractor. Mid-eclipse is at the foot of the perpendicular to the lunar orbit at the shadow center.

16.3.3.3 The Position of the Terminator of the Earth's Shadow on the Earth's Surface.
In order to visualize which terrestrial regions affect the appearance of the eclipsed Moon, the terminator may now be graphed in a world map, preferably in Mercator projection. Tables in the *Astronomical Almanac* give times of sunrise and sunset depending on date and geographic latitude. (It is not necessary to use the *Almanac* for the current year as the differences between years are negligibly small.) Converted into angles, the tabulated times give the geographic longitudes of the shadow terminator at 0^h UT and at the given latitude, counting longitude east of Greenwich positive, west negative. By another instant H of UT the terminator is shifted westward by H degrees. It can thus be graphed for any instant during eclipse.

Meteorological and geographic features have a certain influence on the illumination of the eclipsed Moon, but generally only along some part of the terminator. In order to determine this "effective" part, imagine an eclipsed point on the Moon defined by the distance γ and the position angle P with respect to the shadow center. An observer at this point would see the Earth with a radius π (Fig. 16.9) while the position of the Sun behind the Earth would be given by the same coordinates γ and P. The effective part of the shadow terminator on the Earth's disk is bounded by the tangents t_1 and t_2. The geographic latitude ϕ of the bisecting point on the terminator follows from

$$\sin \phi = \cos \delta_\odot \cos P, \qquad (16.12)$$

and the bounding points of the effective arc have the latitudes

$$\sin \phi_{1,2} = \cos \delta_\odot \cos(P \pm \Delta P), \qquad (16.13)$$

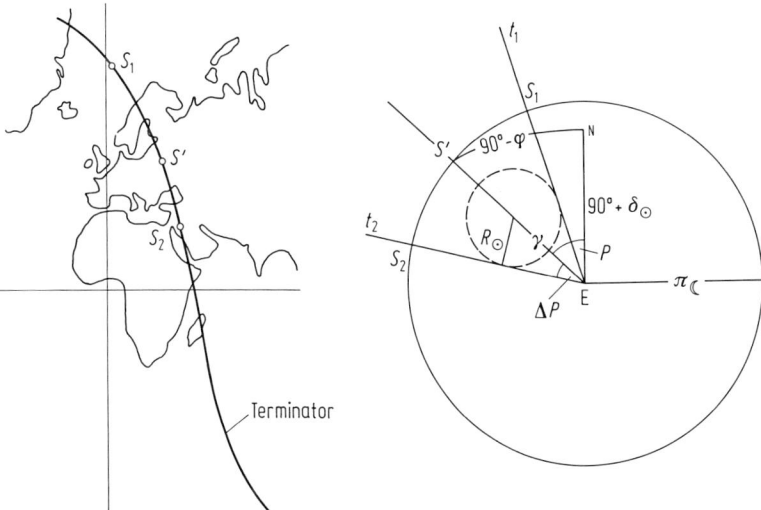

Fig. 16.9. Position of the terminator on the Earth's surface.

where

$$\sin \Delta P = \frac{R_\odot}{\gamma}. \tag{16.14}$$

Thus, the effective terminator of the shadow on the Earth can be located for any selected point on the Moon and for any eclipse phase (say, contacts or mid-eclipse). For contacts, points at the lunar limb will be chosen. Any point with a distance $\gamma \leqq R_\odot$ will be affected by the full 40 000 km length of the terminator.

16.3.3.4 Graphical Determination of the Transit of Craters. The same graph as used in Sect. 16.3.3.2 serves to approximately determine the moment at which a certain crater or other suitable feature passes through the boundary of the umbral shadow. The lunar disk is drawn on tracing paper at the same scale as before and the selenographic coordinate system ξ, η (see Sect. 15.2.4) graphed through the center and also the position of the η-axis relative to celestial north, which is given in almanacs as the position angle of the axis. Graphed on the lunar disk are then the formations of the Mucke–Rükl system [16.9] or other suitable points. This would be the view of the Moon as seen from the Sun if, at the time of eclipse, the selenographic longitude $\lambda_\odot$ and latitude $\beta_\odot$ of the subsolar point both equaled zero. During an eclipse, $\beta_\odot$ is indeed negligibly small (the constancy of the tilt of the lunar axis against the ecliptic has long been known as Cassini's second law), but the influence of the libration ℓ must be incorporated into $\lambda_\odot$. The selenographic coordinates are corrected using the relations

$$\begin{aligned}\xi' &= \cos b \sin(\ell - \lambda_\odot),\\ \eta' &= \eta.\end{aligned} \tag{16.15}$$

The transparent graph of the Moon, correctly oriented toward north, is then shifted along the lunar path, and the instants of passage of the craters through the edge of the umbra (1.01σ) are readily found. *The Astronomical Almanac* gives the colongitude $90° - \lambda_\odot$.

This method serves only as a guide when planning an observing program aimed at determining the increase of the shadow (Sect. 16.4.4.4). In this case, higher precision is not advised, as it may bias the observer.

The positions in the system of Mucke and Rükl [16.9] are preferred here over the formerly used "Berliner System." Their system contains 70 bright and rather point-like craters of suitable albedo, which can be more readily identified and photometrically measured (see Appendix Table B.13 in Vol. 3).

16.4 Enlargement of the Earth's Shadow

16.4.1 General Considerations

The fact that during a lunar eclipse the Earth's shadow appears somewhat extended compared with geometric expectations has been known for about three centuries. The problem centers, in principle, on the difference between the theoretical value σ of the umbral radius from Hipparchus's formula (16.1) and the somewhat larger σ_1 derived from observations of transits of craters and other lunar formations. Ephemerides are computed with an agreed-upon value of increase (Sect. 16.3.3.1), but the task remains to accurately determine σ_1, which may change from one lunar eclipse to the next, or even along the circumference of the umbra during a single lunar eclipse.

From a physical standpoint, the concept of the "edge of the umbra" and the value σ have no real meaning, as the depth D changes as a continuous function of distance γ from the center, increasing from the edge of the penumbra ($D = 0$) to the center ($D \approx 4$). But the density D changes rapidly in the vicinity of σ. As explained by Kühl's contrast theory [16.10], the eye then places a line of separation at approximately the distance where the $D = f(\gamma)$ curve shows zero curvature (inflection point); this position seems to be related to a layer of meteoric dust about 120–150 km above the Earth's surface.

16.4.2 Determination of the Enlargement of the Shadow by the Observation of the Transit of Craters Through the Terminator

The instant of passage of a particular crater or other suitable surface feature through the boundary of the umbral shadow can be determined at every lunar eclipse, provided the crater's position relative to the center of the Moon is known and the position of the latter relative to the the center of shadow has been calculated. Thus the distance of the crater from the shadow center at the instant of transit is determined, that is, the quantity σ_1. Its ratio with the geometrically expected σ gives the enlargement of the shadow.

Apart from the cooperation of the local weather, some other technical prerequisites include the following:

1. For observations, a small- to medium-sized telescope (aperture 5–15 cm) with just enough magnification (30–100×) to produce a sufficiently clear image and with a field of view that is large enough to make a survey, and an appropriate selection of craters.
2. A coworker who can keep a record, that is, read the time to an accuracy of 1–2 seconds, and record the instants of transits, comments given by the observer, etc. The latter thus is relieved of the writing, is not bothered by lighting conditions, and can concentrate fully on the observations. As lunar eclipses are none-too-frequent, enlisting the help of an assistant is always a good idea.
3. A prepared approximate ephemeris (see Sect. 16.3.3.4). This expedites observations substantially, forewarns the observer of the crucial times, and familiarizes him with the positions of the relevant craters. Also, the loss of time when searching for suitable craters is avoided. The predicted times—announced by the assistant—need not be more accurate than about one minute and will thus prevent any bias on the part of the observer. The necessary precision of the time readings can easily be achieved nowadays with a wristwatch possessing a large second hand or digital seconds display.

Corrections of times are carried out before and after eclipse according to time signals broadcast on the radio or telephone. The actual precision of time readings of transits is usually lower. For small and bright craters, one reading of transit through the shadow boundary should suffice; larger craters (if included in the program at all) are read for the transit of both walls and the estimated center, which is often characterized by a central peak.

The actual determination of transit is to some extent a subjective estimate. Precision and homogeneity of the series are impaired by straining to follow the crater too far into the shadow. Hoffmeister [16.11] proposes that, in cases of doubt, one should determine for each crater three instants: T_1, when the crater begins to disappear, T_2, when it in highest probability passes the edge, and T_3, when it is certain that the transit has been completed. The weighted mean of these three instants is then taken using the relation

$$T = (T_1 + 2T_2 + T_3)/4. \tag{16.16}$$

16.4.3 Reduction of Measurements to Determine the Enlargement of the Shadow

The reduction of measurements need not be left to professionals but can be performed using the method of Kozik [16.12], which is adapted to the modern ephemerides of the Sun and the Moon. The coordinate system chosen has its origin at the Earth's center, and the unit of length is the Earth's equatorial radius. The axis $+z$ is the shadow axis, directed along the line joining Sun and Earth, axis $+y$ directed toward the north ecliptic pole and $90°$ away from $+z$, and the $+x$-axis perpendicular to the y–z-plane in the eastern direction. In this system, the Moon has the coordinates

$$x_{\mathrm{C}} = \frac{\cos\delta_{\mathrm{C}} \sin(\alpha_{\mathrm{C}} - \alpha_1)}{\sin\pi_{\mathrm{C}}},$$

$$y_{\mathrm{C}} = \frac{\sin(\delta_{\mathrm{C}} - \delta_1)}{\sin\pi_{\mathrm{C}}} + 0.008\,726(\alpha_{\mathrm{C}} - \alpha_1) x_{\mathrm{C}} \sin\delta_1, \tag{16.17}$$

where α_1, δ_1 are the equatorial coordinates of the point opposite the Sun (z-axis), and $(\alpha_{\mathrm{C}} - \alpha_1)$ is in degrees.

The selenographic coordinates (λ, β) are converted into rectangular ones:

$$\begin{aligned} x_0 &= r_{\mathrm{C}} \cos\beta \sin\lambda, \\ y_0 &= r_{\mathrm{C}} \sin\beta, \\ z_0 &= r_{\mathrm{C}} \cos\beta \cos\lambda, \end{aligned} \tag{16.18}$$

and transferred into the Kozik coordinate system by translation and rotation:

$$\begin{aligned} x &= x_{\mathrm{C}} + a_x x_0 + b_x y_0 + c_x z_0, \\ y &= y_{\mathrm{C}} + a_y x_0 + b_y y_0 + c_y z_0, \end{aligned} \tag{16.19}$$

with the coefficients

$$a_x = -\cos \lambda \cos p - \sin \lambda \sin p \sin \beta,$$
$$b_x = \sin p \cos \beta,$$
$$c_x = \sin \lambda \cos p - \cos \lambda \sin p \sin \beta,$$
$$a_y = \cos \lambda \sin p - \sin \lambda \cos p \sin \beta, \quad (16.20)$$
$$b_y = \cos p \cos \beta,$$
$$c_y = -\sin \lambda \sin p - \cos \lambda \cos p \sin \beta,$$

where λ, β are the selenographic coordinates of the subsolar point and p the position angle of the lunar axis projected onto the x–y plane I through the center of the Moon.

All quantities in the preceding formulae are in Universal Time (UT) and can be interpolated from the *Almanac* data for full hours and, further, for the moments of observed crater transits through the edge of the shadow. The distance from the shadow center then follows from

$$r_0 = (x^2 + y^2)^{1/2} \quad (16.21)$$

and the position angle, reckoned from the equator, from

$$\tan \psi = \frac{y}{|x|}. \quad (16.22)$$

The theoretical radius of the shadow follows from the dimensions of the geoid,

$$r_C = 1 - \operatorname{cosec} \pi_{\mathrm{C}} \tan(R_\odot - \pi_\odot) - 3.376 \times 10^{-3} \cos^2 \delta_\odot \sin^2 \psi. \quad (16.23)$$

A comparison with the quantity r yields the enlargement factor of the shadow. A large number of very precise readings compiled into groups at intervals of, say, every $10°$ of the angle ψ, indicate a distinct flattening of the shadow which exceeds the theoretical value. Therefore, upon constructing a graph of r_0 versus ψ, the coefficients m and n can be obtained using the relation

$$r_0 = m - n \sin^2 \psi. \quad (16.24)$$

The constants m and n characterize the size and ellipticity of the shadow for the given lunar eclipse. The (sometimes asymmetric) shadow silhouette and the enlargement depend on the lunar parallax and on the flattening of the Earth and its atmosphere. Measurements taken over several decades, particularly by Bouška, yielded enlargements from 1/36 (2.8%) to 1/67 (1.5%) compared with the geometric value; this is yet to be explained physically; the present chapter retains 1/85.

16.5 Photometry of Lunar Eclipses

16.5.1 Principles

Ideally, photometry of lunar eclipses aims at the determination, preferably by plotting isophotes, of the overall behavior of the depth of Earth's shadow in all parts of the plane I through which the Moon passes (in approximation) during the course of an eclipse. The practical execution then consists of measuring the brightnesses of certain lunar details at various phases during the eclipse as well as outside of eclipse. The ratio of brightness i during eclipse to i_0 outside eclipse yields the depth of shadow according to the relation

$$D = -\log \frac{i}{i_0}. \tag{16.25}$$

It may also be given as the differences Δm, expressed in stellar magnitudes:

$$\Delta m = 2.5 D. \tag{16.26}$$

The measurements shown in Fig. 16.7 demonstrate that shadow depths up to $D = 5$ can occur, which means that the measurements ought to bridge intensity ratios up to $1:10^5$ in a reliable fashion, and this is indeed a delicate task. It requires an accurate calibration of the photometric equipment and adaptation of sensitivity over this range. The quantities i and i_0 cannot be measured simultaneously, but only with an interval of one hour or more. Over this time, the atmospheric extinction is likely to change, first because of variable meteorological conditions, and second because of the changing position of the Moon in the sky. The elimination of the influence of the variable extinction is thus an important problem inherent in any photometric method applied to lunar eclipses.

The scattering of light which occurs in the Earth's atmosphere and within the optical system also interferes with the measurement of i and i_0. When, immediately after the beginning of the partial phase, measurements are made of a point on the Moon in the umbra, then the remaining part of the Moon, with little penumbral dimming of light, illuminates the optical surface of the instrument (dust, diffraction, etc.) as well as the column of air between the Moon and observer. This results in unwanted scattered light being superposed on the rather weak light of the detail measured in the umbra, and causes a spurious diminution in the depth of the shadow.

Near the edge of the umbra, the shadow depth varies rapidly with distance from the center, and thus also with time. Consequently, a small area on the Moon must be measured in quick succession.

Finally, the measurements are to be made using glass, gelatin, or interference filters in a sufficiently narrow wavelength range (monochromatic light) since, as theory has demonstrated, the shadow depth depends rather strongly on the color of the light (Sect. 16.3.2).

Blue filters in particular should be checked for transmission of small amounts of red light. This red leakage may seem at first glance to be harmless enough, and is so for normal photometry or photometry in the penumbra. In the umbra, however, with the blue light weakened by about a factor of 100 compared with red light, any traces of red light leaking through the filter may seriously contaminate the measurements.

Data on average values of the shadow depth are also yielded by global photometry of the eclipsed Moon. This method is to some extent even simpler with respect to the allowance for scattered light and extinction. But it cannot, of course, supply the richness of detail obtained by photometry of selected lunar formations. The following section presents the various techniques, from estimates of the overall magnitude up to precise measures of individual regions. It will illustrate the intrinsic difficulties, but will also convey the assurance that careful work will lead to scientifically interesting results.

16.5.2 Global Photometry

16.5.2.1 Visual Estimates by Danjon's Method. It is known from numerous experiences that lunar eclipses may be bright or dark, gray or colored. Among the historical records of cases where the Moon disappeared entirely for a shorter or longer time span, there are reports of no fewer than 4 such eclipses during Kepler's time (17th century), and other similar cases were reported in later centuries. The eclipse of 1964 June 24 also belongs to this category. On the other hand, the dimming was so weak in some other cases that the observers expressed doubts about the reality of the event.

When Danjon [16.13,14] was collecting and processing observations which refer to brightness and coloring of the eclipsed Moon, he introduced a a five-step scale for the classification of these eclipses (see Table 16.2).

This scale is based upon the simultaneous gradation of brightness and color, which relies upon observed data as well as on theoretical reasoning. Theoretically (Sect. 16.3.2), the eclipsed Moon should, assuming ideal atmospheric scatter, display a distinctly reddish tint. Any contamination of the atmosphere due to condensation and volcanic or meteoric dust will cause a further diminution of sunlight. Unlike the very selectively absorbing Rayleigh atmosphere, however, this effect should be practically neutral. For these reasons and the fact that the human eye lacks color perception at low light levels, extremely dark eclipses will display less color than bright ones.

The many experiences made with the Danjon classification scheme appeared quite satisfactory for the observing data which covers the past three centuries. Danjon was eventually able to find a relation between the brightness of a lunar eclipse and the solar activity as represented by the 11-year cycle. At the beginning of the cycle, lunar eclipses are dark ($L = 0$ to 1), but in the further course of the cycle they brighten ($L = 1$ to 2), and are usually brightest ($L = 3$ to 4) shortly before the new minimum which begins the next cycle. The minimum of solar activity corresponds in lunar

Table 16.2.

Step L	Description
0	Very dark eclipse, the Moon is almost invisible especially near mid-totality;
1	Dark eclipse, gray or brownish tint, detail difficult to distinguish;
2	Dark red or rust-colored eclipse, with a very dark spot at shadow center, rather bright outer portions of shadow;
3	Brick-red eclipse, shadow often bounded by a yellowish zone;
4	Copper-red or orange-red, very bright eclipse with a bluish, very bright umbral zone.

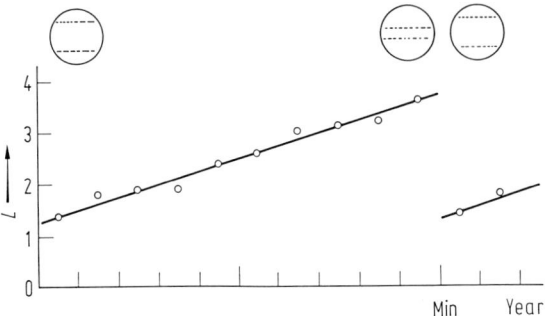

Fig. 16.10. The Danjon relation between eclipse darkness and positions of sunspot activity.

eclipses to a sudden drop in brightness (Fig. 16.10), while the activity maximum does not affect the brightness curve at all.

This relation suspected by Danjon may have connections to lunar luminescence (Sect. 16.6.4). This normal behavior of eclipse brightness is, however, disrupted by, for instance, volcanic eruptions, which pollute the atmosphere with large quantities of fine volcanic ash, or by the influence of meteoric dust (Sect. 16.4.6.2).

It should be noted that not all investigators have distinctly verified the Danjon relation, and many do not give it credence at all. For instance, Bernheimer [16.15] suspects the following influences as playing a role: (1) the position of the center of the Moon relative to the shadow center (the northern part of the Earth's shadow is brighter than the southern part), and (2) a seasonal effect (darker eclipses in winter) with the ecliptic longitude of the shadow at opposition. Also, Sekiguchi [16.16] has failed to find any correlation which supports Danjon's conjecture.

Although the interpretation may be controversial, Danjon's classification scheme by itself should continue to be practiced, especially since it is so easily applied. During the eclipse it is particularly applicable for the study of the total phase, the brightness of which need not be symmetric with respect to mid-eclipse. It is therefore recommended to continue the estimates for the duration of totality. During the partial phase, the light from that part of the Moon which is still in the penumbra and thus much brighter, is a hindrance. Hence estimates of $L = 0$ to 2 of the umbrally eclipsed portion may not be real. Estimates of $L = 3$ or 4 are more significant, as the brightness contamination in this case acts in the opposite sense. Lunar eclipses of this class are truly very bright.

16.5.2.2 Silver-Ball Photometry. Very satisfactory total visual photometry can be obtained with a *silver-ball photometer*, as described by Richter [16.17] and others. It consists primarily of a colorless glass ball possessing a smooth and highly reflective inner surface, similar to that of the popular ball ornaments which are used to adorn Christmas trees. A diameter of about 60 mm has been found satisfactory. (It is possible, though less favored, to use two different balls, e.g., of 40 and 80 mm diameters.)

The Moon reflected at the convex surface appears much reduced in size and very nearly starlike. The ball is attached to the end of a long pole and held in the vicinity of a suitably chosen star. The position of the eye is then varied until the reflected images of the Moon and the star are matched in brightness. Since the human eye is sensitive to even minute brightness differences, quite satisfactory precision can be achieved using this technique. The distance between ball and eye is found using a tape measure, and is recorded together with the time and comparison star.

Some practical hints for the application of this method follow. The evaluation will be less problematic if reference stars with approximately K-type spectra have been selected; the colors of such stars are very nearly that of the eclipsed Moon. The comparison stars should suffice to bridge a substantial interval of brightness. Thus, faint and bright comparison stars should be selected, perhaps even an object as bright as the planet Jupiter, if in a convenient location in the sky and if its brightness at that time is known. The observer should beware of variable stars.

Should the Moon become so faint that the ball must be held close to the eye, readings become unreliable because then the lunar image ceases to be starlike. Note also that the observer faces the direction more or less opposite to that of the Moon in the sky, and it is in this region that comparison stars must be chosen. To keep the correction for extinction small, stars and Moon should, on the other hand, be at about the same altitude; unfortunately, this prevents the use of the same stars throughout the entire eclipse. There is no ideal solution. One must find the best compromise between the stated requirements.

Owing to the simple photometric behavior of the silver ball, the observations obtained by it are readily reduced. According to Richter, the amount B of reflected lunar light obeys the proportionality relation,

$$B \sim \frac{ik\rho^2}{4\Delta^2}, \tag{16.27}$$

where i is the received light intensity from the Moon, k the reflection coefficient of the ball (usually ≈ 1), ρ the radius in meters of the ball, and Δ its distance in meters from the eye. Converted into the logarithmic form of magnitudes, it becomes

$$m_{\mathrm{C}} = m_* - A - 5\log\Delta, \tag{16.28}$$

where m_* is the visual or photovisual magnitude of the comparison star, and where

$$A = 2.5\log\left(\frac{4}{k\rho^2}\right) \tag{16.29}$$

is a constant characterizing the ball. For instance, for a 60-mm ball with perfect reflectivity ($\rho = 0.03$ m, $k = 1$), $A = 9.11$ magnitudes.

When the extinctions E depending on zenith distance are allowed for—for instance with the tables by Müller [16.18] for low-lying or mountaintop stations—the final reduction formula reduced to the zenith (indicated by superscript 0) is obtained. The brightness of the Moon in magnitudes becomes

$$m_{\mathrm{C}}^0 = m_*^0 + E_* - E_{\mathrm{C}} - A - 5\log\Delta. \tag{16.30}$$

A light curve thus obtained for the eclipse of 1982 January 09 by Haupt [16.19] is shown in Fig. 16.11.

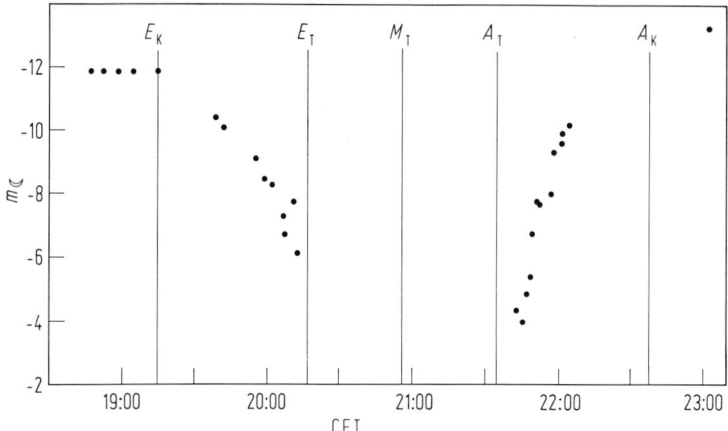

Fig. 16.11. Change in total brightness of the Moon during the eclipse of 1982 January 09. E_K = entrance umbra, E_T = entrance into totality, M_T = middle of totality, A_T = emergence from totality, A_K = emergence from umbra. After Haupt [16.19].

16.5.3.3 Photoelectric Global Photometry. Direct photoelectric photometry of the entire Moon is naturally more precise than the silver-ball method, but requires, first, that the focal image of the Moon be small enough to fit entirely onto the photocathode, and second, that the large brightness differences that occur can be safely bridged. If necessary, one can resort to objective stops, whose light reduction is calculated from their free surfaces or is calibrated from light sources. Much more difficult is the comparison with stars and the correction for extinction.

The aforementioned problems are indicated, for instance, by measurements by Morton [16.20]; the brightness and color of the sky at a position 20° north of the Moon change during an eclipse. Figure 16.12 gives an idea of the precision attainable photoelectrically. It shows an asymmetric pattern particularly in the *B-V* color. Light curves representing total brightness are generally expected to be asymmetric. Only if the lunar disk were of uniform albedo and the isophotes in the Earth's shadow circular would the light curve become symmetric relative to mid-eclipse, as Figure 16.13 a illustrates.

The actual concentration of lunar maria in the northern hemisphere and of highlands in the south causes some predictable asymmetry of the light curve. The isophotes in the shadow also are not circular, but their unpredictable deformations contribute to the asymmetric light curve. In practice, two categories may be distinguished. If the light curve is asymmetric in the sense relating to the albedo distribution on the Moon (Fig. 16.13 b), nothing can be deduced about the shape of the isophotes, and moreover the case has little interest. More revealing is the alternative case when the observed asymmetry of the light curve runs in conflict with the distribution of lunar albedo, and the phenomenon indicates rather substantial deformation of the isophotes in the Earth's shadow (Fig. 16.13 c).

The enormous recent progress in electronic methods opens a broad field to both professional and non-professional observers. Area detectors such as CCDs are being

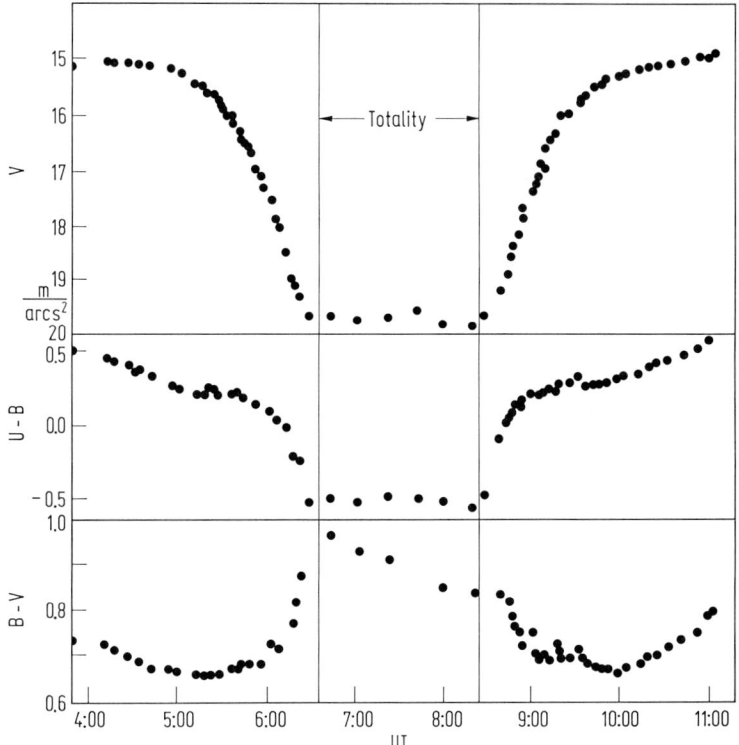

Fig. 16.12. Brightness and color changes of the sky during the eclipse of 1982 July 06. After Morton [16.20].

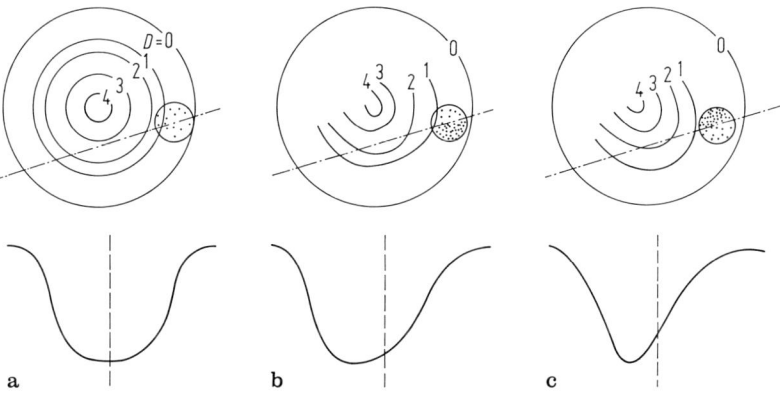

Fig. 16.13a–c. Asymmetry of light curves in total photometry of the Moon.

16.5.3 Detail Photometry of Individual Features on the Moon

General comments regarding the photometry of small areas and details on the lunar surface have already been given (Sects. 16.3.2 and 16.5.1). Here, some specific hints on their execution will be mentioned (for results, see also Sect. 16.6.4).

With a large telescope, small craters and other prominent features are masked out. They are continuously measured, for instance, with a photoelectric photometer and selected color filters while the shadow passes over them. The values thus obtained are graphed against the penetration into the shadow, and the measured shadow density can subsequently be compared with the computed one. The resulting agreement or potential deviations may provide useful information for the rather complex examination of theories on the upper atmosphere of Earth. As a simple example, measurements of the crater Tycho by Widorn [16.22] on 1953 January 30 are shown in Fig. 16.14. Allowing for foreground and scattered light, the transition from penumbra into umbra ($\gamma = 1$) was particularly well recorded. In this case, satisfactory agreement with theory was found.

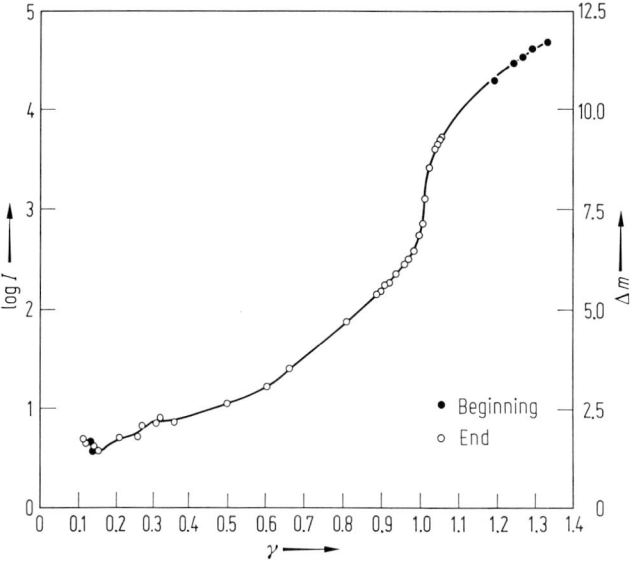

Fig. 16.14. Shadow depth ($\log i$ or Δm) of the crater Tycho as a function of the distance γ from the shadow center (in units of the geometric umbral radius) during the eclipse of 1953 January 30. After Widorn [16.22].

16.5.4 Photography of the Eclipsed Moon

It is nowadays quite common for amateur astronomers to record a series of single exposures of a lunar eclipse on 35-mm or movie film. Because of the availability of a wide variety of photographic emulsions with high speeds, it is now feasible to take very short exposures, thus permitting the continuous monitoring of an eclipse with either a fixed or tracking camera. Many variations can easily be imagined. For example, the Moon can be made to appear as a trail of varying width (= brightness) with the shutter held open during the total phase. Another good plan is to capture, in several consecutive exposures, the rising Moon as it enters eclipse against the terrestrial backdrop. In this case, the position of the camera and the intervals between exposures are chosen so that as many as possible (or at least some) non-overlapping pictures are secured on one film frame. Since the Moon moves through its own diameter in about two minutes when it is near the equator, an interval of about 5 minutes should in most cases be sufficient to give a reliable separation of the resulting images. Obviously, a stable tripod is essential for this kind of work.

Of course, telescopic exposures will also be instructive; Figure 16.1, for example, was made using a telescope with a focal length of 2 m. Because developments in photographic techniques are quite rapid, it is difficult to give recipes which will not soon become superseded. The exact procedure to follow will depend entirely upon the planned program. Owing to the short exposure times, image blur due to the Moon's motion need not be feared, even though the telescope usually tracks at the sidereal rate.

With some experience, it should be possible to obtain color photographs which cover wide ranges of true brightness and color. Nevertheless, the possibilities of present-day image processing have so far scarcely been applied to this task, let alone exhausted it. In the future, color photographs with suitable calibration (which for known and random photographic effects is always the most delicate problem), may permit a more accurate classification of elipses over their course than the estimates of the Danjon type.

Time-lapse photographs of eclipses on movie film (cinephotography) are very instructive when obtained using the correct, previously tested exposure time and with good tracking. Careful preparation is a must. In any event, amateurs have played a major role in and contributed substantially to the progress of this area of eclipse photography.

16.6 Scientific Conclusions from Photometric and Astrometric Results

The purpose of this section is to outline the scientific benefits to be reaped from careful observations of lunar eclipses. The most detailed treatment of such problems is by F. Link [16.7], who investigated in particular the Earth's upper atmosphere by means of such observations. As is generally the case in physics and astronomy, research starts with plausible model assumptions; in other words, the photometric course of a lunar eclipse is calculated under certain simple hypotheses (e.g., an uncontaminated atmosphere with the usual scattering and refractive properties). The theory also gives the depth and color of the shadow under normal conditions as a function of the distance from the shadow center. The observed values are then confronted with the calculated ones, and explanations must be found for any deviations which may have occurred. Actually, several possible causes for abnormal patterns have been investigated, some of which will be mentioned in the following sections.

16.6.1 Cloudiness at the Earth's Terminator

It is easily seen that the presence of clouds at the Earth's terminator, i.e., where the solar rays are refracted into the umbra, will have an effect on the eclipse phenomenon. This effect changes with the continuous rotation of the Earth, as new zones with different meteorological features come into play. Seasonal and geographic effects are displayed by even the cloudless Rayleigh atmosphere: at the poles, refraction is stronger and the cloud cover lower; the result is a greater brightening of the shadow center. The change of cloudiness often effects an irregular deformation of the isophotes.

16.6.2 Volcanic Eruptions and Meteoric Dust

As previously indicated, it is the refractive properties of the atmosphere, particularly at heights of between 5 and 25 km, which have the greatest influence on the illumination of the eclipsed Moon. Apart from clouds and the ozone content (see below), dust particles especially can considerably change the refractive index. Volcanic eruptions such as that of Krakatoa in 1883 were soon identified as having caused a series of unusually dark lunar eclipses. When the observed brightnesses during the eclipse were compared with those of model calculations corrected for other effects, it was found that all major volcanic eruptions have caused extreme eclipse darkness. Keen [16.23] showed that this effect arises because volcanic ash is injected into the statosphere and distributed all around the Earth by atmospheric circulation. Some noted examples are the eruptions of Mt. Agung in 1963 and of El Chichón in 1982. A strong eruption may render the Moon almost entirely invisible at a subsequent eclipse. It can take months to years for the fine dust to settle and therefore for eclipse brightness patterns to return to normal.

The character of a lunar eclipse can also be affected by the presence of meteoric dust, which is located originally at great heights (100 to 150 km). By contrast with volcanic dust, which can darken the umbral shadow, meteoric dust contributes to an *enlargement* of the shadow. There is indeed a relation between the percentage enlargement of the shadow and the annual pattern of meteoric activity. Major meteor streams also appear to cause a diminution of brightness of the eclipsed Moon which, however, returns to normal after a moderately short time (1 to 3 months).

16.6.3 The Ozone Problem

The spectral distribution of the visible radiation falling onto the eclipsed Moon is substantially influenced by atmospheric ozone. The ozone generates spectral absorption bands, which in turn are sensitive indicators of the total amount of ozone in the layer at 20–25 km altitude. The theory of a pure or a haze/dust-affected atmosphere is compared with the actual observations, and conclusions on the ozone content as well as on the instantaneous distribution of ozone over the Earth can be drawn. This distribution is confirmed by other measurements (balloons, satellites, etc.), and it also varies with the seasons and with geographic latitude (Paetzold in [16.24] and other papers). The ozone absorption is, according to Paetzold, also responsible for the often-seen, narrow greenish-tinted zone located at the edge of the umbra, which is discussed, for instance, by the Berlin Lunar Observers [16.21].

Even in this time of continuous satellite monitoring of the atmosphere and in particular of the "ozone hole," discussion of past lunar eclipses may still, in retrospect, provide some information on the long-term changes of atmospheric ozone content.

16.6.4 The Solar Influence: Luminescence on the Moon

The examination of Danjon's relation must at this point be resumed, insofar as it attempts to relate to the appearance of lunar eclipses with solar activity (Sect. 16.5.2.1). Its rationale was: (1) The corpuscular radiation from the Sun (i.e., the solar wind) does not propagate rectilinearly, and therefore can get into the umbra, where it causes luminescence (i.e., additional brightening) when incident on the lunar surface; (2) it is assumed that this is all the more the case when the activity zone is located at a low heliographic latitude. This discontinuity in both latitude of solar emission and lunar brightness occurs just at times of activity minima, as is illustrated in Fig. 16.10. It may be objected that, just at these times, the spot activity in any event is very low, and that coexistence of high-latitude and low-latitude spots may last for several years. Thus, while the explanation of the effect through the spots themselves does not appear plausible, there could still be the influence of the *solar wind* which manifests itself in the suggested way, i.e., by luminescence.

Penumbral eclipses (which are, by the way, the event of a partial solar eclipse as seen from the Moon) are of interest in obtaining evidence of luminescence. In these cases it is found that a large part of photospheric sunlight is blocked by the Earth; this is true to a much lesser extent, however, for the high-energy ultraviolet radiation coming from the corona. It is precisely this radiation which excites luminescence and makes the umbra substantially brighter than would be expected from theory. Also, the photometry of individual formations described in Sect. 16.5.3 also reveals events of brightening compared with theory, and thus gives quantitative data regarding luminescence.

References

16.1 Oppolzer, T. von: *Canon der Finsternisse* (reprint), Dover, New York 1962.
16.2 Mucke, H.: *Der Sternenbote* **30**, 2–15 (1987).
16.3 van den Bergh, G.: *Periodicity and Variations of Solar and Lunar Eclipses*, T. Willink, Haarlem 1955.
16.4 Mucke, H., Meeus, J.: *Canon of Solar Eclipses – 2003 to 2526* (2nd edn.), Astronomisches Büro, Vienna 1992.
16.5 Meeus, J., Mucke, H.: *Canon of Lunar Eclipses – 2003 to 2526* (3rd edn.), Astronomisches Büro, Vienna 1992.
16.6 Paetzold, H.K.: Mondfinsternisse und das Studium der Erdatmosphäre. *Die Sterne* **28**, 86–91 (1952).
16.7 Link, F.: *Eclipse Phenomena in Astronomy*. Springer, New York 1969.
16.8 Duffet-Smith, P.: *Practical Astronomy with your Calculator* (2nd edn.), Cambridge University Press, Cambridge 1981.
16.9 Mucke, H., Rükl: *Lunar Maps for Eclipse and Occultation Observers*, Astronomisches Büro, Vienna.
16.10 Kühl, A.: Über den Einfluß des Grenzkontrastes auf Präzisionsmessungen. *Phys. Zeitschrift* **29**, 1-34 (1928).
16.11 Hoffmeister, C.: Zur Beobachtung der Mondfinsternisse. *Die Sterne* **29**, 166 (1953).
16.12 Kozik, S.M.: Kontur zemnoj teni pri Lunnom zatmenii. *Bull. Tashkent Astr. Obs.* **2**, 79 (1940).
16.13 Danjon, A.: Sur une relation entre l'éclairement de la Lune éclipsé et l'activité solaire. *Comptes rendus, Paris* **171**, 1127–1129 (1920).
16.14 Danjon, A.: Nouvelle détermination de la période solaire basée sur la loi d'éclairement des eclipses de Lune. *Comptes rendus, Paris* **171**, 1207–1210 (1920).
16.15 Bernheimer, W.E.: Kernschattenantritte und Klassifikation der totalen Mondesfinsternis, 1931 September 26. *Lund Observatory Circular No. 4*, 72 (1931).
16.16 Sekiguchi, N.: Photometry of the lunar surface during lunar eclipses. *Moon Planets* **23**, 99–107 (1980).
16.17 Richter, N.: Photometrische Beobachtungen der Gesamthelligkeit des Mondes im Verlauf der totalen Verfinsterung vom 2. zum 3. März 1942. *Zeitschrift für Astrophysik* **21**, 249–253 (1942).
16.18 Müller, G.: in E. Schönberg: Theoretische Photometrie, Handbuch der Astrophysik, Vol. II/1, pp. 264–267, Springer, Berlin 1929.
16.19 Haupt, H.: Verlauf der Gesamthelligkeit des Mondes während der totalen Finsternis am 9.Jänner 1982. *Mitt. Univ.-Sternwarte Graz Nr. 87 = Anzeiger der math.-nat. Klasse der Österr. Akad. d. Wiss.* **119**, 65–66 (1982).
16.20 Morton, J.C.: Sky Brightness and Colour Changes During the 1982 July Lunar Eclipse. *Observatory* **103**, 24–25 (1983).
16.21 Berliner Mondbeobachter, Protokolle der Sitzungen der Berliner Mondbeobachter, serially appearing at the Wilhelm-Foerster-Sternwarte, Munsterdamm 90, D-12169 Berlin.
16.22 Widorn, T.: Die Helligkeiten im Erdschatten nach lichtelektrischen Beobachtungen am Astrographen. *Mitt. der Universitäts-Sternwarte Wien*, Bd. 6, Nr. 10. = *Sitzungsberichte der math.-nat. Klasse der Österr. Akad. d.Wiss.* **162**, 88–92 (1953).
16.23 Keen, R.E.: Volcanic Aerosols and Lunar Eclipses. *Science* **222**, 1011–1013 (1983).
16.24 Paetzold, H.K.: Die optischen Erscheinungen bei einer Mondfinsternis. *Zeitschrift für Astrophysik* **30**, 282–292 (1952).

17 Occultations of Stars by the Moon

W.D. Heintz

17.1 The Moon as an Astronomical Clock

A *lunar occultation* is a small-scale eclipse event—less spectacular but far more frequent than ordinary eclipses—which occurs when the Moon passes in front of a star or planet. The sudden "switching off" or "on" of a star, particularly at the dark lunar limb, is noteworthy as it directly illustrates the motion of the Earth's companion. In addition, the accurate timing of these events has for many years enabled a large number of observers to participate with surprisingly modest instruments in the solution of basic research questions. The task is to empirically connect the astronomically measured universal time (UT), the international atomic time, and the dynamical time, a strictly uniform measure of time on which the calculations of the motions of celestial bodies are based. The definition of time touches upon some very deep yet subtle problems in astronomy and physics ([17.1] and Chap. 2), including general relativity theory. Occultations of stars form a direct link between the Moon as a celestial clock and the system of space coordinates represented by stars.

The motion of the Moon is quite fast (about $13°$ per day) and thus easy to measure, making it suitable as a chronometer. On the other hand, its motion is affected by a multitude of perturbations, the largest effects of which were observed long ago by Hipparchus. Achieving agreement between orbit theory and actual observations has presented one of the most formidable challenges in celestial mechanics. Observed eclipses and stellar occultations from past centuries and even from antiquity have been of great help in separating the long-term changes of the Moon's orbit from those in the Earth's rotation and the time measurement. Despite the low precision of the old data, the large time difference to the present is invaluable.

Of the lunar coordinates with respect to the Earth, the distance from the Earth to the Moon is now known to centimeter accuracy from laser reflectors. Not so satisfactorily precise is the lunar position in declination or ecliptic latitude; this fact has stimulated an interest in "grazing" occultations (Sect. 17.4).

The Moon travels with a speed of 1 km s^{-1}, the equivalent of its own diameter in about one hour. This corresponds in the sky to $0\rlap{.}''5$ per second of time. Because of the Earth's rotation, which is in the same west-to-east sense, the Moon's apparent motion is somewhat diminished by diurnal parallax (while the Moon is above the horizon). Thus, the duration of central occultation of a star may as much as about 70 minutes (at midlatitudes) or even more (in tropical latitudes).

The largest displacement of the Moon's orbit is caused by the retrogression of the nodes; for a given star, there can be one or two series of occultations which recur in the 19-year cycle. The brightest objects within 6° from the ecliptic which can be occulted by the Moon (at 5° orbital inclination and 1° parallax) are four first-magnitude stars (Aldebaran, Regulus, Spica, and Antares), a few stars of second and third magnitudes, and, of course, the planets.

As everywhere in astronomy, the demands on observational precision have increased greatly. The systematic errors affecting the timing of occultations should be controllable to within 0.1 seconds of time, the random uncertainty should not exceed on the average ± 0.25 s (though at reappearances and under other impeding conditions, larger errors are often inevitable), and binoculars cannot be considered sufficient. Nevertheless, a small telescope, a few accessories, some experience, and care are all that are needed to carry out useful work. In particular, one observer should perform a large number of timings. This not only is helpful in the sense of experience gained, but also helps during evaluation to judge the systematic consistency of the data. Isolated observations are virtually useless; the quantity of work is of as much value as the precision obtained at each individual event. In geographic coordinates on the Earth, $1''$ corresponds to about 30 m. The position of the observer (including elevation above sea level) needs to be known to at least this accuracy, lest the large influence of the Moon's parallax cause noticeable systematic errors.

In addition to the direct errors of observation, error sources of other kinds, but of the same order, limit the overall accuracy. These are, in particular, the uncertainty of the star's position and the uneven limb of the Moon. Small errors of proper motions cause all computed positions of stars to lose accuracy in the course of time; thus, the *Smithsonian Astrophysical Observatory Catalogue* (SAOC) is currently becoming outdated, as did all earlier catalogues, and positions of fainter stars from other sources have even larger random errors of several arcseconds. As the revisions of these catalogues progress and the corrections for the lunar profile become better known, these error values are being gradually reduced.

The desired control of the lunar motion through numerous observations and from widely distributed locations is supported by central offices by supplying instructions, reports, and special prediction lists of events to active observers. Also, the results of measurements (differences between observed and calculated contact times) can be obtained later after processing. The contact address in the U.S. for observing forms, instructions, and the *Occultation Newsletter* is H. DaBoll, 6 N. 106 White Oak Lane, St. Charles, IL 60175, U.S.A.; the U.S. Naval Observatory is equipped to compute lists of predictions, and working groups supported by amateur organizations also exist in other countries. The data are collected and evaluated centrally at the International Lunar Occultation Centre, Hydrographic Dept., Tsukiji-5, Tokyo 104, Japan.

The aspiring occultationist may first view and measure a few events according to published predictions, and find out if the task is to his/her liking, and if the instruments—what few that there are—work up to expectation. Then he/she should seek out advice and also the proper forms. The predictions available and the methods of observations are outlined in the sections that follow.

17.2 Predictions

The algorithms for predicting at what time and for which region a star will be occulted are basically the same as those for solar eclipses but much simplified since the diameter of the eclipsed object, its parallax, and its motion during eclipse are significant for the Sun (and also for the planets; see Sect. 17.5) though negligible for stars. The calculation still requires a computer or at least a programmable calculator as it is applied to so many potential cases: the formulae in Sect. 3.6 in Vol. 1 are solved for equal time intervals and interpolated; $(x - \xi)^2 + (y - \eta)^2 = 1$ is the condition for the contact. Within 24 hours, the occultation "shadow" cylinders of as many as 100 stars of magnitude +9.0 or brighter may sweep across the Earth's surface, and the Moon's passage through the Pleiades or Hyades may present the observer with a string of bright-star occultations in a single night. The positions of zodiacal stars from the SAOC and AGK catalogues are machine-readably stored, as are the parameters of the lunar motion; the computer can then obtain the apparent star positions, and quickly filter out the occultation candidates for every lunation and every terrestrial region.

The *ingress*, or disappearance, and *egress*, or reemergence, of a bright star is an exciting event to witness; such occurrences are previewed in many astronomy magazines. Some calendars, such as the *Observer's Handbook* (published by the Royal Astronomical Society of Canada) and *Kalender für Sternfreunde*, list selected predictions with contact times t_0 for certain reference stations (western longitude λ_0, latitude φ_0) along with coefficients a, b which permit the computation of the expected instant t (in minutes of time) at another location (λ, φ) by means of the relation

$$t - t_0 = a(\lambda - \lambda_0) + b(\varphi - \varphi_0), \tag{17.1}$$

usually with adequate accuracy to about 500 km. Standard lists include NZC stars to magnitude +2 for contacts at the dark limb but brighter stars only for events in daylight or twilight, near the horizon, or around full moon.

The letters "D" and "R" denote disappearance and reappearance; the position angle of contact relative to the center of the Moon (counted in degrees from north) is particularly useful for reappearance observations.

The parallactic displacement of the Moon for a given site and time can be calculated in good approximation as follows: geocentric latitude φ' and radius vector r' of the location are found as in Sect. 2.2.1. With the declination δ, the hour angle t, and the horizontal parallax π of the Moon at that instant (taken from the *Astronomical Almanac*), one obtains

$$\begin{aligned} 15\cos\delta\Delta\alpha &= -\pi r' \cos\varphi' \sin t, \\ \Delta\delta &= -\pi r'(\sin\varphi' \cos\delta - \cos\varphi' \sin\delta \cos t). \end{aligned} \tag{17.2}$$

The central offices in charge of predictions also offer calculations for individual sites and include fainter stars. To make the work worthwhile, the observer should aim for observations of at least 60 contacts per year. With a telescope reaching to magnitude +8 or +9 (of good visibility near the Moon!), this requires minimal labor, as one night often presents two or more occultations. If the observer's coordinates are good enough, these predictions should be correct to 1 second of time, unless the star position from the catalogue used or the lunar profile causes larger deviations. A revised version of the NZC was prepared by the US Naval Observatory in 1990.

17.3 Timing of the Contacts

Assuming good transparency, the NZC stars to magnitude $+7.5$ near the Moon and their occultations are seen distinctly enough even in a small telescope (often the available optics will reach to magnitude $+9$ and fainter). The magnification employed is usually only $50\times$ to $100\times$. Low power is especially useful for reappearances, so that the position angle can be easily estimated in the field of view, and the region of the expected egress conveniently viewed. Also, the scintillation will, at so low a power, interfere with the perception of the contact only in very poor seeing. A higher power eyepiece is advised in hazy sky with much scattered light, and also for ingresses of bright stars at the limb.

If the observer has not already determined the coordinates of his telescope—a must for lunar occultation work—the commercially available ordnance survey maps on a scale 1:25 000 easily permit a reading to the requisite ± 30 m.[1] The height above sea level is read with corresponding precision from the contour lines, and the height of the telescope above ground level is added.

The time is provided by a radio which receives one of the continuous time signals on long-wave or short-wave (see Table 2.8 in Vol. 1), and the time of contact is determined using a stopwatch, cassette recorder, or by the "eye–ear" method. If no continuous time signal is received, a chronometer should be compared with the signal at the full hour before and after the contact, and its rate determined.

The telescope need not track, so long as the setting is prepared such that the star (or point of egress) does not move to the edge of the field by the expected time of contact. After preparation, the observer should return to the eyepiece about 2 minutes before the event in order to await the contact with a relaxed eye which has not yet been fatigued by bright moonlight.

The stopwatch is most frequently employed to measure time intervals. The observer usually starts the watch at the contact and then, after turning up the radio volume, stops it at the next full minute of the time signal. The watch should have $0\overset{s}{.}1$ beats; it is used fully wound, but, for protection against wear, it should be allowed to run down at least partially following use. This procedure requires the determination of the clock rate, clock errors, and the reaction time of the observer.

In mechanical action clocks, the clock rate often changes with temperature owing to the thickening of the lubricating oil. The watch should be tested occasionally by extended time runs of 30 to 60 minutes, and, on each observing night, brief comparisons with the time signals will aid in finding irregularities. Some watches lag or jump when put into motion, catch slightly at certain positions, or are incorrectly read because the dial is slightly eccentric. They may run slightly differently depending on whether they are held horizontally or vertically, or display irregularities in very cold weather. It is worth checking the watch in various ways in order to find its little quirks; even good makes are not guaranteed to be free of errors to the amount needed here.

---

1 For enhanced legibility, the cartographer may sometimes "generalize" the map so that, for instance, streets in cities and towns are drawn slightly wider than scale, but this should not exceed 1 mm.

The "personal equation" includes the reaction time of the observer, plus the systematic delay of the timing instrument. The reaction time differs individually according to experience, and also to disposition at the moment; it is also longer for an "unannounced" contact than for an "expected" event (from a time-signal tone). Depending on procedure, only the difference between the reaction times or else the full amount of the personal equation enters the result. The latter can be determined, for instance, by timing the first tone of a signal after a longer interruption, or by covering part of the dial with tape so that the hand is invisible, and timing its reappearance from under the tape; the amount by which the hand runs into the visible sector gives the personal equation.

The reading can be checked (and perhaps corrected) by reading for some seconds after the event which tenth of a second on the dial coincides with the signal tone, before the watch is stopped at the full minute.

Example: The time interval is $41^s\!.9$; but the 7/10 second has been determined as coinciding with the signal before stopping the watch. If the personal equation has been found for (unexpected) contacts as $0^s\!.4$ and for signal tones $0^s\!.2$, then the two readings agree and the contact occurred $42^s\!.1$ before the full minute. The "one-second delay" (operation of watch exactly 1 s after contact) purports to avoid the variability of reaction time, but the success is doubtful and the method seldom used.

As the clock rate (and other clock errors, if any, known to be variable), are determined after every completed observation (i.e., under the same conditions as during the measurement), many observers also habitually add a test of the personal equation. (All these checks may sound tedious but they soon become routine, and are easily completed within the few minutes between making the measurement and filling in the observing record.) If the observer feels that he/she has reacted too slowly or quickly, the personal equation can be modified accordingly. The amount of correction that has been subtracted from the recorded time is stated in the observing report. If nothing is stated, the evaluation will assume a correction of $-0^s\!.3$.

An estimated error range of, for instance, $\pm 0^s\!.2$, is needed as the uncertainty in recording contact times. Note that this is the *random* error, not to be confused with the correction for the personal equation, a *systematic* error. A realistic error estimate helps during the processing to distinguish a very dependable measurement from one impeded by unfavorable circumstances.

For grazing and other especially important contacts, errors of $\pm 0^s\!.5$ are tolerable, but more than this would give the measurement a very low weight.

Example: An occultation timed with stopwatch, chronometer, and an hourly time signal yields the following results:

Chronometer reading (when stopping watch)	$21^h 48^m\ 0^s\!.0$ UT
Chronometer correction (from time signals 21^h, 22^h)	$+\ \ \ 0.4$
Stopwatch (from contact to reading)	$-\ 52.6 \pm 0.2$
Watch correction (stopwatch too slow)	$-\ \ \ 0.2$
Personal equation	$-\ \ \ 0.3$
	$21^h 47^m\ 7^s\!.3 \pm 0.2$

Very similar to the stopwatch method and even more accurate is the operation of a chronograph, where the signal is recorded onto sprocketed paper tape against time signal marks by electrical contacts. Few observers, however, will have access to such a device.

The portable cassette recorder provides a reliable way of recording the tones of a permanent time signal and the voice signal of the instant of contact. The personal equation is also readily determined, for instance, as the voice reaction to a noise. One should make certain that the time signal does not have a long interruption exactly at the time of contact.

The *eye–ear method* compares the *visual* contact with the simultaneously *audible* time signal tones (or the ticks of a chronometer), and interpolates the visual signal into the intervals of the audio in tenths of seconds; this was the only time recording method available in the era before chronometers and stopwatches. The coordination requires some experience, but can be brought to good precision; trained observers have an error of estimation of less than $\pm 0.^s2$. The method can be practiced using transits of stars across a wire. The personal equation is also largely eliminated; the remaining correction for the response to the sudden event of contact may be $-0.^s1$, but rarely more. Attention is needed to count the seconds correctly, either until the instant of contact, or from it to the next full minute mark or clock reading; errors cannot be hunted down afterwards! The observer could, for instance, follow the second count with "null-null-null" until contact, and thereafter count "one-two-three."

Over 90% of observed occultations are ingresses. Therefore, the data on the lunar motion during the waning phase are much weaker than for waxing. This is because egresses at the dark limb occur in the less-favored early morning hours when the reaction time may be a bit longer; they can be more easily missed as the observer watches the wrong part of the limb. Thus, the observer often is less satisfied by sensing a lower reliability of the result. Observing egresses is also more likely to be jeopardized by clouds since the reappearance of a star from a cloud may be confused with that at the lunar limb, whereas at ingress the ability to to distinguish between fading and sudden extinction is better owing to anticipation. In spite of this, more visual monitoring of reemergence should be encouraged, especially since photoelectric recording still presents substantial problems.

As a preparation, the region of expected reappearance is viewed according to the given position angle. No micrometer or degree circle is needed; an eyepiece crosswire, oriented by a star transit to the N–E–S–W directions, or even four small notches in the eyepiece diaphragm, will suffice to estimate the predicted position angle from the center of the Moon. With the telescope's polar axis adequately adjusted, a star set on before ingress will—at least in declination—not run much off the center of the field of view until reappearance. In this case, however, the telescope cannot be used for anything else for about an hour.

Contacts at the bright lunar limb are more difficult to observe, require more experience, and are deemed reliably measurable only for bright stars to be included in the list of predictions. The observer should be certain to see the real, instantaneous contact and not an illusion by irradiation; any doubts are recorded in the report sheet. Higher magnification and—for reappearances—a more accurate location of the point of egress will help.

For a visual double star, the times of ingress or egress can often be recorded for each component. Even when the pair is close and not separated by the telescope, the duplicity is often revealed in a distinctly "slow" or 2-step contact. The difference of contact times is given by

$$\Delta t = \frac{\rho \cos(p - P)}{\mu \cos(m - P)}, \tag{17.3}$$

where p, ρ are the polar coordinates of the pair (Sect. 26.1 in Vol. 3), μ and m the amount and position angle, respectively, of the apparent lunar motion, and P the position angle of the contact at the lunar limb. Noticeably delayed or stepped contacts which last 1^s thus can occur even for close pairs ($\rho < 0.''5$). Relative positions in doubles are usually well known, and the theoretical difference of contact times can be allowed for in processing. Predictions usually mention duplicities, if known. Indications of a slow contact and its estimated duration will be recorded in the observing sheet. As a sidenote, stopwatches with *two* hands are available which permit the recording of quick, consecutive events.

In order to have all data in time for processing, the Occultation Centre in Tokyo should receive reports no later than 6 months after the first observing night. The coordinates of the telescope and the time signal used should be checked for correctness. The observing forms also have columns for telescope, observer, method, estimated precision, atmospheric conditions, etc. with certain code numbers. The observer performs no reductions other than the corrections mentioned in the example above.

17.4 Grazing Occultations

The marginal contacts along the northern and southern lunar limbs were long neglected as being of inferior precision, but have of late regained significance because of their usefulness in determining lunar declinations. Among other things, this helps to check the position of the vernal equinox (the zero-point in right ascension) in the sky exactly; ultimately, the accuracy of all stellar positions and proper motions depends on how precisely the vernal equinox and its precessional rate are known.

The observing program thus should not avoid occultations which are near-grazing at the observed point. A small expedition to the "grazing zone" can even be arranged. Special predictions are issued to identify the zone of grazing contact and thus the best position in geographic latitude to be accurate to a few hundred meters. The program is not restricted to bright stars; otherwise, these events would only very rarely occur sufficiently close to home. But the observation is impeded as a grazing contact always occurs, if not at the bright limb, then in its vicinity. To eliminate a major fraction of the interfering moonlight, a focal diaphragm (i.e., stop) may be used during bright lunar phases. Multiple disappearances behind and reemergences from mountains at the lunar limb should be anticipated; the profile is dramatically mountainous, especially in the polar regions. A cassette recorder together with a battery-operated radio will be found useful.

Such excursions will be particularly promising as a group project. Many cases of grazing contacts have been successfully observed by amateur groups with 10 or more observers using small, portable telescopes. The observers place themselves along the direction of the lunar azimuth, perhaps near a country road with low traffic (but not near a high-voltage power line which would interfere with short-wave radio reception). A tape measure is brought along with the ordnance survey map to record their positions precisely. The tape recorders are placed where multiple contacts are expected; stopwatches can operate in the zone of deeper occultation, and one observer outside the expected grazing zone is delegated the unrewarding but important task of reporting that no occultation has occurred.

As is all too well known, many things can go wrong on an eclipse expedition. As only one failure at a vital point can ruin many hours of careful preparation, the functioning of instruments, organization, and electricity is rehearsed in daytime. Even then it is essential that all observers, especially those who carry important equipment or parts, be present. Nighttime observations in the open may also attract mosquitoes, dogs, police patrols, and other operating deterrents. Also, bystanders who are curious about the "UFO" being sought are apt to stand so as to block the telescope view or to trip over electrical cables.

17.5 Occultations of Planets

Owing to the brightness of planets and also to the longer durations of ingress and egress of their disks, these celestial events are spectacular to view but occur more seldom, and are also less important for research than the occultations of stars. Near-grazing occultations of Venus with durations of ingress of half an hour, and also some partial occultations, have been observed.

The contact times in predictions hold for the center of the planetary disk. The duration of contact may be computed as follows:

The planetary radius ρ is found in the *Astronomical Almanac*, as are the hourly motion of the Moon and daily motion of the planet. Both of these are converted into arcseconds per minute of time and then subtracted from one another. The differences are n_α and n_δ, and the time interval of ingress and egress at the selenocentric position angle P becomes

$$\Delta t = \frac{2\rho}{|n_\alpha \sin P - n_\delta \cos P|}. \tag{17.4}$$

To this end, the lunar parallax can be neglected, but P needs to be known. For Saturn, the ingress of the ring system takes $2\frac{1}{2}$ times longer than that of the planetary disk itself.

Scale drawings of the motion of the planet relative to the Moon appear occasionally in astronomical magazines, and can be prepared for a given location with the aid of the ephemerides data. The lunar positions are computed, corrected for parallax (see Sect. 3.6) for a few full hours before and after occultation, and the position differences of planet (interpolated per hour) versus Moon graphed. (The lunar parallax is included only to first order and that of the planet not at all; but this error is irrelevant except perhaps in grazing contacts.) The radii are graphed, as are the phases of illumination, with their position angles; then, beginning, end, and position angle of contact can be read to good approximation.

All measured contact times are measured to full seconds, as the uncertainty for extended objects is invariably much larger.

17.6 Photoelectric Registration

In the area of astronomical photometry, the advent of rapid data storage systems has been an immense help in recording processes which occur on very short time scales; occultations of stars are a particularly rewarding application. Photoelectric recording of disappearances at the dark limb has increased considerably over the years. With the widespread availability of personal computers (PCs), diode arrays, etc., the operation of an occultation photometer is now within reach of the smaller observing stations. Several variants have been tested and described [17.3–6]; the observer will have to acquire some knowledge of electronics to construct the equipment and those parts which may not be commercially available.

By virtue of the *Fresnel diffraction* of light at the lunar limb, a wave pattern of measured light intensity appears briefly at the contact. With significant amplitude, it lasts usually less than $0\overset{s}{.}1$, and thus is not visible to the eye. It is fed point by point into an on-line computer with a time resolution of milliseconds and then evaluated for the geometric instant of disappearance by the equations of wave optics.

The precise position of the Moon is not the only item of interest. A registration curve can be used to identify very close binary stars, and large stellar diameters can even be determined. A star of finite diameter blurs the diffraction pattern compared with a point source, and a double star will in this procedure—depending on the separation and brightness difference of the components—cause a delayed, or two-step, decline in the light. In this way, separations below $0\overset{''}{.}01$ and diameters down to as small as $0\overset{''}{.}003$ have been measured [17.7]. Irregularities of the lunar limb contribute to the distortion of the registration curve; in particular, obstacles of sizes in the range 10–20 m interfere most strongly with the diffraction pattern, while substantially larger mountains displace it only *in toto*. For this reason, double-star and diameter measurements are based on repeated registrations of occultations of the same star, or simultaneous ones from different stations; the McDonald Observatory (Texas, USA) in particular specializes in this technique. At large telescopes, allowance must even be made for the size of the aperture, showing how high the time resolution is: at the west side of the tube the star ingresses a few milliseconds earlier than on the east side. Simultaneous observations in two or more colors may permit the separation of the geometric entrance of the lunar profile from the wave-mechanical—and thus wavelength-dependent—influence of the light source upon the diffraction pattern. A detector can be inserted which subtracts an amount of current corresponding to the scattered moonlight from the signal, so that the signal from the star is not drowned out. In order to make do with limited storage capacity, the registration is programmed "self-triggering": before contact, the storage contents are overwritten until the incipient diminution of light, when the detector stops the erasure and thereafter also the registration itself. About one-half of the currently evaluated events are observed photoelectrically, but only from a few stations.

Why is it that here a rather simple arrangement of equipment (compared with other scanners and interferometers) can be used and yet achieves a superior resolution not attainable even with large telescopes? It is primarily because the "natural" image scanning by the lunar limb occurs before the light reaches the Earth's atmosphere and the telescope, where it is scattered. For a similar reason, the solar corona is much more easily viewed during natural eclipses than when looked at through a coronograph. Most astronomers are displeased with the Moon because its overwhelming brightness fogs photographic plates and impedes photoelectric measures. But, at least for the present task, it proves its worth as an extraterrestrial, movable diaphragm; the drawback is, however, its restriction to stars near the ecliptic.

Photoelectric reappearance observations are few in number. The range of current to be recorded, which is set before ingress, may meanwhile have shifted substantially with changing transparency, so that subsequently the registration fails. In order to block a large part of moonlight, a narrow stop is also needed, and without a computer-operated telescope (which also allows for the change of refraction), it is difficult to place the small field of scan onto the point of reappearance. At the 60-cm refractor of the Hamburg-Bergedorf Observatory, reappearances have been recorded photoelectrically by offset guiding on a distant star.

The author thanks M. Lukac of the U.S. Naval Observatory, Washington, for reading through the manuscript.

References

17.1 Mulholland, J.D.: *Publ. Astr. Soc. Pacific* **84**, 357 (1972).
17.2 Winkler, G.M., Van Flandern, T.C.: *Astronomical Journal* **82**, 84 (1977).
17.3 Nather, R.E., Evans, D.S.: *Astronomical Journal* **75**, 575, 583, 963 (1970).
17.4 de Vegt, Ch.: *Astronomy and Astrophysics* **8**, 85 (1970) and **48**, 245 (1976).
17.5 Morbey, C.L., Fletcher, J.M.: *Publ. D.A.O. Victoria* **14**, 271 (1974).
17.6 White, N.M.: *Publ. Astron. Soc. Pacific* **89**, 238 (1977).
17.7 Evans, D.S.: *Sky and Telescope* **54**, 164 (1977).

18 Artificial Earth Satellites

R. Kresken

18.1 Introduction

In the years between the launch of *Sputnik 1* in 1957 and the last manned landings on the Moon in 1972, the observation of Earth satellites was a popular activity among amateur astronomers. Many observers were organized in international programs like the famous "Moonwatch" to provide observational data for the tracking of satellites. The announcements in the press of favorable passages of bright satellites enabled everyone to watch those silently moving points of light and to take part in the upcoming space age.

Since that time, the public interest in the observation of satellites has steadily diminished. Today, after thousands of launches, only special events and unusual manned flights are mentioned in the press. Under suitable conditions, it is easily possible for a single observer to see several satellites simultaneously.

All satellites that can become bright enough to be observed by amateur astronomers are catalogued and tracked by several ground stations all around the world. Nevertheless, they are still rewarding objects for visual observers, and, more than thirty years after the first attempts by amateurs, the precise measurement of their slow movement has lost nothing of its fascination.

18.1.1 The Population of Satellites in Space

To minimize the risk of collisions and accidental re-entries, the population of man-made objects is surveyed and catalogued as extensively as possible. In the United States, this work is done by the US Space Command (USSPACECOM). Its very sophisticated phased array radars can detect objects as small as 10 cm in low Earth orbit (LEO) and 1 m in geosynchronous orbit at a distance of approximately 36 000 km. The total of more than 7000 catalogued objects (as of December 1990) can be divided into two main categories: *operational payloads* and *space debris*. The first category comprises all satellites that are active or at least under control. Depending on their mission, these satellites are subdivided into several classes:

– scientific and technological satellites;
– satellites for special applications, such as telecommunications, Earth observation, navigation, and meteorology;

- satellites for specific military use, such as photographic reconaissance, electronic detection, and anti-satellite satellites (ASATs);
- manned space stations and spacecraft.

Everything else—and therefore the vast majority of all artificial objects in Earth orbit—are summarily categorized as *space debris*. In this category, we have thousands of objects that are big enough to be observed by visual means or radar. such as deactivated satellites, spent rocket stages and fragments of satellites and rockets. In addition, there are also small and very small particles such as instrument covers, paint flakes, screws, and exhaust particles of solid-fuel engines. These objects are too tiny to be under surveillance from Earth, but they can cause a considerable threat to manned and unmanned spacecraft.

18.1.2 International Designations

All spacefaring nations are obliged to announce each satellite launch to the international Committee on Space Research (COSPAR), which assigns to all satellites and fragments an official designation, based on the year of launch. As an example, the European communications satellite *Olympus 1* bears the designation 1989 53 A, because it was placed in orbit with the fifty-third launch in 1989. The letter "A" indicates that it was the main payload of that flight. The letters B, C, and so on are assigned to other objects that are placed in orbit with a launch, such as upper and inter-stages or fragments. The letters I and O are omitted to avoid confusion with the digits 1 and 0.

If one launch produces more than 24 pieces, the sequence continues after Z with AA, AB, AC, and so on. The destruction of the astronomical *Solwind* satellite by an ASAT missile in 1985 was such a case. That incident caused the spacecraft to break up into 251 known pieces.

It should be mentioned here that COSPAR used another system of designation before 1963. It was also based on the year of launch, but instead of numbers, it utilized a Greek letter. This was followed by an index which was assigned to the different objects with decreasing visual magnitude. For example, the faint *Sputnik 1* was designated 1957 α_2, the bright upper stage was registered as 1957 α_1.

18.2 Satellite Orbits

18.2.1 Undisturbed Motion

To a first approximation, the orbital motion of an Earth satellite can be regarded as a point mass moving in the gravitational field of a homogeneous sphere. The satellite's mass is negligible.

Under these assumptions, the motions can be described by Kepler's laws. The resulting ellipse has a point on it nearest to Earth called the *perigee*, and one farthest from Earth called the *apogee*. These extremum points are jointly referred to as the *apsides*, and the line which connects them as the *line of apsides*. That point at which

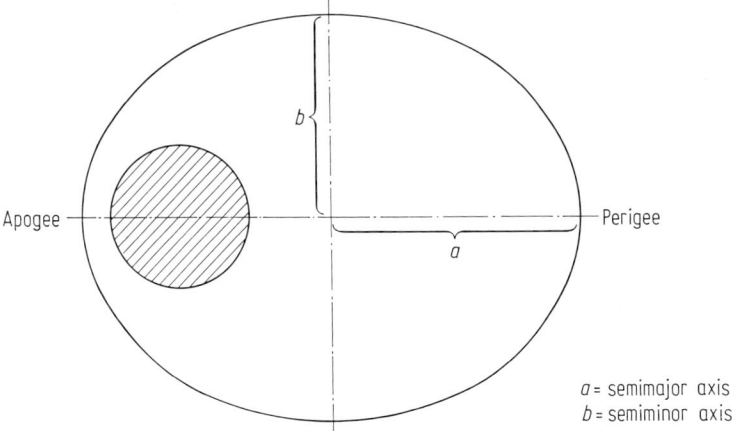

Fig. 18.1. The apsides of an Earth satellite orbit.

the satellite intersects the equatorial plane of the Earth from south to north is called the *ascending node*.

The position and the velocity in the orbit can be calculated from six quantities, known as the *orbital elements*. They are commonly expressed in the so-called Keplerian form, similar to the one used to describe the orbits of comets and planets around the Sun. It uses the following values:

- The *inclination i*, which is the angle between the orbital plane and the equatorial plane of the Earth. Satellites that move westward instead of eastward have a *retrograde* orbit with an inclination of more than 90°.
- The right ascension of the ascending node Ω.
- The *argument of the perigee* ω.
- The mean motion n in revolutions per day.
- The *eccentricity e*. This quantity measures the departure of the orbit from a circle. It is defined as

$$e = \frac{\sqrt{a^2 - b^2}}{a}, \tag{18.1}$$

where a and b are the semimajor and semiminor axes, respectively.
- The *mean anomaly M*. This angle describes the position of the satellite in its orbit at a given instant.

18.2.2 Orbit Perturbations

If the motion of a satellite around the Earth followed Kepler's laws exactly, its orbital elements would remain constant. In reality, there are certain mechanisms that disturb the ideal two-body motion and cause the elements to vary with time.

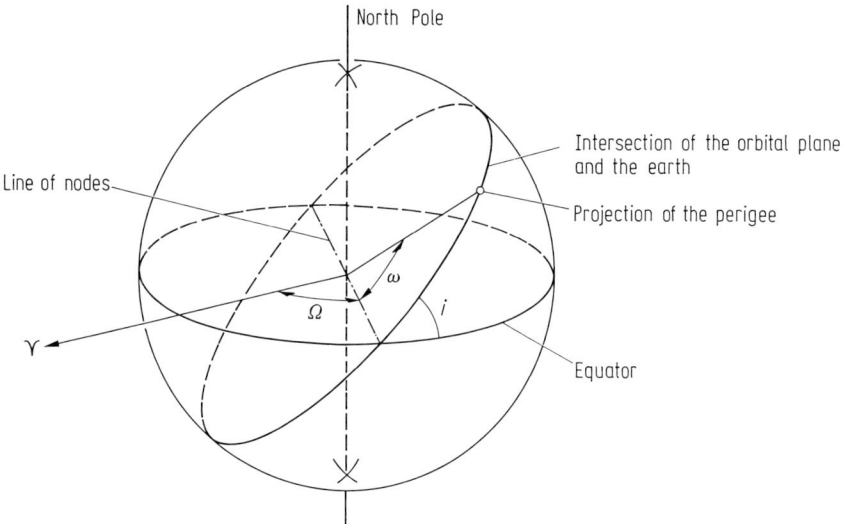

Fig. 18.2. The orbital elements of a satellite.

18.2.2.1 Perturbations by the Gravitational Field. The deviation of the gravitational field surrounding the Earth from that of a homogeneous sphere gives rise to some important effects on orbiting satellites. The two most important perturbations of this kind are due to the oblateness of the Earth.

One is a precession of a satellite's orbital plane. The magnitude of this effect can be expressed in terms of the variation of the right ascension of the ascending node in degrees per day. It can be calculated as follows:

$$\frac{\Delta\Omega}{\text{day}} = -\frac{9°\!.964}{(a/R_e)^{7/2}(1-e^2)^2}\cos i, \tag{18.2}$$

where R_e is the equatorial radius of the Earth (6378 km), a the semimajor axis of the orbit, e the eccentricity, and i the inclination. For $i < 90°$, the line of nodes moves continuously westward, and for $i > 90°$, eastward. For a so-called *polar orbit* with an inclination of $90°$, the motion of the nodes vanishes.

The other effect is a rotation of the line of apsides in the orbital plane. Its velocity depends on the inclination i, the semimajor axis a, and the eccentricity e, and is given by

$$\frac{\Delta\omega}{\text{day}} = \frac{4°\!.982}{(a/R_e)^{7/2}(1-e^2)^2}(5\cos^2 i - 1). \tag{18.3}$$

The fact that this quantity vanishes for the so-called critical inclinations of $i_1 = 63°\!.43$ and $i_2 = 116°\!.57$ is of great importance for some special satellite applications.

It should be mentioned that the above calculations are based on a homogeneous, oblate and rotationally symmetric Earth, which is still an idealization, even though a far better one than the homogeneous sphere.

Artificial Earth Satellites 173

The real mass of the Earth, the geoid, is an inhomogeneous and asymmetric body which causes perturbations that cannot be predicted exactly by simple formulae. Conversely, observations of the evolution of satellite orbits yield indispensible data for determining the precise shape and mass distribution of the Earth.

The gravitational field surrounding the Earth is also disturbed by the gravitational attraction of the Sun and Moon. The resulting perturbations are very difficult to predict, and depend strongly on the relative orientation of the orbit and the attracting bodies. As a general rule, near-equatorial orbits are much less affected than near-polar orbits.

If an orbit has a large eccentricity, the gravitational force of the Sun and the Moon can cause substantial changes in the perigee height.

18.2.2.2 Non-gravitational Perturbations. For satellites in low-Earth orbits, *atmospheric drag* is the most important non-gravitational perturbation. It is due to free collisions of a spacecraft with atmospheric molecules that are abundant even at these altitudes. This interaction causes the orbital energy of the satellite and therefore its semimajor axis to decrease. This leads to the paradoxical phenomenon that the period shortens and the satellite speeds up.

In an orbit with a notable eccentricity, atmospheric friction is a factor mainly near the perigee. Therefore, the drag changes with the orbital position of the spacecraft and tends to decrease the eccentricity of the orbit.

The density of the upper atmosphere is highly variable. It depends on the degree of solar activity, the season of the year, the latitude, and the change from day to night.

As a result of the energy lost, any satellite with a perigee height of less than a few thousand kilometers will sooner or later decay in the atmosphere. A first approximation of a satellite's lifetime in revolutions or days, respectively, can be derived from Fig. 18.4, which is based on the assumption of a mean atmosphere. It reflects

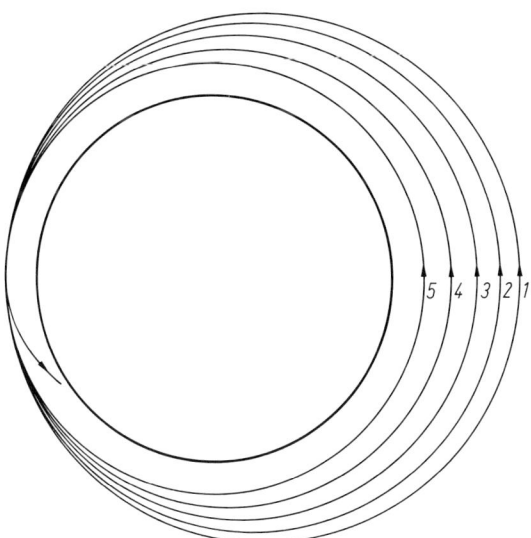

Fig. 18.3. The evolution of an orbit under the influence of atmospheric drag.

the strong influence of the so-called *ballistic coefficient* c_b, which can be determined as follows:

$$c_b = \frac{m}{c_d A}. \tag{18.4}$$

Here m is the satellite's mass, A the cross-sectional area perpendicular to the velocity vector, and c_d the drag coefficient. It is very important to note that neither A nor c_d is constant for non-spherical satellites. These quantities depend on the attitude of the spacecraft and can therefore change rapidly, particularly during the last days in the lifetime of a decaying satellite. Typical values for c_b lie between 25 and 100 kg m^{-2}.

Since Fig. 18.4 is calibrated with the ballistic coefficient, one has to know its magnitude to determine the approximate lifetime in revolutions N. When N is known, the lifetime in hours N_h can be calculated by multiplying with the orbital period:

$$N_h = N \frac{2\pi}{3600\sqrt{\mu}} a^{3/2}, \tag{18.5}$$

where $\mu = 398601.3$ km^3 s^{-2}.

Another perturbation arises from the solar radiation pressure. The same force that forms the well-known comet tails acts as a disturbing force on any satellite that is illuminated by the Sun. Although it is almost negligible for a typical spacecraft in low Earth orbit, radiation pressure can have a significant impact on a satellite in a higher orbit, especially one which, like the balloon satellites of the 1960s, possesses a low density. The radiation pressure tends to increase the eccentricity of the orbit, while leaving the semimajor axis approximately constant. Thus, the perigee height decreases continuously, which can in turn cause a rapid decay due to atmospheric friction.

18.2.3 Classes of Orbits

Nowadays, there are artificial Earth satellites in a multitude of orbits. Nevertheless, it is possible to distinguish several classes of orbits which reflect the requirements of some special satellite applications.

18.2.3.1 Polar orbits. For satellites with Earth observation or reconnaissance missions, it is essential to overfly almost every region of the Earth in a low altitude. This is possible only in a low Earth orbit with an inclination of approximately 90°. If the inclination is exactly 90°, the orbital plane does not precess around the Earth.

In a retrograde orbit with an inclination of approximately 97°, the ascending node moves with a rate of 0°.984 eastward. This compensates for the angular velocity of the Earth's orbital motion around the Sun; the orbit is *heliosynchronous*. This is essential if the satellite requires permanent illumination of its solar arrays or if its mission demands illumination of the overflown areas. As an example, a heliosynchronous orbit with an altitude of 200 km has an inclination of 96°.32.

18.2.3.2 Geostationary orbits. The tasks of modern telecommunications, such as television broadcasting and the provision of telephone links, demand an uninterrupted connection between the ground stations involved. To fulfill this, most communications satellites are placed in a *geostationary orbit* (GEO), i.e., a circular orbit that lies

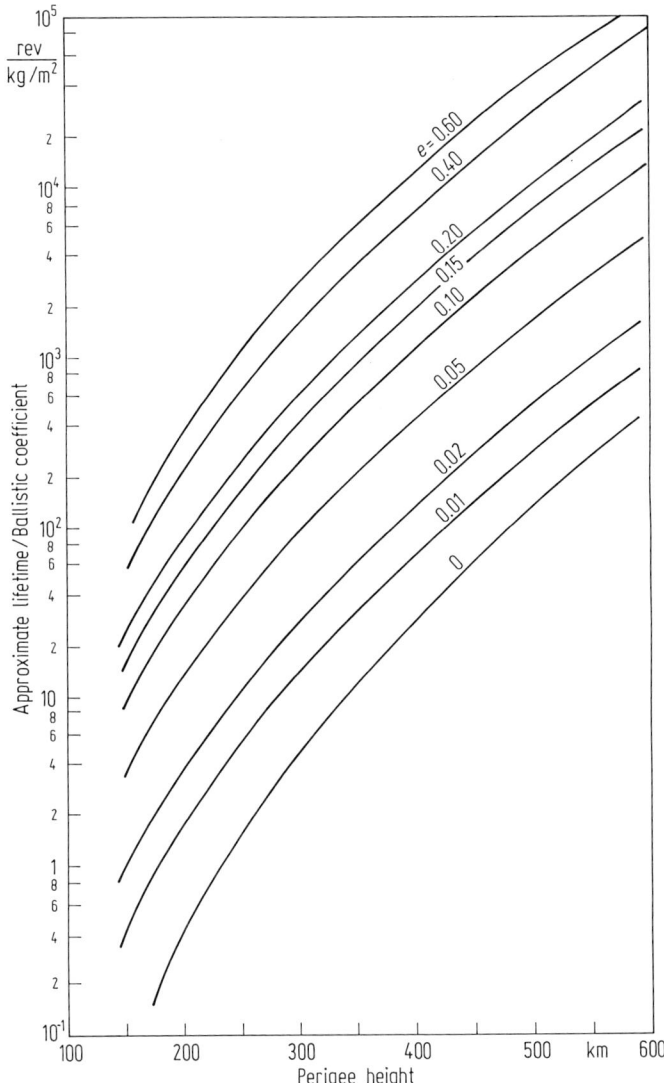

Fig. 18.4. Diagram for the determination of the approximate lifetime of an Earth satellite from its orbital eccentricity e and perigee height h_p. Adapted from *Spacecraft Attitude Determination and Control*, J. Wertz, ed., D. Reidel Publ. Co., Dordrecht 1980.

in the Earth's equatorial plane. Owing to its orbital radius of approximately 42 164 km, the orbital period is $23^h 56^m 04$. Therefore, the satellite remains virtually fixed with respect to the Earth.

18.2.3.3 Molniya orbits. A disadvantage of the geostationary orbit is the fact that ground stations in extreme northern or southern latitudes cannot be reached. To connect such stations, Russia places communications satellites in so-called *Molniya orbits*.

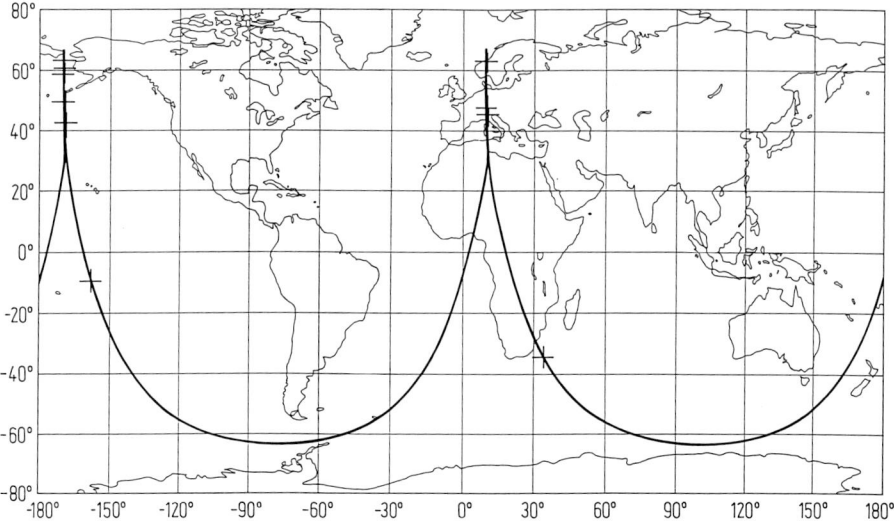

Fig. 18.5. Groundtrack of a Molniya satellite (ESA/ESOC).

Such an orbit has a period of $11^h58^m02^s$, which is one-half of a sidereal day, and the following elements:

$$a = 26\,555.4 \text{ km},$$
$$\text{Perigee height} = 1000 \text{ km},$$
$$\text{Apogee height} = 39\,356.7 \text{ km},$$
$$\text{Argument of the perigee} = 270°,$$
$$\text{Inclination} = 63°\!.4.$$

Such an orbit allows a satellite to stay for more than 8 hours in northern latitudes. The critical inclination of $63°\!.4$ (see Eq. (18.3)) is crucial because it avoids a rotation of the line of apsides, which would soon render the orbit useless.

18.3 Conditions of Visibility

As with any other celestial object, an artificial Earth satellite has to fulfill two basic conditions to be observable: it has to be above the observer's horizon and it has to exceed a certain minimum brightness to be photographed or recognized. This depends on the brightness of the sky background and on the sensitivity of the eye or the photographic emulsion. Since very few satellites have a built-in light source, it must reflect sunlight to be visible. In this section, discussion will center on how to predict the passage of satellite and how to assess its illumination.

Fig. 18.6. A Space Shuttle Orbiter with the Spacelab in its cargo bay. It is usually launched from Cape Canaveral with an orbital inclination of 28°5, but the rare European Spacelab missions require an inclination of more than 50°. With a maximum magnitude of between −1 and −4, it is easily visible to the naked eye (MBB/ERNO).

18.3.1 Influence of the Inclination

The inclination i of the orbital plane of a satellite is critical with regard to its visibility, because it limits the area on Earth that can be overflown. The point on Earth exactly under the satellite, the so-called *subsatellite point*, always moves between the latitudes $+i$ and $-i$. Therefore, the area from which the satellite is visible does not extend much further.

To gain the most from the Earth's rotation, most satellites are launched eastward. Thus, the geographic latitude of the launch pad equals the orbital inclination of the satellite. For Cape Canaveral, the latitude is 28°5, and, as a result, many American satellites are unobservable from mid-northern latitudes. The situation for the Russian spacecraft is much better, because they are launched from different sites with considerably higher latitudes.

18.3.2 Accurate Predictions

To observe a special satellite, it is imperative to have a more or less precise prediction or *ephemeris* for it. Nowadays, it is most efficient to obtain these predictions with a home computer and a program that is capable of calculating positions from a set of orbital elements. The problem is that these elements do not remain constant for a long time. No program in use by amateurs can predict the perturbations mentioned in the last section with sufficient accuracy for more than several weeks. Additionally, active satellites are occasionally prodded into new orbits by ground controllers. Thus, a source of current orbital elements is absolutely indispensible.

In the United States, amateurs can receive these elements and predictions for equator crossings in the form of the *NASA Prediction Bulletin*, which is available from the NASA-Goddard Space Flight Center, Control Center Support Section, Code 513.2, Project Operations Branch, Greenbelt, MD 20771. In Europe, they can be provided by institutions in England (see References).

18.3.3 Simplified Predictions

Many amateurs may be interested in predicting the reappearance of a satellite based upon their own direct observations. This can be performed with a minimum of calculation if a circular orbit is assumed. This is justified, because most satellites, especially those in low orbits, indeed have a near-circular orbit. The method presented here is recommended only for satellites that become relatively bright, because they are unlikely to be confused with fainter ones during a predicted passage.

To be able to predict the reappearance, one has to obtain three observational data:

- the azimuth A of the satellite trajectory,
- the apparent maximum elevation above the horizon, and
- the apparent angular velocity in the culmination point.

From the observer's latitude φ and the azimuth A, which is, as usual, measured astronomically from the south through the east, the orbital inclination i can be calculated as follows:

$$\cos i = \sin A \, \cos \varphi. \tag{18.6}$$

To determine the height of a passing satellite, one must proceed as follows. First, the satellite must be tracked until it reaches its maximum elevation above the horizon. This angle must be measured or estimated. Then, to ensure that the satellite moves perpendicular to the line of sight, its angular velocity must be measured when it still has the maximum elevation. This can be performed best with a pair of binoculars with a known field of view and a stopwatch.

With these two data, the height of the satellite above the ground and the distance to the observer can be determined from the diagram in Fig. 18.7.

With the satellite's height H, it is easy to determine its orbital period P, measured in minutes, from the equation

$$P = \frac{2\pi}{\sqrt{\mu}}(R+H)^{\frac{3}{2}}, \tag{18.7}$$

where $\mu = 398\,601.3$ km^3 s^{-2} and $R = 6378$ km. With these data, it is possible to predict when and where to look for the satellite.

The satellite will cross the observer's latitude once during each orbital revolution, but it will be visible only when the orbit has roughly the same orientation with respect to observer and under similar conditions of illumination. Thus, it may be observable

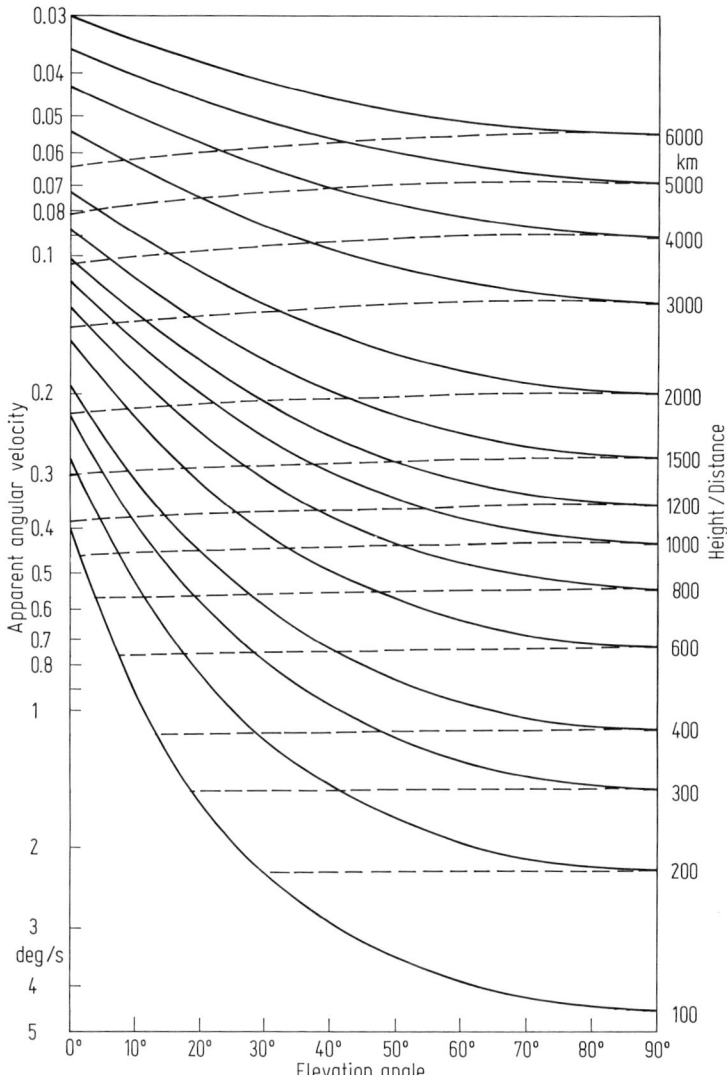

Fig. 18.7. Diagram for the determination of the height and distance of an observed satellite. The *solid lines* correspond to the heights, the *dashed lines* to the distances indicated on the right.

again during its next pass after P minutes or after approximately 24 hours. In the latter case, the precise time of reappearance can be calculated by multiplying the orbital period P with the number of complete revolutions per day.

For predictions from one day to the next, the decrease in period caused by air drag can be neglected, but it can render long-term predictions for satellites in low orbits

Fig. 18.8. The European Remote Sensing Satellite *ERS-1* was launched in 1991. It has a sunsynchronous circular orbit at a mean altitude of 758 km with an inclination of 98°.5 (ESA).

inaccurate. After a week, such a satellite can arrive more than 10 minutes earlier than predicted.

To know where to look for the satellite during a future passage, it is important to consider the motion of the orbit with respect to the Earth's surface. This motion is a consequence of two mechanisms: the rotation of the Earth and the precession of the orbit. The Earth's rotation contributes a westward motion of 360°.986 per day or 0°.250 6847 per minute. The daily motion $\Delta\Omega$ due to the precession can be obtained from Fig. 18.9. It is based on Eq. (18.2) with $a = R + H$ and $e = 0$.

Now it is possible to calculate the total relative displacement ΔL for a passage that occurs ΔT minutes after the observed one:

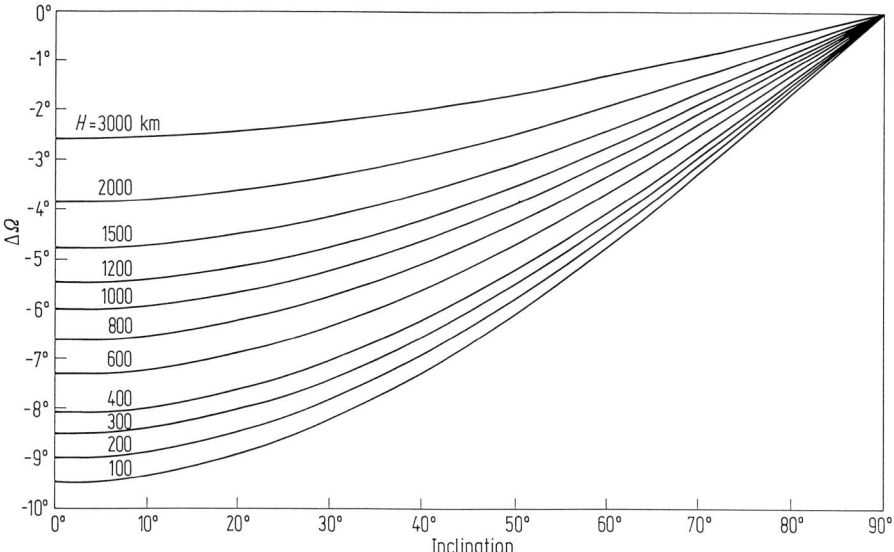

Fig. 18.9. The daily drift $\Delta\Omega$ of the orbital plane for circular orbits as a function of the inclination i and the altitude H.

$$\Delta L = \left(0\overset{\circ}{.}250\ 6847 - \frac{\Delta\Omega}{1440}\right)\Delta T. \tag{18.8}$$

If a satellite is expected to reappear after ΔT minutes and the quantity ΔL is an integer multiple of 360°, the apparent track on the sky will be about the same. If it is some degrees more or less, the track will be shifted to the west or to the east, respectively. Since the angle ΔL is measured from the center of the Earth, the apparent displacement of the satellite will always be more than this amount, depending on its altitude.

18.3.4 Geostationary Satellites

Finding geostationary satellites on the celestial sphere is a very specific problem. Because of their circular orbit with an inclination close to 0° and a period of 23^h56^m, they remain virtually fixed in the sky. But as a consequence of their great distance, they never become brighter than magnitude 10 and are therefore difficult to find. The best way to find such a satellite is to calculate its equatorial position for a given instance and to point the telescope with respect to the stars.

The position of a geostationary satellite is specified by the geographic longitude λ_s of its subsatellite point. Longitudes west of Greenwich are considered positive. To calculate its position for an observer with the geographic longitude λ_0 and latitude Φ_0, it is useful to define a coordinate system that rotates with the Earth and has its

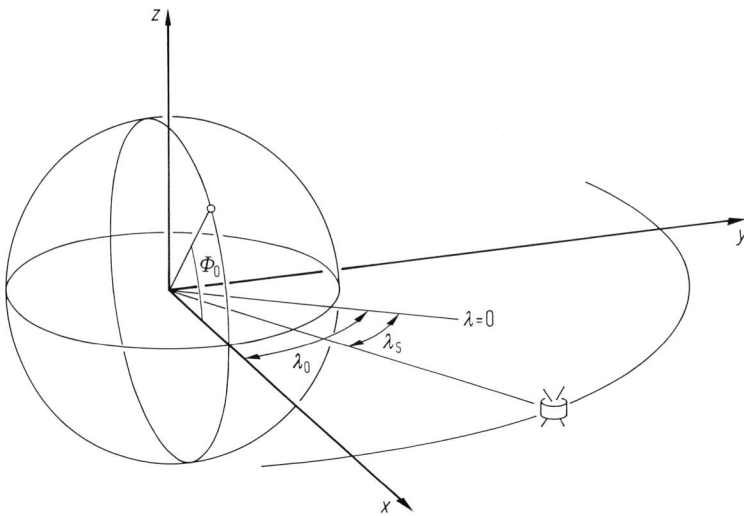

Fig. 18.10. The coordinate system that underlies the calculations for the determination of the apparent position of a geostationary satellite.

origin in the Earth's center. The positive z-axis passes through the North Pole, the x-axis intersects the equator at the point with the same longitude λ_s as the observer. The y-axis passes through the equator at the point 90° further east. R is again the Earth's radius, r_s is the radius of the geostationary orbit (42 164 km).

The rectangular coordinates of the observer are given by:

$$x_0 = R \cos \Phi_0, \tag{18.9}$$
$$y_0 = 0, \tag{18.10}$$
$$z_0 = R \sin \Phi_0. \tag{18.11}$$

With the definition $\Delta\lambda = \lambda_0 - \lambda_s$, the coordinates of the satellite are

$$x_s = r_s \cos \Delta\lambda, \tag{18.12}$$
$$y_s = r_s \sin \Delta\lambda, \tag{18.13}$$
$$z_s = 0. \tag{18.14}$$

Hence, the coordinates of the satellite with respect to the observer are

$$x = r_s \cos \Delta\lambda - R \cos \Phi_0, \tag{18.15}$$
$$y = r_s \sin \Delta\lambda, \tag{18.16}$$
$$z = -R \sin \Phi_0. \tag{18.17}$$

The distance r between the observer and the satellite can now be calculated:

$$r = \sqrt{x^2 + y^2 + z^2}. \tag{18.18}$$

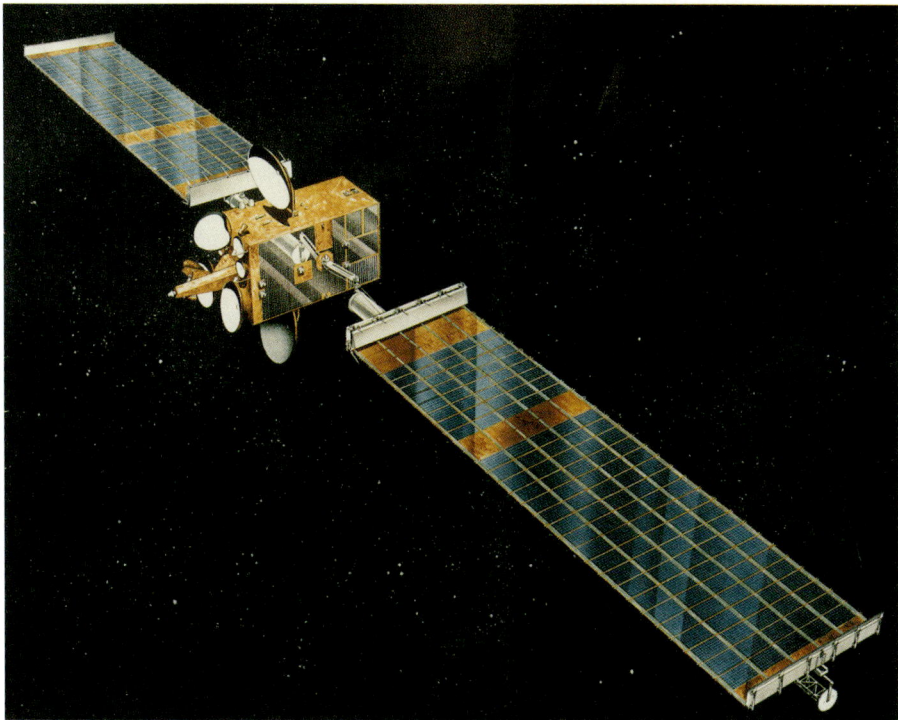

Fig. 18.11. The geostationary communications satellite *Olympus 1*. It is positioned at the longitude of 19° west (ESA).

From these rectangular coordinates, the hour angle u and the declination δ can be obtained:

$$\tan u = \frac{-y}{x} \tag{18.19}$$

and

$$\sin \delta = \frac{z}{r}. \tag{18.20}$$

The right ascension α is given by the classical formula

$$\alpha = \Theta - u, \tag{18.21}$$

where Θ is the local sidereal time at the observer's longitude.

The equatorial coordinates α and δ refer to the equinox of the date. To plot that position on a good star atlas, it has to be transformed to the standard equinox of the chosen atlas by applying the annual precession in both coordinates:

$$\Delta\alpha = 3\overset{s}{.}073 + 1\overset{s}{.}336 \sin\alpha \tan\delta, \tag{18.22}$$
$$\Delta\delta = 20\overset{\prime\prime}{.}04 \cos\alpha. \tag{18.23}$$

Table 18.1. Geostationary satellites that are visible from North America (*left*) and Europe (*right*). The first column gives the nominal longitude (west), the second column the name, and the third column the COSPAR designation of the satellite. The actual and the nominal positions can differ by up to 0°.5. All of these satellites have inclinations of less than 0°.1 and should become brighter than magnitude 12.

Longitude (West)	Satellite	COSPAR Designation	Longitude (West)	Satellite	COSPAR Designation
137.0	Satcom C1	90100A	30.0	Hispasat 1A	92060A
135.0	Satcom C4	92057A	19.0	TV-Sat 2	89062A
128.0	ASC-1	85076C	19.2	TDF 1	88098A
125.0	Galaxy 5	92013A	19.2	TDF 2	90063A
125.0	GStar 2	86026A	8.0	Telecom 2A	91084A
123.0	SBS 5	88081B	5.0	Telecom 2B	92021A
111.0	Anik E1	91026A	− 3.0	Telecom 1C	88018B
107.2	Anik E2	91026A	− 5.0	Tele-X	89027A
105.0	GStar 4	90100B	− 13.0	Eutelsat 2 F-1	90079B
103.0	Gstar 1	85035A	− 13.4	Italsat 1	91003A
101.0	ASC-2	91028A	− 16.1	Eutelsat 2 F-3	91083A
99.0	Galaxy 6	90091B	− 19.2	Astra 1A	88109B
99.0	SBS 6	90091A	− 19.2	Astra 1B	91015A
93.0	GStar 3	88081A	− 19.2	Astra 1C	93031A
90.9	Galaxy 7	92072A	− 21.6	ECS 5	88063B
87.0	Spacenet 3R	88018A	− 23.6	DFS 1	89041B
70.0	Brasilsat 2	82026B	− 28.6	DFS 2	90063B
62.0	TDRS 3	88091B	− 30.7	DFS 3	92066A
53.0	Intelsat 5a F-1	88040A	− 31.2	Arabsat 1C	92010B

18.3.5 The Brightness Behavior of Satellites

To assess the brightness behavior of a satellite during a passage, it shall be idealized here as a simple sphere that is illuminated by the Sun.

Two general kinds of reflections can be distinguished: *mirroring* and *diffuse reflection*. In the case of mirroring reflection, which can be realized with a polished metal coating, the luminous intensity depends on the illuminance E of the incident sunlight ($E = 150{,}000$ Lx), the reflection coefficient C, and the radius of the sphere r_s. It is given by

$$I_{\text{mir}} = \frac{ECr_s^2}{4}, \tag{18.24}$$

and is independent of direction. In the case of diffuse reflection, the luminous intensity depends also on the phase angle α, which is the angle subtended at the satellite between the Sun and the observer. The luminous intensity can be calculated as

$$I_{\text{diff}} = ECr_s^2 \frac{2}{3}\varphi(\alpha), \tag{18.25}$$

where

$$\varphi(\alpha) = \frac{\sin\alpha + (\pi - \alpha)\cos\alpha}{\pi} \tag{18.26}$$

is the *phase law* and the angle α is expressed in radians.

The most famous example of a diffuse reflecting satellite is the Moon. For a spacecraft, diffuse reflection can be realized with a dull coating. For phase angles smaller than approximately 90°, a diffuse reflecting satellite is brighter than a mirroring reflecting one with the same reflection coefficient.

The phase angle α may be calculated from

$$\alpha = 90° + h - \zeta, \tag{18.27}$$

where h is the altitude of the Sun. It is negative when the Sun is below the horizon. The angle ζ is given by

$$\tan \zeta = -\tan z \, \cos d, \tag{18.28}$$

where z is the zenith distance of the satellite and d the difference in azimuth between the Sun and the satellite. The angle ζ is positive if d is greater than 90°.

The apparent brightness of a satellite depends strongly on its distance from the observer. If the distance is doubled, the light flux from the satellite is reduced to one-quarter of its initial value. This corresponds to a difference of 1.5 magnitudes. In general, if a satellite has the apparent magnitude m_d at a distance d, its magnitude m'_d at a distance d' can be obtained by the relation

$$m'_d = m_d + 5 \log(d'/d). \tag{18.29}$$

For a diffuse reflecting satellite, a decrease in phase angle is favorable only as long as it is still illuminated by the Sun. This fact causes a considerable limitation to optical satellite tracking. When the sky becomes dark enough for such observations, the Earth's shadow rises quickly above the observer. The approximate shadow height H_s in the zenith can be calculated from the relation

$$H_s = R \left(\frac{1}{\cos h} - 1 \right), \tag{18.30}$$

where R is again the radius of the Earth (6370 km) and h the altitude of the Sun. A satellite that passes through the zenith has to be higher than H_s in order to be illuminated by the Sun and thus become visible. It is important to note that the Earth's shadow has the shape of a cone with a diffuse boundary. This is a consequence of the finite angular diameter of the Sun.

At the end of the astronomical evening twilight, when the Sun is 18° below the horizon, the height of the shadow directly over the observer exceeds 300 km. All satellites in lower orbits (i.e., the brightest ones), have already penetrated the shadow cone. The situation depends strongly on the season. In winter, the useful time for observations in midnorthern latitudes is less than one hour. In summer, the situation is much more favorable. Depending on the observer's latitude, it may be possible to observe satellites throughout the night.

The observation of geostationary satellites is much less hindered by eclipses. For most of the year, the entire geostationary orbit is sunlit, since the orbit is tilted out of the Earth's shadow. It is only when the Earth approaches the vernal and the autumnal equinoxes that parts of the orbit move into the shadow. Thus, eclipses are possible only from February 27 to April 12, and from August 31 to October 16. The duration of an eclipse never exceeds 72 minutes.

18.4 Optical Observations

Today, any potential object for amateur satellite observations is tracked regularly by ground stations all around the world. Consequently, one might ask what is the point in conducting these amateur activities. The answer lies in the fact that most of the regular professional observations are carried out with radar techniques with limited precision and in time intervals that are just short enough to facilitate low-precision orbit determinations.

18.4.1 Accurate Positional Observations

The result of a satellite observation is a set of three data: right ascension, declination, and the instant of time. Since the equatorial coordinates are topocentric, it is important to know the precise geographic position of the observing site.

The first step to an accurate satellite observation is to obtain a prediction for the satellite from one of the institutions listed at the end of this chapter. This is unavoidable since there is no sense in observing an unidentified satellite. To be able to find the satellite during its passage, it is best to plot its predicted track in a small star atlas or, if the satellite is anticipated to be very bright, on a detailed planisphere.

Since many satellites are usually too faint to be visible to the naked eye and since a moderate magnification is absolutely necessary to attain an acceptable positional accuracy, observations are usually carried out with telescopic aid. A pair of binoculars with an aperture of 50 to 60 mm and a magnification between $7\times$ and $10\times$ is the ideal choice for this task.

To record the time, an accurate stopwatch is needed. It should have a digital display, readable to 0.01 seconds, and a "lap time" facility.

As mentioned previously, air drag can cause the orbital period to decrease with an unpredictable rate. Thus, it is advisable to start watching for the satellite several minutes before its predicted arrival. If the satellite arrives earlier than predicted, it is likely to deviate from its predicted track. This is mainly a consequence of the Earth's rotation. Depending on the satellite's height, the actual apparent track may be several degrees east of the predicted one for every minute the satellite is early. If it arrives late, the track will deviate to the west.

Once the satellite has been found, the observer has to wait until it passes between two stars that lie nearly perpendicular to the apparent track and are separated by less than one degree. When the satellite crosses the line joining them, the stopwatch must be started. In the same instant, he or she has to estimate the position of the satellite as shown in Fig. 18.12.

The next step is to identify and to mark the two stars in a detailed atlas.

Finally, the watch is stopped against the signal of a radio or telephone time service. If it has a split-time function, the accuracy of the time recording can be improved by checking the fraction of a second at several preceding time signals.

The evaluation of such an observation requires the precise positions of the two reference stars. These can be taken from a good star catalogue. With these data and

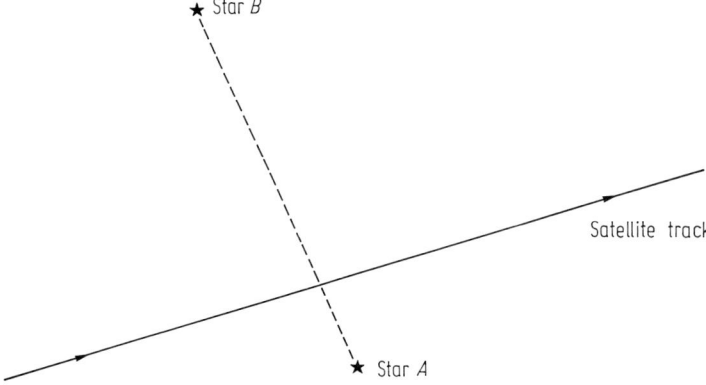

Fig. 18.12. Position measurement of a satellite. When the satellite crosses the imaginary line between stars A and B, the observer estimates the fractional distance to them (e.g., 1/4 AB). Simultaneously, the stopwatch must be started.

the observed fractional distance to each of the stars, an accurate position for the satellite is found by simple linear interpolation.

Another technique that is easier and more accurate can be applied if the satellite has a very close approach to a star. At the moment the satellite appears to coincide with the star, their positions can be assumed to be identical and so a second reference star is not necessary. The time is recorded as described above. An actual occultation is very unlikely. Such an event would last for only a few milliseconds and would therefore be undetectable.

18.4.2 Photographic Observations

Another way to observe satellites is to photograph them. Although it is quite easy to record a spacecraft on film, it is very difficult to achieve photographs that allow precise position measurements. In any event, the evaluation of such a photograph is very laborious and demands expensive equipment. For this reason, only a brief description of how to photograph the passage of a satellite will be given here.

A 35-mm camera with a manual exposure and several f/stop settings and an attachable cable release is quite adequate for photographing satellites. The camera lens should have a low f-number to be able to record a fast-moving object. In addition, it should provide a wide field of view and a distinct satellite trail on the photograph. A standard 50-mm lens fulfills these demands and can be regarded as a good compromise. Of course, only high-sensitivity films should be used. The f/stop must be fully opened.

To achieve the best possible results, the satellite should be photographed when it is approximately opposite to the Sun. In this position, the sky background is darkest and the satellite has a small phase angle.

Because of the rapidly rising shadow of Earth, most satellite observations are carried out before the end of the astronomical evening twilight. Thus, the brightness of the sky background has to be taken into account to avoid overexposures. On the other hand, exposure times have to be long enough to produce sufficient reference star trails.

18.4.3 Light Variations

Most artificial satellites have a shape that deviates considerably from the idealization as a sphere. Cylindrical rocket stages are quite numerous, and many payloads are prismatic or box-shaped. Most of these are equipped with attached solar panels and antenna reflectors. Consequently, their brightness depends not only on the distance and phase angle, but also on their altitude with respect to the observer.

As an example, the tumbling rotation of many spent upper stages can easily be observed as a periodic light variation with an amplitude of several magnitudes. The period of this rotation is influenced by external torques due to atmospheric drag and the interaction of eddy currents with the Earth's magnetic field. As an example, the rotational period of the satellite 1958 δ_1 increased steadily from 15.0 s on 1958 June 02 to 18.4 s on October 02. An amateur observer can easily estimate the tumbling period of a satellite from visual observations. These provide valuable data for the research on the mechanisms that cause the rotational period to change.

But it is not only pieces of uncontrolled space debris that exhibit light variations. Even active spacecraft with a controlled altitude are expected to show light variations because they are usually stabilized with respect to the Earth's limb. Consequently, the orientation to the observer changes in the course of a passage. This effect and the change in phase angle often causes an easily perceptible light variation.

Sometimes satellites exhibit very conspicuous glints of light. These are caused by specular reflections from glazed surfaces on solar arrays or radiators. There have been reports of an increase in brightness of more than 10 magnitudes! Soviet Molniya satellites are especially likely to produce such flashes. These satellites remain for hours in high northern declinations and are sunlit almost throughout the night. In the past, several of these flashes have been misinterpreted as originating from a then-unknown class of variable stars. The famous Perseus and Aries flashers are such cases. In addition, satellite flashes seem to be the explanation for some remarkable phenomena on lunar photographs.

Modern three axis stabilized geostationary satellites such as *Olympus* can also cause bright specular reflections. Such satellites are equipped with a control mechanism that keeps their solar arrays oriented to the point on the celestial equator that has the same right ascension as the Sun. Thus, sunlight can be reflected towards the observer when the satellite's topocentric declination equals that of the Sun. The gain in brightness due to this reflection would reach its maximum when the satellite and the Sun have a difference in right ascension of 12^h (180°), but this always coincides with an eclipse. Therefore, the spacecraft is brightest immediately before and after the shadow passage. Even then, the increase in brightness can be high enough to render the satellite visible in binoculars or very small telescopes.

An orbiting satellite is perpetually being hit by micrometeorites and small particles of space debris. The satellite *LDEF* (Long Duration Exposure Facility) was retrieved in 1990 by the space shuttle *Columbia* after it had spent nearly six years in orbit for the singular purpose of being exposed to the space environment. Upon examining the satellite, engineers found an average of 40 000 impacts per square meter! Such a degradation causes reflective surfaces to become duller and paint coatings to flake off. This results in a perceptible loss in brightness of the satellite after a certain duration in space.

18.4.4 Changes in Color

When a satellite enters the Earth's shadow, it does not vanish abruptly. Rather, it passes through a penumbral phase, when, as seen from the satellite, some portion of the Sun is occulted by the Earth. As the spacecraft progresses further into the shadow, the intensity of the direct sunlight decreases continuously, but there is a growing amount of sunlight that has been refracted and reddened in the Earth's atmosphere. In the shadow core, the direct sunlight is completely obstructed by the Earth. Conversely, the egress of a satellite is accompanied by a change in color from white to red. There have also been reports of a green tinge due to the ozone layer in the upper stratosphere.

18.5 Professional Observing Techniques

In the early days of spaceflight in the late 1950s and early 1960s, there were only a few specially equipped groundstations to track satellites. Almost all observations were carried out with visual astronomical instruments. Nowadays, there are several very sophisticated observational techniques at the disposal of the tracking stations.

18.5.1 Photographic Techniques

As was mentioned earlier, a camera suitable for satellite observations must have a very fast f-ratio and a wide field of view. Therefore, Schmidt or modified Schmidt systems are almost ideal for this task. The famous Baker-Nunn camera was such a modified Schmidt type with a focal length of 510 mm and a focal ratio of $f/1$.

There are camera mounts for three different operational modes:

- The camera remains fixed during the exposure, and the satellite and the reference stars produce trails on the film. This method has the grave disadvantage that only the brightest satellites can be photographed.
- The camera pursues the satellite. With this technique, even very faint satellites can be photographed. The disadvantage of this method is that it requires a high-precision prediction for the track of the satellite and a very special three- or four-axis gimbal system. This operational mode is used in the *GEODSS* (Ground-Based Electro-Optical Deep-Space Surveillance System), a very sophisticated system that is used by the US Air Force to track satellites at altitudes of more than 5000 km.

It employs optical telescopes with apertures of 1.02 m that use vidicon tubes to record images.
- The camera has an equatorial mount and follows the stars. Again, only the brightest satellites can be recorded, but the reference stars produce sharp points. This makes the evaluation of the photograph easier.

18.5.2 Laser Techniques

The laser distance-measuring technique can be applied only to satellites that are fitted with a special retro-reflector. The principle of this method is very simple: the satellite is illuminated with a short laser pulse, which is reflected to a telescope with a sensitive photomultiplier tube. The distance can be calculated from the time taken by the beam to go out and come back.

In practice, this technique causes some very difficult problems. To illuminate the satellite, the laser beam has to be pointed very accurately. This demands a very good prediction for the satellite. Since the beam travels at 300 000 km s^{-1}, a high-precision time measurement is crucial. Today, the laser technique facilitates distance measurements with a precision of several centimeters.

18.5.3 Radar

A radar tracking system "illuminates" the sky with a narrow beam of radio waves and detects the feeble echo caused by a satellite crossing the beam. The direction to the satellite is then given by the the direction of the antenna, and its distance is known by the time delay of the echo. A radar signal is reflected by any metallic body; a special retro-reflector is not necessary. Since clouds are translucent for the radio waves used, a radar system works in all kinds of weather.

The positional accuracy of a radar tracking system is inferior to that of any other professional system; it is equalled by most visual observers using binoculars.

References

18.1 King-Hele, D.: *Observing Earth Satellites*, Van Nostrand-Reinhold, New York 1983.
18.2 Williamson, M.: *The Communications Satellite*, Adam Hilger, Bristol 1990.
18.3 Wertz, J. (ed.): *Spacecraft Attitude Determination and Control*, D. Reidel Publ. Co., Dordrecht 1980.
18.4 Roy, A.E.: *Orbital Motion*, Adam Hilger, Bristol 1982.
18.5 Turnill, R.: *The Observer's Spaceflight Directory*, Warne, London 1978.
18.6 King-Hele, D.: *The R.A.E. Table of Earth Satellites 1957–1980*, Macmillan Press, London 1981.
18.7 King-Hele, D.: Observing Artificial Satellites. *Sky & Telescope*, May 1986.
18.8 Welch, D.L.: Observing Geosynchronous Satellites. *Sky & Telescope*, June 1986.
18.9 Taylor, G.E.: Geostationary Satellites. *Sky & Telescope*, June 1986.

Satellite predictions and current orbital elements can be obtained from:

- NASA Prediction Bulletins, Project Operation Branch (Code 513), NASA-Goddard Space Flight Center, Greenbelt, MD 20771, USA.
- British Satellite Prediction Center, Earth Satellite Research Unit, University of Aston, St. Peter's College, Saltley Birmingham, England.
- Satellite Section, The British Astronomical Association, Burlington House, Piccadilly, London WIV ONL, England.

With current elements, predictions can be calculated using the software package "Spacebirds – A Computer Program for Predicting Naked-Eye Visibility of Artificial Satellites," which is available from Sky Publishing Corp., P.O. Box 9111, Belmont, MA 02178, USA.

19 Observations of the Planets

G.D. Roth

19.1 The Purpose of and Tasks for Planetary Observations

19.1.1 The Amateur Observer and the Planets

The 19th century and the first half of the 20th century have had a distinguished history of amateur observers who achieved scientific renown as planetary observers. These include C.F. Capen, T.A. Cragg, and W.H. Haas (USA), M.B.B. Heath, B.M. Peek, and T.E.R. Phillips (UK), G. Farroni and G. Teichert (France), S. Cortésie and M. du Martheray (Switzerland), and F. Kimberger, W. Löbering, W. Sandner, and W.W. Spangenberg (Germany), to name just a few. Despite the fact that nowadays astronomical research is based primarily on large-scale astronomical equipment and the exploration of the solar system on space missions, the amateur interest in planetary observations remains [19.1]. The ambitious observer will attempt to employ physical measuring methods in addition to merely visual observations, and will also seek cooperation with other observers, since joint efforts often lead to valuable results.

19.1.2 Observational Tasks

With the dawn of the space age in 1957, new life was breathed into planetary research. In some respects, considering the results of the recent flyby missions to as far out as Neptune, one can even speak of the "rediscovery of the planets." Such findings include the exciting verifications that planetary surfaces show craters like those on the Moon, and that Saturn is not the only planet to possess a ring system. A better knowledge of the solid surfaces of planets and moons in the solar system supplies new ingredients for interpreting the origin of these bodies and of the entire system, not to mention important deductions regarding planetary geology. The amateur devoted to planetary observations should be acquainted with these new developments. Easy-to-read presentations and recent updates appear regularly in numerous readily accessible magazines [19.1,2].

The results of the large programs, such as the *International Planetary Patrol Program*, should be studied. The observer will therefy not only keep informed of the foci of research, but will also receive hints concerning the evaluation of his own observations. Even the traditional topographic survey—still the center of most amateur work—gains more weight when processed according to the state-of-the-art methods.

The *Voyager* flybys at Jupiter and Saturn were accompanied by worldwide organized simultaneous observing programs for assiduous amateur observers, namely the *International Jupiter/Saturn Voyager Telescope Observations Program*. The national working groups have reported on this and other joint efforts [19.3].

There are several observational tasks which, when performed with care and perseverence, may aid the scientific exploration. They include, for instance:

(a) Long-term monitoring of planets with respect to the appearance of special phenomena (Martian dust storms, bright or dark markings in the atmospheres of Jupiter and Saturn) [19.4];
(b) Position measurements of individual objects (e.g., clouds on Mars, the Great Red Spot on Jupiter) on planets in order to derive rotation periods or other motion patterns;
(c) Spectrophotometry, visual as well as photographic (observations with filters of known spectral transmission);
(d) Total and detailed photometry, visual and photographic (e.g., intensity estimates of bright zones and dark bands on Jupiter, estimates of brightnesses of minor planets);
(e) Special observations, e.g., of planetary satellites, occultations, or transits of planets across the solar disk.

The following pages will outline these tasks in connection with the individual planets to which they apply.

When performing a serious observational project, however, it is always of fundamental importance that the observer pay attention not only to the personal reliability and precision of the equipment but also to the relevant literature and care in the processing of the results. It is advisable to join a working group or a planetary section of an astronomical association (see the listing in Sect. A.3 in Vol. 1). Knowledge of the literature also permits highly interesting comparisons, for instance, between photographs taken from space and early drawings, such as have been made by R. Koppmann:

"Considering ... the minimal means with which evidently excellent work has been done over the last century, the results are sufficient motivation, especially for amateur observers, for continuing to observe planets in spite of all the recent space missions, and also to publish such observations even though they may appear to be minor or seemingly incredible" [19.5].

In no way, however, should the interested planetary observer be rated as merely an assistant to science. It is advantageous that most of the planets reveal to the observer more than just a point of light, and the viewer can thus become acquainted with cosmic landscapes and directly observe natural processes. In this view, planetary observations have great experience value, a fact which should be kept in mind especially by those who teach astronomy in high schools, community colleges, etc. The astronomical experience is enhanced by systematic observations.

19.2 Observing Equipment

19.2.1 The Telescope

As successful planetary observations depend on very good image definition, even with medium and higher magnifications, an instrument of long focal length is favored (aperture ratio $f/15$ or larger). In principle, there is no preference between refractors and reflectors (see Chap. 4, Vol. 1), although in reflectors, the silhouetting effect of the secondary mirror limits the definition. The disadvantage of silhouetting can be surmounted by choosing a larger aperture.

The high-quality telescope should be accompanied by similarly good eyepieces and by a stable mounting (Chap. 5, Vol. 1). For medium and high magnifications, orthoscopic and eyepieces of similar quality are preferred. Eyepieces containing thick meniscus lenses, which diminish the off-axis imaging errors, have proven useful. The Astro-Planokular of Zeiss, for example, and also the wide-angle eyepieces of Baader, Meade, and Tele-Vue deserve attention. Binocular adaptors (supplied by, e.g., Baader, Munich, Celestron, Torrance, and Carl Zeiss, Jena) have also been found useful [19.6].

Tables 19.1 and 19.2 provide guidelines for answering the often-asked question as to which apertures and which magnifications are actually useful. Planetary observations require a minimum power so that the observer can ascertain more than just a bright dot. The magnifications necessary to see the planetary disk in the telescope at the same apparent angle as the full moon viewed with the unaided eye are provided in Table 19.1. The experienced observer M. du Martheray has compiled experiences concerning the choice of an optimum power for planetary observations into a graph [19.7], on which Table 19.2 is based.

Not to be underrated are personal qualities and the experience of the observer, as well as the quality of the telescope and the atmospheric conditions. Figure 19.1, which is based on a study by H. Wichmann [19.8], considers different atmospheric conditions on the magnification.

The experiences of numerous observers in both Europe and the U.S. emphasize the fact that useful powers can be employed with telescopes of small and moderate

Table 19.1. Magnifications for which the planets would appear the same angular size as the full moon in the sky.

Planet	Apparent Diameter	Magnification
Mercury	$4''\!.8 - 13''\!.3$	$280\times$ at elongation $(6''\!.5)$
Venus	$10'' - 64''$	$70\times$ at elongation $(25'')$
Mars	$4'' - 25''$	$70\times$ at opposition
Jupiter	$31'' - 48''$	$40\times$ at opposition
Saturn	$15'' - 21''$	$100\times$ at opposition
Uranus	$3'' - 4''$	$500\times$ at opposition
Neptune	$2''\!.5$	$750\times$ at opposition

Table 19.2. Optimal magnifications for planetary observations.

Objective Aperture (mm)	Magnification for Observing Detail							
	Mercury	Venus	Mars	Jupiter	Jupiter's moons	Saturn	Uranus	Neptune
75	150	150	175	150	150	175	175	—
135	200	225	275	200	375	250	300	300
250	300	300	325	275	400	350	350	350
300	350	350	350	300	500	375	450	500

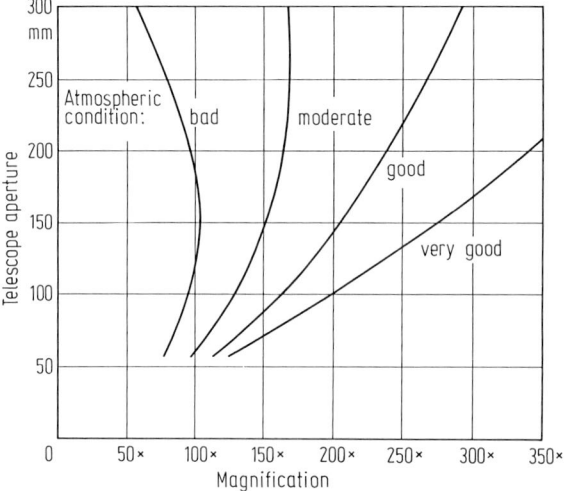

Fig. 19.1. Effect of atmospheric conditions on the useful magnification for planetary observations as dependent on aperture size.

apertures (100 mm to 150 mm) even under fair or poor atmospheric conditions, and thus these apertures can be considered "optimal" for amateur observers over a wide range of atmospheric circumstances.

19.2.2 Accessories

In order to reach more diversified physical methods of observing and measuring, certain accessories will be indispensible for the amateur observer. Visual as well as photographic observations may be upgraded by the use of additional instruments, the most important of which are:

(a) color filters (Chap. 6, Vol. 1),
(b) the micrometer (Sect. 4.7.5, Vol. 1),

(c) the photometer (Sect. 4.7.6 and Chap. 8, Vol. 1), and
(d) the spectroscope (Sect. 4.7.7 and Chap. 7, Vol. 1).

Certainly the color filters are the easiest to work with. They can be advantageously combined with small instruments to make observations of planetary surfaces in various well-defined selected spectral ranges [19.9]. As a rule, these ranges include the red (e.g., Schott RG2 and RG5), yellow (e.g., Schott GG10) and blue (e.g., Schott BG12 and BG23). The individual filters are framed so as to cap the eyepiece [19.10]. A micrometer can be used to measure, for instance, the diameter of the crescent of Venus, to determine the time of dichotomy and the cusps (Sect. 19.5.2.2), or to measure the positions of striking details on Mars (clouds, polar caps, etc.) and in the atmospheres of Jupiter and Saturn (bright and dark features, widths of zones and bands); see Sect. 19.5 for further details. With a photometer one can, in particular, perform *area photometry* of various regions on the planetary surface (Chap. 8, Vol. 1) and *point photometry* to obtain the total integrated brightness of a planet or minor planet; see Sect. 19.5.3.4. The spectroscope enables the observer to find the rotation period of a planet via the Doppler effect and to gain information regarding the planetary atmosphere [19.11]. The observer may even contemplate the use of a CCD camera [19.12], a device particularly suited for photometric tasks (cf. Chap. 8).

19.3 Visibility of the Planets

19.3.1 Apparent Diameter, Phase, and Oblateness

The conditions of visibility for the individual planets differ greatly. Information on their visibility during a year can be found in various astronomical calendars (see also Appendix Table B.15 in Vol. 3). The orbit and size of a planet, as well as the location of the observer, determine its visibility. Specifically, the determining factors are:

(a) rising-culmination-setting of the planet as viewed at the observing site;
(b) right ascension and declination of the planet;
(c) apparent diameter (see Fig. 19.2);
(d) apparent brightness;
(e) phase angle.

The outer planets (Mars, Jupiter, Saturn, Uranus, Neptune, and Pluto) are closest to the Earth at *opposition*, when they are visible all night long. The inner planets (Mercury and Venus), on the other hand, appear either in the evening or morning sky (at, respectively, the times of *greatest eastern* or *greatest western elongation*). The position of the ecliptic relative to the horizon and the seasonally varying duration of twilight, depending on the geographic latitude of the observer, influence the visibility of the planets, in particular the inner ones. But also for the outer planets, the observing conditions are most favorable when the planet's opposition occurs during the winter, when the ecliptic band is high above the horizon. At lower geographic latitudes, the ecliptic is steeper with respect to the horizon, and visibility conditions are better. Even for a planet with a substantial orbital eccentricity, the altitude above the horizon

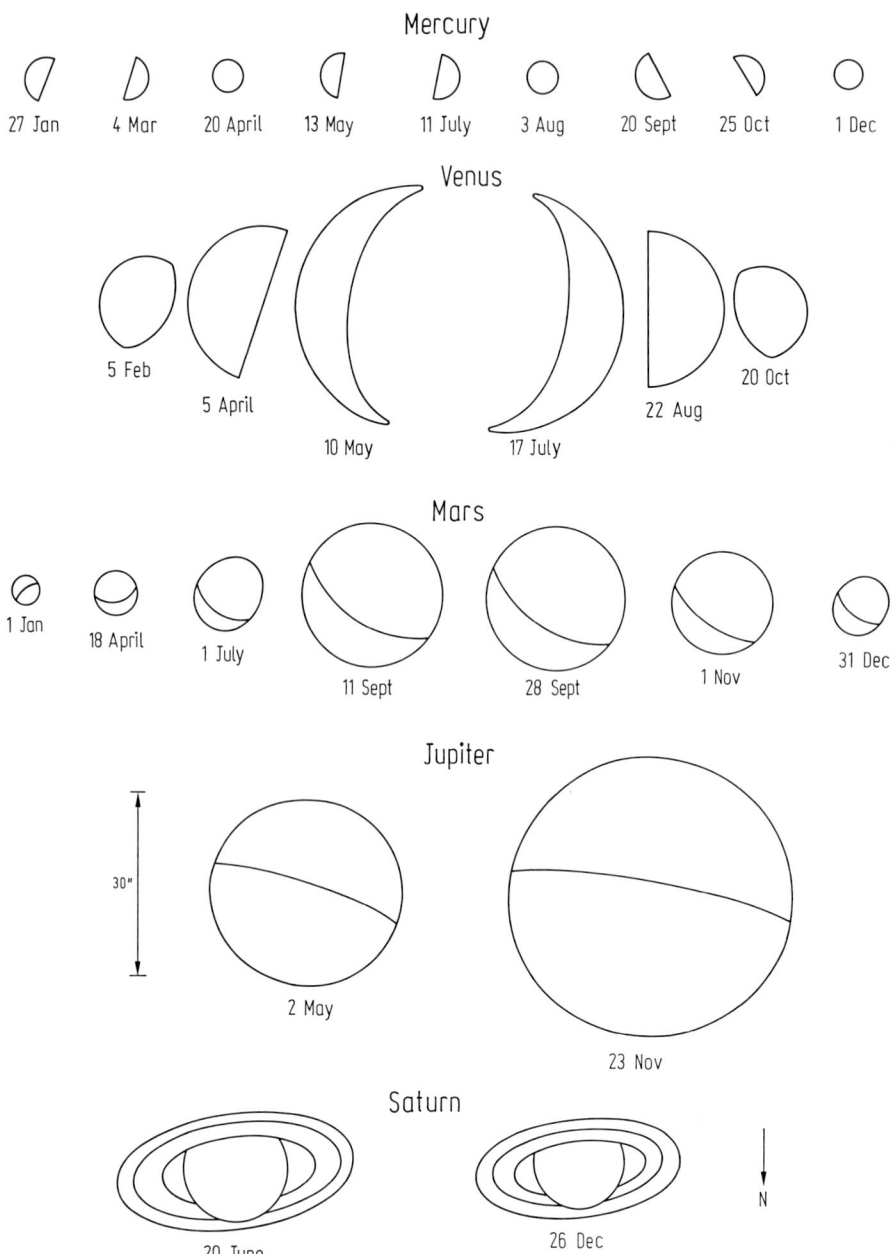

Fig. 19.2. Appearance of the planets during 1988. From G.D. Roth, *Taschenbuch für Planetenbeobachter*, 3rd edn., Sterne und Weltraum, Munich 1987.

outweighs the perhaps less-favorable orbital conditions (i.e., nearness to the Sun for Mercury, small apparent diameter for Mars).

The *phase angle* of the planet (which is listed in astronomical almanacs) will have a significant influence on the observations, as it determines the degree of apparent illumination of a planet. At a phase angle of 0°, the disk of the planet is fully illuminated, and corresponds to the full-moon phase, while at 180° it is invisible (corresponding to new-moon phase). For the inner planets, the phase angle varies from 0° to 180°, while for the outer planets, it plays a significant role only for Mars (where it is 46° when one-eighth of the planetary disk is invisible). Some almanacs give, instead of the phase angle, the degree of illumination of the planetary disk.

In addition to the changes of apparent diameter and illumination, the observer must also be concerned with the position of the rotation axis and the oblateness, information on which is to be drawn from the annual almanacs.

19.3.2 Atmospheric and Environmental Influences

The effects of Earth's atmosphere on astronomical observations is detailed in Chap. 23. For planetary observations, the atmosphere over the observing site should be as free from turbulence as possible. The medium and high magnifications required for planetary observations are vitiated by atmospheric unsteadiness. Haze, on the other hand, is usually less of a problem as it is often associated with very stable conditions.

In order to objectively rate the atmospheric conditions at the observing site, the observer should use an estimated scale and record the result in the observing protocol. Two important data are estimated: the *seeing*, or image stability, which depends on the magnification employed, and the *transparency*, which is best estimated without a telescope.

Seeing Scale

Seeing 5 — excellent; even at high magnification, the image of the planet in the eyepiece is steady and sharp.

Seeing 4 — good; general impression as in 5, but with very occasional turbulence.

Seeing 3 — fair; the turbulence is higher over time, but there are sufficiently long time intervals during which moderate magnification can be used to obtain a useful overall impression of the planetary disk.

Seeing 2 — poor or mediocre; the turbulence interferes perceptibly; only momentarily can details be seen.

Seeing 1 — very poor or useless; it is impossible, even at low power, to see a sharp image of the disk.

Transparency Scale

T5 — very clear sky; 5th-magnitude stars visible.

T4 — clear sky; 4th-magnitude stars visible.

T3 — hazy sky; visibility reaches to 3rd-magnitude stars.

T2 — thick haze; only 1st-magnitude stars are seen; planetary observations marginally possible.

T1 — haze, fog, clouds so thick that the planet is intermittantly visible with the naked eye; in the telescope the image is too dim to perceive details.

The atmospheric turbulence is best seen on a sharply focused image of a bright star. The image wiggles around its central position. This "turbulence disk" has a radius of $5''$ or more in bad seeing. In quiet air, the disk size diminishes to less than $0\rlap{.}''5$ [19.13].

Atmospheric turbulence is caused not only by climate and weather at the observing site, but also by other circumstances in the vicinity of the telescope which set into motion the layers of air directly overhead (e.g., heating, smoke updrafts, airplanes). Artificial light sources are not as detrimental to visual planetary observations as they are for other kinds of astronomical work. Therefore, opportunities exist even for planetary observers who live in densely populated areas. However, localized urban heating does favor the generation of small turbulence cells.

19.3.3 Personal Qualities

In the final analysis, the perception of planetary details depends on the approach of the individual observer. Here, experience gained from long-term, well-planned observing runs plays an important role. Every observer should be aware of the following rules:

(a) Always observe with rested eyes.
(b) Avoid light glare (adapt to darkness, dim all light sources needed in the vicinity).
(c) Do not overmagnify images.
(d) Practice monocular vision, perhaps (at the beginning) by covering the unused eye with the hand or a patch. Note, however, that the experienced observer does not close the unused eye while observing, but merely subconsciously suppresses its signal. Binocular adaptors may be tried.
(e) Telescope operation, such as guiding, must be performed effortlessly in the dark.
(f) Several breaks should be taken between observations to avoid eye fatigue.
(g) No observion should be performed without sketching, as this forces the eye to carefully examine every detail.

W.W. Spangenberg, who over several decades became acquainted with the personal idiosynchrasies of numerous observers and the accompanying physiological effects, notes:

"Observing experience is generally acquired and maintained through extensive and continuous activity. Interruptions thereof, even if of short duration, can lead to a loss of practice" [19.14].

Details on planetary surfaces are often not easily perceived and interpreted. Except for rare occasions, the ubiquitous atmospheric turbulence during observation impedes the rendition of fine detail. There is a substantial risk that errors will enter the drawings. It may be advantageous with regard to observing quality to specialize on just one planet [19.15].

Statistics show that about one-half of active amateur observers have already reached an age where diminished sensory efficiency can be expected. Diminished vision acuity is not the only result of inevitable aging process. As Spangenberg showed, the posture at the telescope is also, with increasing age, "a genuine physiological problem which affects both vision and perception" [19.16].

19.4 The Representation of Planetary Observations

19.4.1 Drawings

When making drawings, the directions on a planet must first be defined: following the IAU decision of 1961, planetary maps have the same "astronautical" orientation used for Earth maps in atlases: north on top, south at bottom, east to the right (in the direction of planetary rotation), and west to the left. Thus the image is upside-down, or rotated by 180° for observing a planet south of the zenith with an inverting telescope (from most northern locations on Earth), or for a southern observer using a standard, non-inverting pair of binoculars. Surface detail, such as the Great Red Spot on Jupiter, is carried by the planet's rotation to the east (to the left in a northern inverting telescope). Formerly, the terms "preceding end" and "following end" were used, where preceding was in the sense of the diurnal displacement of the planet due to Earth's rotation as well as to the shift due to the planet's rotation. In any event, observers are advised to avoid confusion by proper labeling of graphs, maps, or photographs.

Stencils for graphs can be obtained from various amateur supply companies. There are circular stencils for Mercury, Venus, Mars, while those for Jupiter include the flattening, and for Saturn both flattening and various inclinations of the rings. The two most important things to be kept in mind for Mercury, Venus, and Mars—phases and axial orientation—will be explained in the respective subsections.

As a basic technique in drawing, every graph should be developed from coarse contours into fine detail. A medium-soft pencil should be used. For the fast-rotating planets Jupiter and Saturn, the coarse contours should be completed quickly lest substantial longitude errors occur.

Apart from general presentations of the surface, drawings may be used to determine positions. In this case, the emphasis is not on innumerable detail but rather on the careful representation of certain selected features, such as bright and dark spots on Jupiter, and their arrangement into a coordinate system (see below). With some practice, carefully drawn detail near the central meridian can yield good results (cf. Sect. 19.5.5.3 on transit and micrometer observations):

"Under particularly good circumstances, standard errors of ±0.3° to ±0.5° (jovicentric) were reached, while under unfavorable conditions (turbulence, personal indisposition, etc.) the deviation reached ±1°." (W. Löbering on observations of Jupiter).

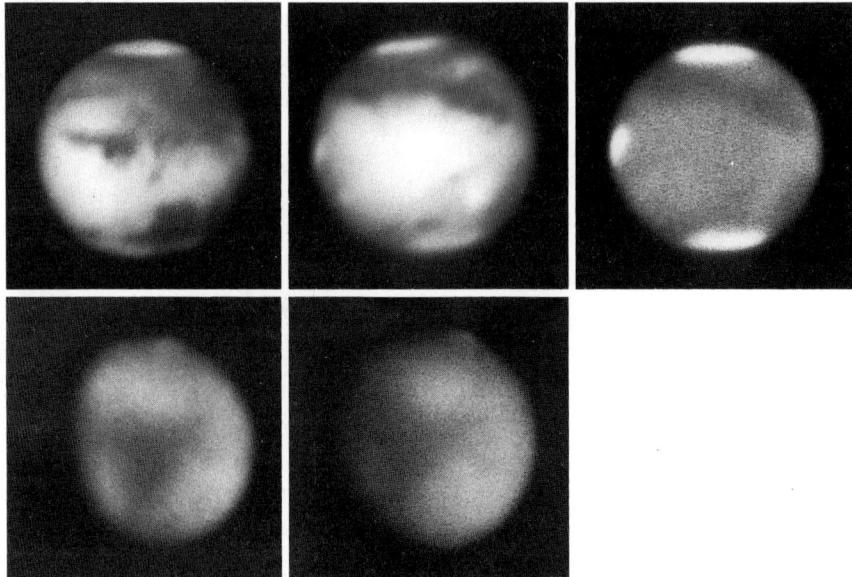

Fig. 19.3 a. *Top left*: Mars, photographed on 1986 July 20, 22^h18^m UT with a 106-cm reflecting telescope at the Pic-du-Midi Observatory by J. Dragesco. Central meridian $11°3$, apparent diameter $23''$, $f/D = 52$, exposure time 1/4 s on TP 2415 with W29 filter. *Top center*: Mars, photographed on 1986 July 12, 04^h52^m UT with a 32-cm Newtonian reflecting telescope by Donald C. Parker, Coral Gables, Florida, USA. Central meridian $186°$, $f/D = 198$, exposure time 5 s on TP 2415, no filter, developed in Rodinal 1:100. *Top right*: Same observer and date, but photographed at 04^h42^m UT using a violet filter (W47). $f/D = 198$, exposure time 2.5 s on ED 200, developed E-6 process. *Lower left*: Mars, photographed on 1973 September 07, 03^h46^m UT with a 25-cm Newtonian reflecting telescope by P. Hückel, Weilheim. $f/D = 108$, exposed 10 s on Kokak high-contrast copy. *Lower right*: Mars, photographed 1973 September 04, 02^h25^m UT with a 25-cm Newtonian reflecting telescope by P. Hückel, Weilheim. $f/D = 72$, exposed 5 s on Copex-Pan.

19.4.2 Photographs

Techniques and photographic materials are detailed in Chap. 6, Vol. 1. Good planetary photographs are also a useful supplement for comparison with the visually obtained sketches (Fig. 19.3). Color filters will show the varying visibility of some surface or atmospheric details on planets as depending on wavelength (Fig. 19.4). Planetary photographs with amateur instruments serve primarily to measure positions of prominent details with the aid of grids. Photographic positions on Jupiter are impeded by the substantial limb darkening across the disk; the true limb of the disk may be difficult to locate. It is therefore advisable to photograph a prominent object near the central meridian in each series, as the position of the central meridian is better determined than that of the limb [19.17].

The use of CCDs has been mentioned in Chap. 4, Vol. 1. The increased speed of a CCD camera is a valuable asset for planetary photography. It is about 75 times faster

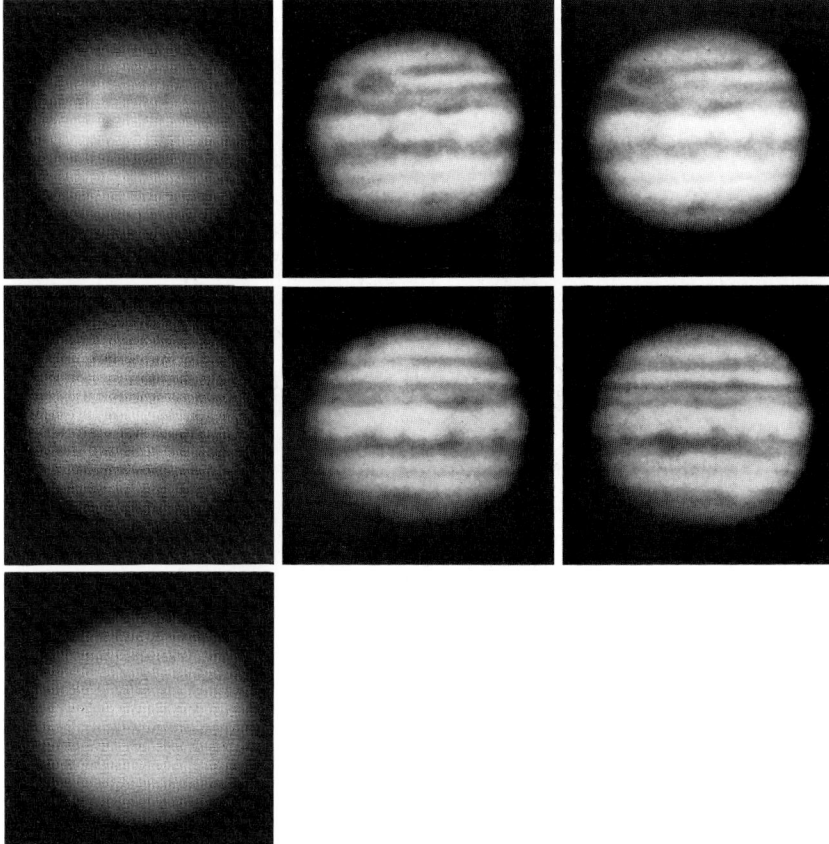

Fig. 19.3 b. *Top row, left*: Jupiter, photographed on 1967 February 24, $21^h 10^m$ UT with a 25-cm Newtonian reflecting telescope by P. Hückel, Weilheim. $f/D = 56$, exposed 3 s on Agepan FF. *Middle and right*: Jupiter photographed with a refractor of 200/4000 mm with an effective focal length of 36 m, using 14/10 DIN, by G. Nemec, Munich. *Middle row, left*: Jupiter, photographed on 1967 March 04 at $19^h 40^m$ UT with a 25-cm Newtonian reflector by P. Hückel, Weilheim, $f/D = 60$, exposed 2 s on Agepan FF. *Middle and right*: Jupiter photographed with a refractor (200/4000 mm) with an effective focal length of 36 m, using 14/10 DIN, by G. Nemec, Munich. *Lower left*: Jupiter, photographed 1966 January 08, $20^h 15^m$ UT with an 11-cm Schiefspiegler by P. Hückel, Weilheim. $f/D = 118$, exposed 1 s on Ilford FP 3.

than the conventional photographic technique and yet maintains the desirable feature of being able to integrate over long exposure times. CCDs are available commercially in different makes, but there is also something to be said for "do-it-yourself" construction. Terry Platt (UK) used a Sony ICX 021 CL CCD chip to build a camera for a 32-cm Schiefspiegler. The camera acts as a photographic emulsion of about ISO 2000°, permitting short exposure times and thus high-resolution images even in inferior seeing [19.18]. Pictures of Jupiter and Mars photographed off the monitor attain the quality

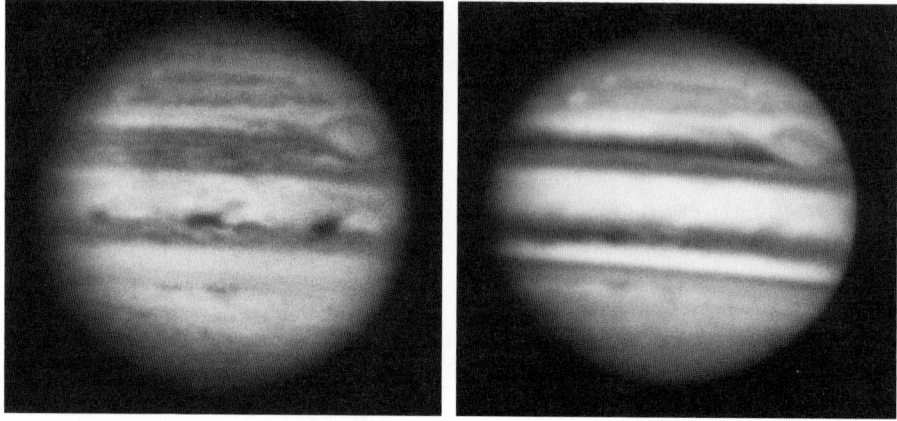

Fig. 19.4. Examples of photographs taken using color filters. *Left*: Jupiter, photographed 1986 July 20, $03^h 14^m$ UT with an 106-cm reflecting telescope on the Pic-du-Midi Observatory by J. Dragesco. $f/D = 50$, exposed 2 s on TP 2415, using a Wratten W29 red filter. *Right*: Same, but taken at $03^h 15^m$ UT at $f/D = 32$, exposed 8 s on TP 2415, using a Wratten W49 blue filter.

of J. Dragesco's photos (Figs. 19.3 and 19.4). Astronomical magazines often report on this rapidly developing technology [19.19].

19.4.3 Maps and Planispheres

In principle, planetographic coordinate systems are always the same: the equator is the basic circle, and its position depends on the orientation of the rotational axis. Meridians intersect both poles at right angles to the equator. The position on a planet is expressed by the *planetocentric longitude L* and the latitude *B* (e.g., "areographic" longitudes and latitudes for positions on Mars, or "jovicentric" ones on Jupiter). Longitudes are counted from a zero-meridian at 0° westward to 360°. In the course of time, the meridians of increasing longitude traverse the central meridian. Latitude is counted from the equator to the poles (0° to +90° north, 0° to 90° south).

It is not difficult to construct such a grid for spherical planets. Normally, the so-called *orthographic projection* is used [19.3]. For Mars, the tilt of the axis with respect to the Earth is allowed for, which converts the latitude circles into ellipses (as the meridians already are). Orthographic horizontal projection permits the construction of a grid of longitude and latitude at the given tilt. The axis orientation may be found in an astronomical almanac or can be computed as follows:

With α and δ as the coordinates of Mars, *A* and *D* those of the Martian north pole, the inclination *i* of the axis, e.g., the areographic latitude of the apparent center of disk (+ to north, − to south), is given by

$$\sin i = -\cos D \cos \delta \cos(\alpha - A) - \sin D \sin \delta. \tag{19.1}$$

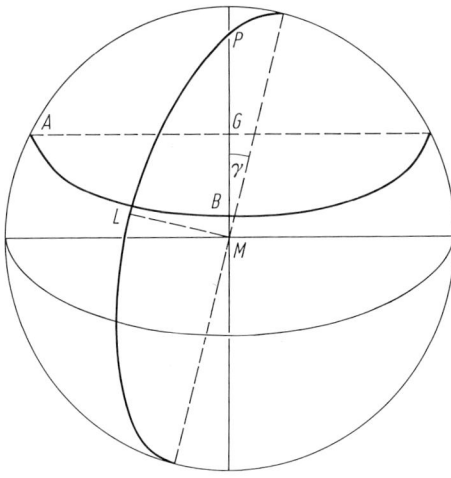

Fig. 19.5. Construction of the grid of a planet in orthographic horizontal projection.

The coordinates of the Martian north pole are $A = 317.61°$, $D = +52.85°$ for 1990, with centennial changes of $+0.77°$ and $+0.42°$, respectively.

The completion of the grid (cf. Chap. 15) then follows (Fig. 19.5): with a graph radius s, meridians become ellipses with semimajor axis s, latitude circles with $s \cos b$. Hence,

$$MP = s \cos i, \qquad MG = s \sin b \cos i,$$
$$GA = s \cos b, \qquad GB = s \cos b \sin i,$$

and if the longitude difference between central meridian and the meridian at L, then

$$ML = s \sin l \cos i, \qquad \tan \gamma = \tan l \sin i.$$

Intervals in l and b need not be chosen small; $30°$ will suffice to adequately read the position of the detail recorded. (The vertical in Fig. 19.5 is the position angle p of the axis, given by $\sin p = -\cos D \sin(\alpha - A) \sec i$.)

In order to compile the readings into a map, cylindrical projection (Fig. 19.6) may be used, as it is suited to represent extended equatorial regions. The projection is true to area everywhere, but true to length only at the equator. Parallel circles are stretched to the length of the equator, and graphed at $x = \sin b$, while the vertical meridians are spaced at $y = \text{arc } l$. The area of a zone from the equator to latitude b is then correctly represented as $2\pi \sin b$.

For Jupiter and Saturn, the oblateness, in addition to the substantial axial tilt, complicates the construction. The definition of longitude is not changed by the flattening, but the planetocentric and *planetographic* latitudes are distinguished (see Fig. 19.7) [19.11].

The preparation of grids helps to determine the positions of surface and atmospheric details, and can in principle be used to process drawings as well as photographs. It also allows the combination of several observations into a synopsis of the surface. The rapid change of observable detail on some planets, particularly Jupiter, requires a

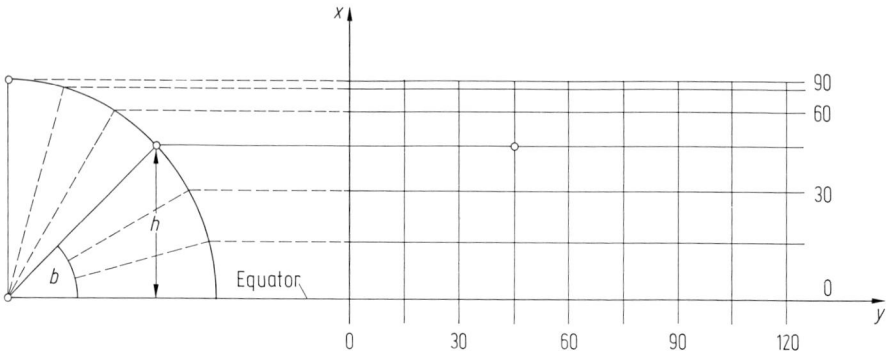

Fig. 19.6. Construction of a map using cylindrical projection.

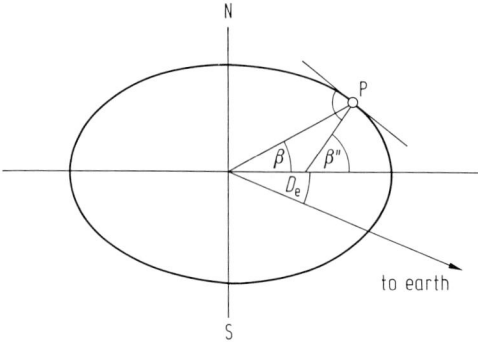

Fig. 19.7. Definitions of planetocentric latitude β and planetographic latitude β'' of an oblate planet.

combination of results of several consecutive nights. To represent extended equatorial regions, a mercator projection is advised [19.20]. A personal computer is a useful tool for storing and processing planetary observations, and it can also expedite the preparation of maps and planispheres, as was reported by H. Haug and C. Kowalec in [19.3]. Examples and programs for PCs and pocket calculators can be found in various astronomical magazines [19.21].

19.5 The Planets

19.5.1 Mercury

19.5.1.1 Visibility. The planet Mercury travels around the Sun in an orbit which lies inside that of Earth. During its synodic period, it shows phases similar to those of the Moon, as it can move between the Earth and Sun, but never into opposition. Its largest angular distance from the Sun (maximum elongation) is 28°. Beginning with superior conjunction, the pattern of visibility is as follows (with several days' variation owing to the eccentricity of the orbit):

Day 0: Superior conjunction, "full Mercury";
Day 12: Mercury appears in evening sky;
Day 36: Greatest eastern elongation, "last quarter";
Day 47: Retrogression of Mercury begins;
Day 53: Mercury disappears from the evening sky;
Day 58: Inferior conjunction, "new Mercury";
Day 63: Mercury appears in the morning sky;
Day 69: Retrogression ends;
Day 80: Greatest western elongation, "first quarter";
Day 104: Mercury disappears from morning sky;
Day 116: Next superior conjunction, "full Mercury."

Astronomical almanacs supply the following data:

1. Right ascension and declination (needed for finding objects during the daytime).
2. Apparent diameter, which varies between $4\rlap{.}''8$ and $13\rlap{.}''3$.
3. Apparent magnitude, varying between $+3.0$ and -1.2 during visibility.
4. Phase angle (0° = full, 180° = new) or phase (1.00 = fully illuminated, 0.00 = unilluminated).
5. Planetographic coordinates of the center of the disk (sub-Earth point), which may be ±7° in latitude off the equator during periods of visibility.

Northern observers at mid-latitudes will have their best opportunities to view Mercury in spring in the evening sky and in autumn during the morning. Better observing configurations can be found near the Earth's equator, owing to the steeper angle of the ecliptic to the horizon, or in the southern hemisphere, as Mercury at times of maximum elongation is farther south than the Sun.

Mercury can also be observed in the daytime sky. Its apparent brightness is highest around superior conjunction. In 1971 June, F. Dorst found the planet with a 6-inch refractor at only $1\rlap{.}°7$ away from the limb of the Sun [19.22]:

"Some sharp, deep-blue gaps occurred in a cloud cover near the Sun but were free of scattered light and permitted the location of the planet with surprising ease .... The disk was easily recognized at a magnification of 78×, and a shading was seen, though somewhat uncertain, of the center disk at 250×."

Low light scatter near the Sun is evidently the most essential condition for such observations.

19.5.1.2 Visual Observations. The crater-covered surface of the planet revealed by photographs taken from spacecraft [19.23,24] cannot be recognized as such by Earth-based observers. Small and middle-sized telescopes show surface detail only in the form of somewhat difficult shadings. The work is also impeded when the planet is observed near the horizon. While it helps to make more frequent observations during the daytime, the best advice is to observe during twilight. The experienced observer W.H. Haas (ALPO) suggests that "the most important factor for success is to observe the planet when the contrasts are strongest; detail appears most prominent when the image is neither too pale in daytime, nor too bright in the evening or morning twilight."

This experience also holds for observations of Venus. The use of filters is advised in order to enhance contrast. The detail on Mercury is—at least in principle—not fuzzy or diffuse, but rather comparable with the distinct outlines on Mars. Apparent "changes" of dark areas and possible brightenings may depend on the phase angle.

The times of greatest elongation and of dichotomy differ by up to ± 6 days, owing to the substantial orbital eccentricity of the planet.

19.5.1.3 Photography. When Mercury appears in the evening or morning sky, it can be easily captured on film using a conventional camera with a telephoto lens. Very nice "constellation photographs" [19.25] may result, showing, for instance, Mercury near the Moon or near Venus. It may also prove worthwhile to try color film.

Photographs taken through the telescope require conditions of low light scatter. Photographs of phases or surface details are very difficult to obtain with the telescopes of the size normally used by amateurs. On the other hand, photographs of an occultation of Mercury by the Moon [19.26] or the rare event of a transit of Mercury across the face of the Sun [19.27,28] are well within the reach of modest equipment.

19.5.1.4 Transits of Mercury. Two more transits of Mercury across the face of the Sun will occur in this century: 1993 November 06 and 1999 November 15. They can be observed either directly or by projection. The tiny, deep-black planetary disk is a difficult object to observe against the brilliant solar light, as certain optical effects may lead to visual errors. For instance, the occurrence of the "black drop" [19.29] will interfere with determining the times of second and third contacts.

19.5.2 Venus

19.5.2.1 Visibility. As with Mercury, the planet Venus travels in an orbit which lies inside that of the Earth; it is the second "inner" or "inferior" planet. Although still always within the proximity of the Sun as viewed from Earth, Venus is much more conspicuous in the morning or evening sky than Mercury; its maximum angular distance from the Sun reaches $47°$. Thus, at the time of optimum visibility Venus can be seen for several hours after sunset as an "evening star" or before sunrise as a "morning star." The planet shows phases just as the Moon and Mercury do. Counting from superior conjunction, the visibility pattern is as follows.

Day 0: Superior conjunction, "full Venus";
Day 35: Venus appears in evening sky;
Day 221: Greatest eastern elongation, "last quarter";
Day 271: Retrogression of Venus begins;
Day 286: Venus disappears from the evening sky;
Day 292: Inferior conjunction, "new Venus";
Day 298: Venus appears in the morning sky;
Day 313: Retrogression ends;

Day 362: Greatest western elongation, "first quarter";
Day 549: Venus disappears from morning sky;
Day 584: Next superior conjunction, "full Venus."

The data supplied by almanacs are the same as for Mercury. The apparent diameter of Venus varies from $10''$ to $64''$, and the apparent magnitude between -3.9 and -4.7. Northern-hemisphere observers have the best evening viewing conditions when eastern elongation falls in the springtime and the planet then remains visible until around midnight.

Venus can be observed over a period of about 7 months at each elongation. The planet reaches greatest brilliancy 35 days after the eastern elongation, and 35 days before western elongation. The high apparent brightness facilitates finding and observing the planet during the daytime. Such observations around maximum elongation are possible with just the naked eye. The observer should stand in the shadow of a house or tree in order to avoid the glare of the Sun. The approximate position of the planet in the daytime sky can be found using the almanac coordinates and a rotatable star map.

An equatorially mounted telescope with circles (see Sect. 5.14.1 in Vol. 1) will facilitate the setting on the planet. It is advisable to have focused the eyepiece on a star beforehand (e.g., on the previous night) in order to immediately secure a sharp image of the planet in the field in the daytime.

Although observation of Venus near conjunction is by no means easy, it should be noted that the planet can, under excellent conditions, be seen with the naked eye about two weeks before or after a superior conjunction [19.30]. If, moreover, binoculars are used, then Venus may be viewed almost up to or just after the time of conjunction. Indeed, F. Dorst observed Venus in August 1971 on the day of conjunction just one hour before the event with 8×30 binoculars.

During inferior conjunction, the apparent diameter of the narrow crescent of Venus reaches about $60''$ from cusp to cusp. Since the planet may be found up to $9°$ north or south of the Sun, the overlap of the cusps can be seen around inferior conjunction with a 2- or 3-inch telescope under reasonably good viewing conditions. Some observers even report that around the time of inferior conjunction the planet can be seen with the naked eye as an "oblong triangle." When Venus is far north of the Sun at inferior conjunction, it is for a few days both a morning and an evening star for northern hemisphere observers.

19.5.2.2 Visual Observations. Concerning planetary detail, the range of visual viewing from Earth is very limited, and not just for amateurs equipped with small-aperture instruments. "The appearance of Venus resembles that of a closed, sunlit stratus layer seen from an airplane high above." This often-cited experience of the long-time observer W. Sandner summarizes succinctly what the observer can expect to see even in the most favorable case. Observers sometimes report subtle differentiation in terms of brighter or darker shadings, but the reality of these results is doubtful. As K. Graff stated over 60 years ago: "Villiger's studies on brightly illuminated spheres showed probably beyond doubt that most drawings of Venus containing bright and dark shadings represent physiological effects and say nothing about the planet itself" [19.31].

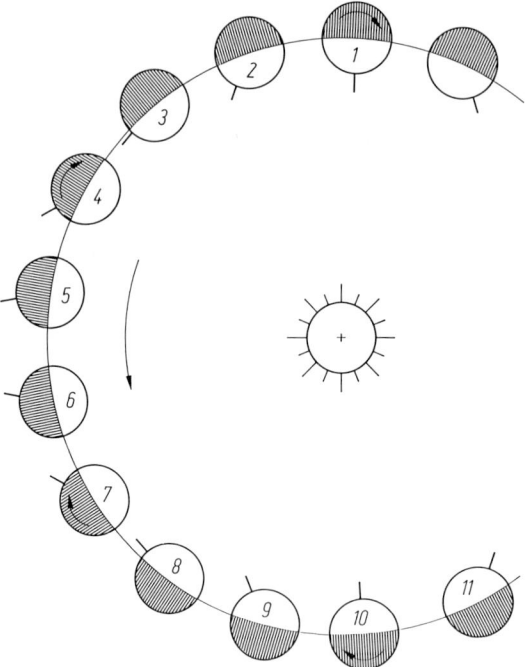

Fig. 19.8. Part of the orbit of Venus during a full Venusian day (between two consecutive noons). The sense of the rotation is retrograde (*clockwise short arrow*), while the sense of the orbital motion is prograde (*counterclockwise long arrow*). Noon occurs at position *1*, sunset at *3*, midnight between *5* and *6*, sunrise shortly before *8* (in the west!), and *10* marks the next consecutive noon. From P. Ahnert, *Kalender für Sternfreunde*, Johann Ambrosius Barth, Leipzig 1981, p 166.

The *Venera* space probes of Russia (former USSR) and *Mariner* and *Pioneer* probes of the USA obtained new information on the atmospheres and surfaces of the inner planets. The solid surface of Venus rotates in 243 Earth days (Fig. 19.8) and thus the rotation is longer than the Venusian year of 225 Earth days. The rotation is also retrograde from east to west, so that on Venus the Sun rises in the west and sets in the east. More than half of the desert-like surface is quite flat. Almost one-fifth lies below zero-level. Elevations reach up to 10 000 m. When the NASA space probe *Magellan* arrived at Venus on 1990 August 15, the first pictures it took of the surface revealed many more craters and other structures [19.32]. The atmosphere is composed of several cloud layers with the upper layer rotating distinctly faster than the lower ones. Virtually all of the incoming solar energy is absorbed near the upper edge of the cloud deck. This triggers heating, which in turn drives the circulation patterns in the atmosphere. Ultraviolet photographs taken by the probes (Fig. 19.9) have confirmed fine structure in the clouds. In particular, eddies in the shape of the letters "Y" or "C" have been repeatedly observed by amateurs.

 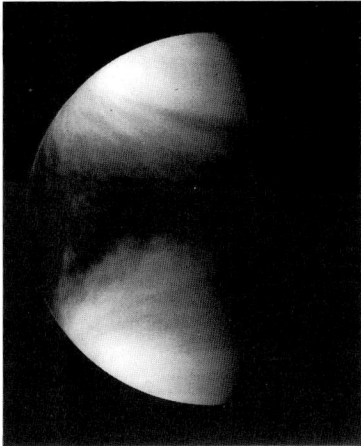

Fig. 19.9. UV photograph of the planet Venus taken by the *Pioneer Venus* spacecraft. *Left*: 1979 January 14 at a distance of 65 000 km; *Right*: 1978 December 30 at a distance of 43 000 km. Photographs courtesy of NASA.

Indeed, the Venusian atmosphere is very dense, nearly structureless in visible light, and conceals the surface from the viewer. Several visual observers working with filters have reported the impression that the structure of the dense atmospheric shell is better recognized. Blue filters apparently offer more detail than yellow and red glasses (see Chaps. 4 and 8, Vol. 1).

The visual observer is left with the following tasks [19.33]:

1. to follow the change of phase and to determine dichotomy;
2. to make observations of the shape of the terminator;
3. to make observations of the extension of cusps of the narrow crescent and of the "ashen light."

Monitoring the phases and the dichotomy shows the difference between computed and observed times owing to the action of the atmosphere. Many observers have studied the dichotomy and its deviation from the geometric prediction.

The phase angle represents the unilluminated arc of the planetary hemisphere toward Earth (Eq. (3.38) in Vol. 1). It is by no means a simple task to estimate the phase correctly; the image of Venus is generally too pale in the daytime, but too bright even during twilight. Moreover, the presence of Venus's atmosphere makes the terminator noticeably less sharp than the limb.

For observations aimed at estimating the phase, it is advised to start the series some time before the expected dichotomy. Paul Ahnert [19.30] advises that the width k of the illuminated part be estimated as a fraction of the diameter. The phase angle φ follows from k from the formulae

$$\begin{aligned}\sin(\varphi - 90°) &= 2(0.5 - k) \quad \text{planet less than half-illuminated,} \\ \sin(90° - \varphi) &= 2(k - 0.5) \quad \text{planet more than half-illuminated.}\end{aligned} \qquad (19.2)$$

Ahnert undertook phase observations spanning several years, and used refractors with 80-mm and 100-mm apertures and magnifications from 100× to 200×. He reported that neither aperture nor magnification had a substantial influence on the estimates [19.34]. His observations should provide helpful hints on the use of color filters:

"2-mm Schott filters were used, OG 4 for yellow, VG 5 for green, and BG 23 for blue. Estimates without a filter and with a green filter agreed closely, the mean difference between observed phase angles being $+0.°01 \pm 0.°84$ from 33 good observations. There was also good agreement between estimates made with yellow and blue filters, but a substantial deviation was found from the result without a filter and with a green filter" [19.34].

Ahnert suspected that the lower image brightness was the primary cause for the phase to be estimated narrower in the yellow and blue filters.

A series of phase estimates is best evaluated in graphical form. The data are entered into a diagram which shows the observed (O) phase and the geometrically computed one (C) with respect to time. It graphs the displacement of observed versus calculated dichotomy due to the action of the Venusian atmosphere (the *Schröter effect*). It should be noted that the time of greatest elongation (E) does not necessarily coincide with the geometric dichotomy (D); the observer must use the latter. With a maximum time difference $E - D$ of 3 days, a mistake in this calculation may result in $O - C$ ($\approx 10^d$) being off by 30%, as W. Kunz [19.35] has pointed out. In addition to visual estimates with or without filter, the filar micrometer (see Sect. 4.7.5) may also be used.

Observations on the shape of the terminator concern its possible deformations. Numerous observers have reported protrusions or indentations at the terminator during the phase cycle. It may here be helpful to use an eyepiece with a crosswire. This task can also be facilitated by the application of various color filters. It may be assumed that most of the reported anomalies in the terminator are not optical illusions but are indeed real effects in the Venusian atmosphere. This becomes particularly clear in the case of the extension of the cusps of the narrow crescent beyond the poles of illumination. Such observations fall around inferior conjunction. The observer may simply estimate the extension using a crosswire eyepiece and recording the data using stencils with position marks.

Micrometer measurements will give more precise results. Near inferior conjunction, the angular cusp extension and its change with the approach of the planet to the Sun

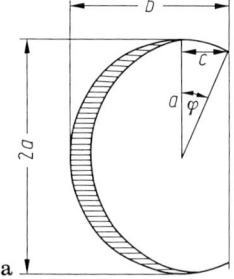 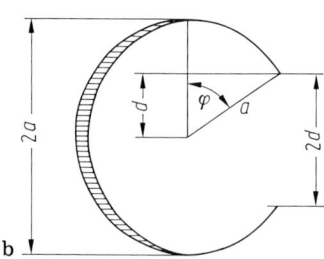

Fig. 19.10 a, b. Measurement of the extension of the cusps of Venus.

are measured according to the scheme of N. Richter [19.36]: for small angles α_p according to Fig. 19.10 a, or for large angles according to 19.9 b, the lengths $2a$ and b, or $2a$ and $2d$ are measured, respectively. Since $c = b - a$, the angular extension is found from

$$\sin \alpha_p = \frac{b-a}{a} \quad \text{and} \quad \cos \alpha_p = \frac{d}{a}. \tag{19.3}$$

Once α_p is known, the observer can calculate the height of the atmosphere of Venus. E. Schoenberg provides the following considerations and formulae.

If the planet Venus is imagined projected onto the center of the solar disk, the dark planetary disk would be surrounded by the bright ring of its atmosphere (although this cannot be seen because of the glare of the Sun). The dark disk appears somewhat reduced in size because solar rays grazing the planet are refracted inward (Fig. 19.11, right side). The outer ring represents the light-permeated outermost atmospheric layer, the inner ring the still-illuminated part which generates the refraction. The left side of Fig. 19.11 then represents the situation before inferior conjunction and the phenomenon of extended cusps due to a high atmospheric layer. In the relation

$$\frac{h}{R} = 1/2 \, \tan^2 V \, \sin^2 m, \tag{19.4}$$

h is the height of the atmosphere, R the radius of the planet, and V the angle of elongation.

One noteworthy phenomenon which is seen around inferior conjunction when Venus displays a very narrow crescent is the secondary or *ashen light*: the unilluminated part of the disk appears in a pale glow resembling the earthshine on the Moon. This phenomenon was first observed by Kircher in 1721 and has since been confirmed by many observers. Unless it is an optical illusion, its cause must again

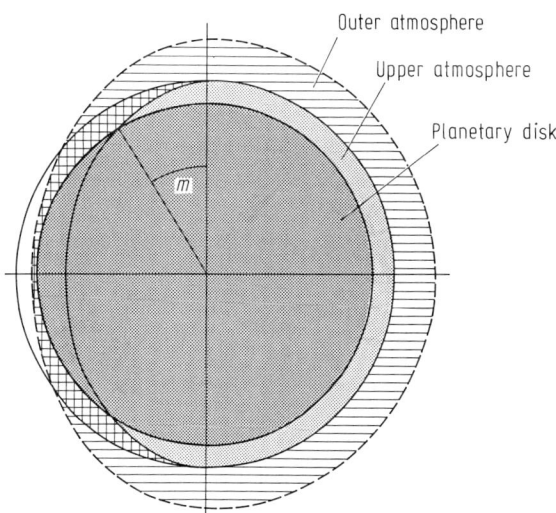

Fig. 19.11. Refraction by the low and high atmospheric layers of Venus at or near inferior conjunction.

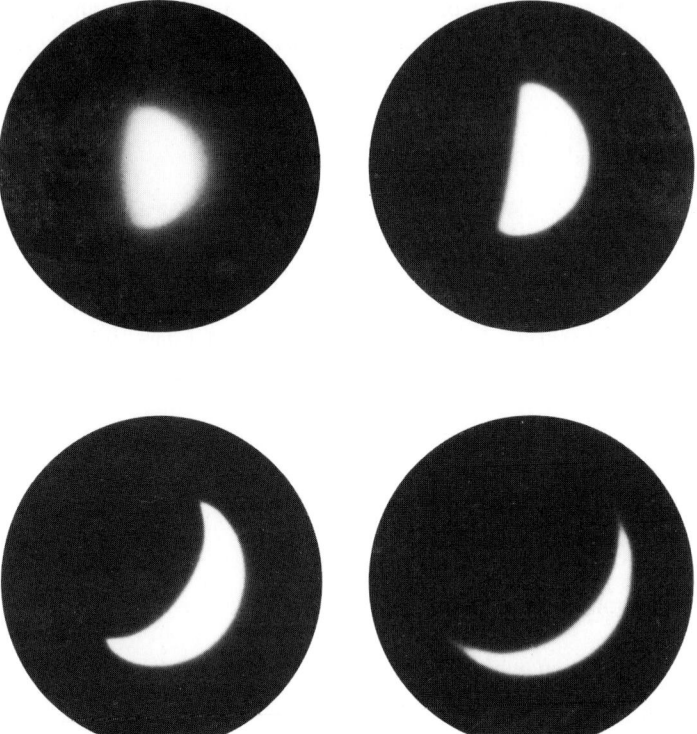

Fig. 19.12. Example of a sequence of Venusian phases captured on film. All photographs are by Ernst Elgass, Munich. Upper left photograph taken with a 5-inch refractor, all others with a 7-inch refractor.

be attributed to the atmosphere of Venus [19.37]. Observers report different colorings ranging from brownish to gray and violet.

As a summary of visual and filter-visual observations, the persevering observer may learn something of the effects of the dense atmospheric shell which surrounds Venus. As is well known, and continuously published, recent research results come from the flybys, flights into the atmosphere, and soft landings of numerous space probes.

19.5.2.3 Photography. Apart from radar scans of the surface, the Earth-based exploration of Venus is restricted to atmospheric and cloud observations in the ultraviolet. The overwhelming brightness of the planet makes it suitable for photographic observations by amateurs. Many observers have obtained interesting "constellation photographs" showing Venus near the Moon, other planets, or bright stars [19.38]. For most photographic work, a single-lens reflex (SLR) camera with exchangable lenses will suffice. Photographs of, for example, Venus and of the lunar crescent require only very short exposure times (between 1/50 and 1/100 second).

Another task is the photographic recording of the phase change (Fig. 19.12) and the determination of the dichotomy. For instrumental requirements and photographic material for this project, see Sect. 6.6.2 in Vol. 1. Good results were secured by amateurs with systems of 5- to 10-inch apertures and with eyepiece projection. For an example, B. Kimmel [19.39] photographed Venus in 1972 January through May. The apparent diameter of the planet increased from $12''$ to $58''$. He used a 10-inch reflector ($f/5$), eyepiece projection, and an SLR camera (Praktica) with 135-mm telephoto lens. He used Kodak Plus-X film, exposure times between 0.02 and 0.1 seconds, and total focal length of the system of 31.8 meters. The photographs show the changes of phase nicely and permit the determination of the dichotomy [19.40].

As early as 1957, cloud motions in the upper layers of Venus's atmosphere were noticed on photographs taken by C. Boyer with a 10-inch scope in ultraviolet light. Numerous photographic series secured using large-aperture instruments, primarily at the Pic du Midi Observatory, have subsequently confirmed the clouds and their motions. Attempts with large (10-inch and above aperture) amateur telescopes should prove rewarding. Along with the ultraviolet filter UG 5, blue filters are also suitable for such photographs. If it is desired to extend the focal length of the system with the aid of a Barlow lens, the transmissivity of that lens in the ultraviolet should be checked. Photographic tests may also be performed using yellow filters, with which clouds in the high atmosphere have been recorded.

A larger telescope is needed to secure detailed photographs from which the extension of the cusps near inferior conjunction can be determined. It may also be worthwhile to use filters such as RG 5 and OG 2. Simultaneous exposures in two colors (e.g., red and orange) of Venus when near inferior conjunction can be used to determine its angular diameter. G.P. Kuiper suspected that the apparent angular diameter depends on the wavelength of light used in photographing.

19.5.2.4 Transits of Venus. The two last transits of Venus took place in 1874 and 1882. Then next ones will be on 2004 June 08 and 2012 June 06.

19.5.3 Mars

19.5.3.1 Visibility. Mars is the nearest of the "superior" planets which travel around the Sun outside the Earth's orbit. During opposition a superior planet can be observed throughout the night. The visibility after a conjunction with the Sun runs on the average as follows:

Day 0: Conjunction with the Sun;
Day 54: Mars appears in the morning sky;
Day 353: Mars begins retrogression;
Day 390: Mars in opposition with the Sun;
Day 427: Retrogression ends;
Day 726: Mars disappears from evening sky;
Day 780: Next conjunction.

An almanac will supply the observer with the following data:

1. Right ascension and declination;
2. Apparent diameter (varying between 4″ and 25″);
3. Apparent magnitude (between +1.50 and −2.52);
4. Phase angle (also phase = illuminated fraction of disk);
5. Central meridian at 0^h UT;
6. Position angle of axis of rotation;
7. Latitudes of Earth and of Sun with respect to the Martian equatorial plane;
8. Distance from Earth in AU;
9. Light travel time from Mars to Earth;
10. Time of transit.

Some calendars list the heliocentric, in addition to the geocentric, coordinates of the planet.

The conditions of observation strongly depend on the position of Mars relative to the Earth. Owing to the eccentricities of both orbits, the distance between Earth and Mars at nearest approach varies from under 56 million km, when Mars is near perihelion at opposition, and 101 million km in an opposition near aphelion. Accordingly, the apparent diameter in a perihelion opposition is over 25″, but only 13.″8 in an aphelion opposition (Fig. 19.14).

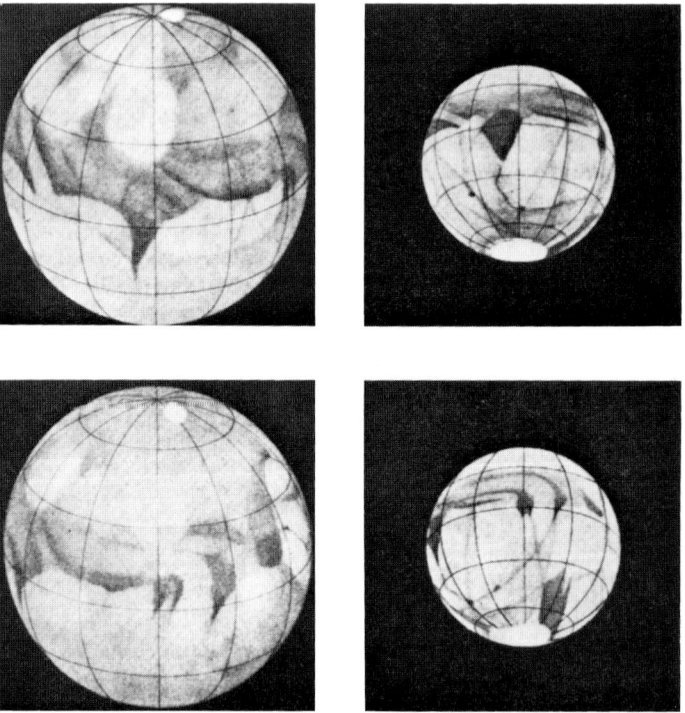

Fig. 19.13. Comparison of apparent sizes and of axis orientations of Mars at oppositions occurring at perihelion (*left*) and aphelion (*right*).

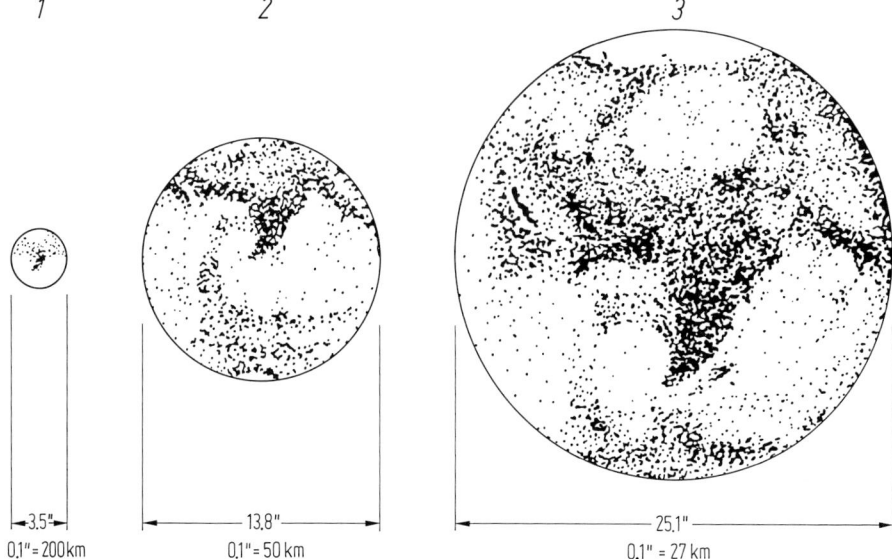

Fig. 19.14. The apparent diameter of Mars as seen during a conjunction at aphelion (*1*), and at oppositions at aphelion (*2*) and perihelion (*3*). From E. Freydank and H. Freydank [19.47], p 9.

Perihelion oppositions occur in late summer, when the planet is situated in the southern part of the zodiac and therefore at an unfavorable elevation for northern-hemisphere observers. Aphelion oppositions which take place after the beginning of a year will find the planet in the northern part of the zodiac. The observer should be aware that an apparent diameter of at least $10''$ is required to see substantial detail.

Since the atmospheric conditions, which improve with altitude, play a major role in the observation, northern-hemisphere observers will find the most favorable oppositions to be those which occur when Mars lies between its aphelion and perihelion positions, such as the one of 1990 November 27 (declination $+23°$, apparent diameter $18''.1$), and again in 2005.

The equatorial plane of Mars is inclined with respect to its orbital plane by $25°12'$. This means that seasons occur on Mars as they do on Earth, and also that the observer sees more of the northern Martian hemisphere at oppositions occurring in springtime, whereas the autumnal oppositions show more of the southern hemisphere (Table 19.3).

Table 19.3. The seasons of Mars.

Heliocentric Longitude	Northern Hemisphere	Southern Hemisphere
88°	Spring begins	Autumn begins
178°	Summer begins	Winter begins
268°	Autumn begins	Spring begins
358°	Winter begins	Summer begins

When the rotation axis of Mars is nearly perpendicular to the line of sight (this happens at oppositions between aphelion and perihelion), the rotation of the planet carries surface details across the disk in straight lines. The axial tilt substantially influences the appearance of most details visible on the planetary surface at oppositions near aphelion and perihelion; see Fig. 19.13.

In addition to the apparent diameter and the tilt of the axis, the observer should also keep in mind the influence of the phase angle. It reaches a maximum of 46° and must be allowed for. Observations made at most one month before and after opposition may neglect the phase angle, and circular stencils may be used for sketches.

The rotation period of Mars is almost the same as that of Earth: one Martian day equals 1.025 Earth days. An observer viewing the planet on consecutive nights at the same time will find very little displacement of surface detail. Only after 38.5 days will the observer see the same central meridian in the eyepiece when observing at the same time. The annual rate of change of the central meridian on Mars is 14°.6. The central meridian must be determined for every observation (on a graph), since the identification of surface detail depends on it (see also Appendix Table B.20 in Vol. 3).

The rotation of Mars can be measured even with a 2-inch telescope, by way of transit observations of prominent dark markings (e.g., Syrtis Major, Solis Lacus, Titanum Sinus). To calculate rotation times, see Wepner [19.41]. W.D. Heintz advises:

"Frequent observations of a few standard points are preferable to scattered and uncertain data from many points. Transit observations should be limited to the times corresponding to 10 days before and after opposition (which does not coincide with the day of nearest approach), as otherwise the phase and the unequal brightness of the limbs will affect the observations with uncontrollable errors."

Daytime observations of Mars are of little significance although they have occasionally been made and surface detail seen.

19.5.3.2 Visual Observations. Images sent back to Earth from the *Mariner 4* spacecraft during its Mars flyby in 1965 gave the surprising result that, instead of "canals," the Martian surface is covered with craters and in many ways resembles that of Earth's moon. This similarity was confirmed by all subsequent space missions to Mars, including the two *Vikings* in 1976 [19.23]. In addition to craters, the spacecraft pictures showed grabens, meandering valleys, and volcanoes. The atmosphere was found to be thinner than expected. The polar caps are composed of both water ice and dry ice.

Despite the astronautical lead in the exploration of Mars, the amateur observer should continue to observe Mars telescopically under a variety of conditions and acquire experience. Some feasible observing tasks include the following:

1. finding the period of rotation (see above);
2. monitoring seasonal changes in the polar caps and the atmospheric cover;
3. observation of bright and dark contours on the surface, and also changes in albedo patterns;
4. observation of atmospheric phenomena: white clouds, ice haze, yellow dust clouds, blue or violet clearing (see below).

As a rule, the most prominent surface feature on Mars is either (depending on the opposition) the northern or southern polar cap. A polar cap provides the observer with an aid in orienting the sketch. In this regard, W. Sandner [19.42] gives the following suggestions:

"It is advisable, at least when using high magnifications, to employ a micrometer wire for fixing the north-south direction; otherwise, if the polar cap and a phase are positioned eccentrically, this may render the perceived orientation incorrect. Next, the contours of dark areas are graphed with a sharp, soft pencil, and this is the moment for which the central meridian is later computed. Significantly brighter regions (clouds, etc.) are usually contoured by dashed lines. Finally, the shading is included in the graph by using a 'stump' to smear out the graphite."

The appearance and extent of the polar cap follows the seasonal changes on Mars (see Sect. 19.5.3.1): the cap develops during autumn as layers of ground fog are allowed to form in its surrounding bright zones. During the Martian winter, the cap grows; the observer may also see cloudlike veils in the vicinity of the polar cap. The cloud layer is often especially prominent in spring, when it has the largest extension. With the approach of summer, the cloud veils disappear, and the appearance and increasing prominence of a dark fringe indicates the shrinking of the polar cap (Fig. 19.15). Owing to the longer winter in the southern hemisphere, more CO_2 becomes frozen near the south pole, and thus a residual southern polar cap remains over the summer.

Observers have reported seeing both caps simultaneously. However, the "cap" of the winter hemisphere actually is an atmospheric cloud layer consisting of CO_2 and H_2O ice crystals and covering the real polar cap. While the caps change in size gradually, the polar cloud veil can change its appearance within a very short time. When observing the polar caps, it is advisable to use a blue filter [19.3].

At the end of the observing season, the observer may wish to measure all of the drawings so that the extent of the polar cap over time can be graphically represented. Observations of the appearance and the evaporation rate of the caps spanning many years may provide useful information regarding possible correlations with solar activity.

The dark areas on Mars are labeled with such terms as *sea* (mare), *bay* (sinus), or *lake* (lacus). But the terms *insula* (island), *fonts* (spring), *pallus* (marsh), *ponts* (bridge), and *regio* (land) also occur in the nomenclature, and often relate to legendary (hellenistic) or biblical names. Noachis, for instance, refers to Noah; Syrtis Major was named after the Syrtis of Libya by the astronomer Schiaparelli, who initiated a major part of Martian nomenclature, including the "canali" [19.43].

The dark features sometimes show greenish or bluish tints, while the extended bright regions (the "continents") have the typical rust-red or salmon-red coloring to which Mars owes its alternate name, "the Red Planet." The telescope reveals N-shaped connections between the dark areas; it was the often straight-line appearance of these features that triggered the furor over "Martian canals." However, observations made decades ago using large telescopes resolved these bands into individual spots and pieces. Whether or not Earth-based observers have seen craters on Mars through the telescope is a question of interpretation of earlier graphs and records [19.44,45].

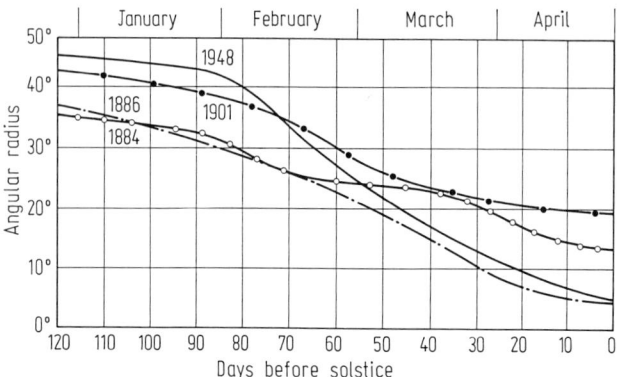

Fig. 19.15. Sublimation of the northern polar cap of Mars in various oppositions. *Abscissa*: days before solstice; *ordinate*: angular radius. After de Mottoni.

A comparison of various maps of Mars shows that there exist several dark regions whose shape changes either very little or not at all (e.g., Syrtis Major, Mare Sirenum, Sinus Sabaeus, Mare Tyrrhenum, Mare Acidalium). Where changes have been repeatedly observed, the cause may be temporary variations due to changes in albedo (resulting from, for example, changes in moisture) or to atmospheric superposition of clouds and veils. This overlying matter can also be dust which has been lifted into the upper atmosphere but which sinks back to the surface after some time. There are always surprises which await the observer from one opposition to the next. In particular, the larger clouds can be seen with only a 4- to 6-inch telescope. There are three basic types of clouds:

1. White clouds, which are commonly visible over the wintery polar regions, but can also be found at mid-latitudes and near the equator. They are cyclonic currents, suggesting the mixing of air masses with different temperatures (see Fig. 19.16).
2. Blue clouds, which strongly reflect blue and ultraviolet light. The Martian atmosphere contains a "violet layer" which significantly limits the visibility of the surface through blue filters. This layer can split open, for instance, in the vicinity of large dust storms, and the observer can seize the opportunity to make visual trials with filters. The *Mariner* and *Viking* missions to Mars have not positively confirmed the existence of the violet layer, and yet the phenomenon impedes Earth-based observations. Temporary clearings have been observed, that is, visibility of dark surface areas, through a blue filter, and this is termed a "blue" or "violet clearing." Roth [19.3] provides a table to grade the blue clearing.
3. Yellow clouds, generally caused by dust storms and which obscure major portions of the atmosphere, as, for instance, occurred in late September of 1971 [19.46].

Any visual observations can be enhanced by the use of color filters. The "classical" Mars filter is the orange filter (e.g., OG 550, 2 mm thick). Red filters (e.g., RG 1 and 2) also may enhance contrasts and are particularly good at showing clouds. Although the transparency in blue light is low, monitoring with a blue filter (e.g., BG 12 and

Fig. 19.16. Cloud formations on Mars in a cyclonic storm. This indicates a mixture of air masses of different temperatures. The cyclone was captured by the *Viking Orbiter 1* on 1978 August 09 at latitude 65°N. At the lower right in the picture is the crater Korolev (196°W, 73°N). NASA photograph. From G.D. Roth, *Taschenbuch für Planetenbeobachter* [19.3], p 254.

23) is worthwhile owing to the violet disturbances mentioned; see Chap. 6. Observers in the U.S. often use the Kodak Wratten filters: W 38 and W 80A are blue filters, W 58 green, W 21, W 23A, and W 25 orange and red.

A micrometer is also helpful for observing Mars (see Sect. 4.7.5) for the following tasks:

1. Measuring the positions of dark features. Such measured points on the surface can serve as a basis for preparing a map. Changes in position or of size or extension of an object during an observing season may thus also be obtained.
2. Position measurements of cloud-like phenomena and the determination of their sizes. This assumes that the cloud to be measured shows reasonably well-defined edges, which as a rule occurs only in small, compact cloud formations.
3. Measuring the polar caps. The most important point is the position of the center of the cap, which does not always coincide with the rotational pole of the planet. Another task is to measure the diameter of the cap and follow its increase or decrease in size. Finally, the observer may measure the width of the dark fringe, if present, and its variations.

Figure 19.17 illustrates the use of a micrometer to determine the areographic longitude and latitude of a feature. With the telescope drive switched off, a micrometer wire is oriented by the transit of a star. In the telescope (which is inverting but not mirror-

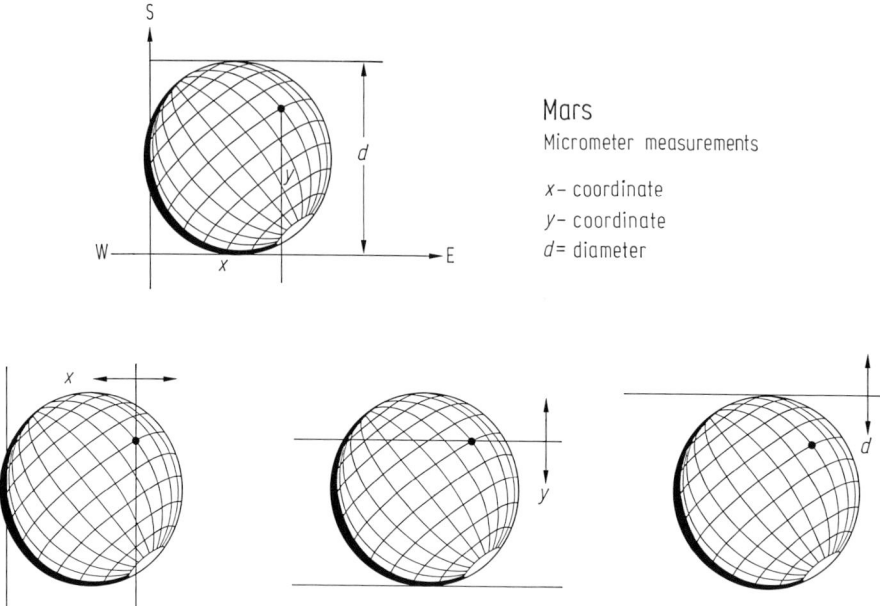

Fig. 19.17. Micrometer measurements of Mars. See explanation in text. From E. Freydank and H. Freydank [19.47], p 21.

imaging), the x-axis points to the right (east in the sky) and the y-axis upward (south). The axes are tangent to the disk and the coordinate origin lies outside the disk. It is unimportant whether or not Mars has a phase at one of the limbs. The phase, position angle, and inclination of the rotation axis must all be considered when computing the coordinates. The apparent diameter of the disk is measured in the y-direction, again disregarding the phase.

To determine the position of an object, the observer measures its x- and y-coordinates. At least five settings are desired in order to form an average. For each setting, the time is recorded to about ±15 s. See also [19.3,47].

19.5.3.2.1 Preparing a Map of Mars. The opportunity to observe Mars daily over at least one month supplies the basis for constructing a surface map of the planet. All observations are recorded in drawings. Owing to the changing orientation of the Martian axis, the stencil must be reconstructed as needed. Only detail near the central meridian (to ±10°) is transferred to the map [19.48]. The old maps by Schiaparelli, Lowell, and Antoniadi have frequently been published [19.49]. An entirely new generation of maps and globes has originated from the flyby missions of *Mariners 6* and *7* [19.50,51].

19.5.3.3 Photography. Apart from the wide-field "constellation photographs," which show a planet as merely a bright point of light against the background stars, Mars, when it is close enough to Earth to display a considerable disk, can also be photographed through a small telescope. When the apparent diameter is 10″ or over, a

6- to 8-inch telescope will readily secure useful photographs of the surface and of prominent features such as polar caps, dark areas, and clouds.

To be sure, planetary photographs cannot take the place of direct visual observations, but they can be used to supplement them. In order to make comparisons between the photograph and a map, the central meridian at the time of the exposure must be determined.

An advisable task for photographic observers is to systematically monitor the planet during the time interval when its apparent diameter exceeds $10''$. The purpose of the monitoring is primarily to secure a record of unusual events in the atmosphere, such as dust storms and prominent bright clouds. Photographs taken in red and blue light are highly desirable, as are ones in ultraviolet light.

Figure 19.3 a demonstrates the image quality obtained with large telescopes at favorable observing sites. The other photographs in the figure demonstrate that good-quality photographs can also be obtained using smaller telescopes. P. Martinez recommends Kodak 2415 emulsion developed with D19 or Acuspeed (125 ASA, $\gamma = 2.9$) as the best solution with respect to contrast as well as to the balance of speed and resolution [19.13]; see also Sect. 6.7 in Vol. 1.

19.5.3.4 Photometry. The drastic change in the Earth–Mars distance between conjunction and opposition results in a variation in the apparent brightness of up to 4 magnitudes. Superposed on this effect is a rotational variation in brightness, and thus an investigation into the visual apparent brightness (via global photometry) of Mars and its variation would be of some interest. E. Pfannenschmidt notes that

"the light curve of Mars, when compared with a map of the planet, shows a light variation with an amplitude of around 18% caused by the presentation of different surface regions. The greatest brightness occurs at that instant when the surface presents the area most covered with bright regions, while the minimum coincides with the appearance of the greatest number of dark areas. Graphing the brightness against rotation phase, the maximum, according to Guthnick and Lau, is at areographic longitudes $117°$ to $120°$, the minumum between $300°$ and $6°$. Red stars are best suited for comparison" [19.52].

Careful observations may also detect atmospheric events and cloud formation. This global photometry proceeds according to step estimates (see Sect. 8.4.2) and can be made with binoculars. For telescopic observations, the area photometer after Gramatzki [19.3] may be used, as it permits more accurate brightness measurements. High precision is demanded for photometry of selected bright and dark areas on the surface. It is simplest to make visual estimates by memory scale, as is used for other planets (see Sect. 19.5.5.3). For global and areal photometries, the photoelectric photometer (see Sect. 8.6.1) can be used, and it is also advisable to employ filters. Changes in apparent brightness also depend on the blue and violet transparency of the Martian atmosphere. When no dark surface areas are visible in the blue filter, the global apparent brightness is measured to be 0.2 to 0.3 magnitudes higher.

19.5.3.5 The Moons of Mars. The two satellites Phobos and Deimos were discovered in 1877 with a 65-cm refractor. Spacecraft have now supplied excellent close-up photographs, especially of the satellite Phobos [19.23], revealing detail of 10 m and smaller.

The apparent magnitude of both moons is around +12. Being so close to the planet, they are very difficult objects to locate. The amateur can at best attempt to find Deimos as it is the moon more distant from Mars. In any event, the observer needs a telescope with an aperture of at least 10 to 12 inches. It may help to mask out the planet by means of a disk or conic diaphragm placed in the focus of the telescope, a set-up similar to that of a prominence telescope.

19.5.4 Minor Planets (Asteroids)

19.5.4.1 Visibility. The minor planets are probably, like the comets, among the oldest bodies in the solar system, certainly older than the Earth. They have partly preserved their original shapes. Astrophysicists hope that the study of organic molecules of minor planets which contain carbon (Type C) will provide clues on the origin of life on Earth.

A "renaissance of minor planet astronomy" is said to be occurring today [19.53]. Observation is not limited to the motion of minor planets; the interest is currently concentrated on the structure and evolution of these bodies. The great majority of minor planets move around the Sun between the orbits of Mars and Jupiter. The number of known minor planets has increased from 1649 in 1960 to 3320 in 1985. Most of them are quite faint; only two of them (Vesta and Ceres) reach an apparent magnitude brighter than magnitude +7 at opposition and only about 70 asteroids have opposition magnitudes brighter than +10. On the average, oppositions recur every 1.2 to 1.4 years (see Sect. 8.6.1.1).

The conditions for the visibility of a minor planet at opposition vary owing to the orbital eccentricity and its inclination against the ecliptic plane; these quantities are often large. Moreover, many minor planets have irregular shapes and thus exhibit light variations with substantial amplitudes. The opposition brightnesses given in almanacs are generally only approximate. Yet an observer working systematically over some years can, assuming a limiting magnitude +13.0, log over one hundred minor planets photographically [19.54]. Even the largest minor planet Ceres (No. 1), which has a diameter of 1000 km and an opposition distance of 1.77 AU, presents an apparent size of only 0.″78. This means that theoretically (see also Sect. 8.6.1.1) an aperture of 23 cm is needed to show even a minute disk.

Minor planets are visible only as points in smaller telescopes; they are also studied photometrically as point sources.

The apparent brightness which determines a minor planet's visibility depends on (1) its distance r from the Sun, (2) its distance Δ from the Earth, and (3) the phase angle α_p.

These geometric data are found along with right ascension and declination in some almanacs, which also give a value for the mean photographic magnitude at opposition. The apparent visual magnitude is on the average 0.6 mag brighter than the photographic one. The following formula determines the instantaneous apparent brightness:

$$B = B(1,0) + 5\log(r\Delta) + \text{Opposition Effect}. \tag{19.5}$$

Since the observation of minor planets is usually restricted to the time near opposition, the phase angle is as a rule quite small and less than 30°. Most minor planets do not approach the Sun closer than about 2 AU. The geometric brightness variation has an amplitude of at most 4.5 mag including a phase effect not exceeding 1 mag. The reason for the small phase effect is that the phase angle for these minor planets does not exceed 30°. The exceptions are minor planets crossing the orbit of Mars (phase angles up to 45°) or even crossing inside the Earth's orbit (phase angles over 90°).

Observations show that for many minor planets a 1° decrease in the phase angle diminishes the brightness by an average of 0.023 mag. Within phase angles less than 8°, observations indicate a brightness increase of about 0.4 mag, which is called the *opposition effect*.

The blue magnitude B of the UBV system by H.L. Johnson (see Sect. 8.3) is most frequently used for observations of minor planets, as it simplifies photoelectric measurements. The formula for the mean opposition magnitude now may be written

$$m_0 = m_r - 5\log[a(a-1)], \tag{19.6}$$

where m_r is the "reduced" blue magnitude $B(1,0)$, m_0 is the mean opposition magnitude $B(a,0)$, and $a\ (= r_0)$ is the semimajor axis (= average radius) of the planetoid's orbit; see Eq. (8.35) in Vol. 1.

Observers will find the comment given by L.D. Schmadel to be of interest:

"The observer and especially the visually observing amateur will find another factor of significance for minor-planet magnitudes, namely the *color*. The color indices $B - V$ and $U - B$ are positive for all objects. That is to say, the brightness increases at longer wavelengths. For example, the asteroid Herculina (532) has $B - V = 0.8$ and $U - B = 0.4$. A B magnitude of $+10.9$ corresponds to $+10.1$ visually, which makes the object accessible to even the smallest telescope" [19.55].

The observer thus may assume that the almanac data are on the unfavorable side when used for visual observations and that the mean visual opposition brightness may be up to 1 magnitude brighter.

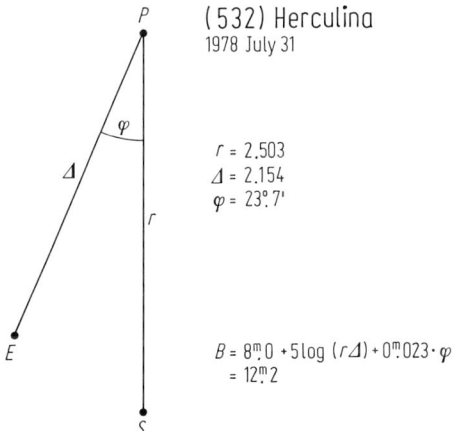

Fig. 19.18. Relation between geometry and apparent brightness. As an example, the minor planet Herculina (532) is shown for the date 1978 July 31. From [19.55].

Table 19.4. Sample ephemeris for the asteroid (532) Herculina at opposition. Here r and Δ are the Sun-planetoid and Earth-planetoid distances, respectively, in AU.

1978		α(1950.0)	δ(1950.0)	r	Δ	φ	B
April	22	$15^\text{h}34.^\text{m}0$	$+6°41'$	2.361	1.440	12°.5	$10.^\text{m}9$
May	02	15 26.1	+7 03	2.372	1.423	10.6	10.9
	12	15 17.2	+7 00	2.384	1.430	10.3	10.9
	22	15 08.4	+6 30	2.397	1.460	11.7	11.0
June	01	15 00.8	+5 34	2.411	1.512	14.1	11.1
	11	14 55.2	+4 16	2.425	1.585	16.7	11.3
	21	14 52.1	+2 42	2.440	1.675	19.1	11.5
July	01	14 51.6	+0 55	2.455	1.780	21.0	11.7
	11	14 53.6	−0 58	2.470	1.896	22.4	11.8
	21	14 57.9	−2 54	2.486	2.022	23.3	12.0
	31	15 04.3	−4 51	2.503	2.154	23.7	12.2

Some amateur almanacs annually publish ephemerides of minor planets which will be observable during a particular year with mean opposition magnitudes $\leq 10^\text{m}$, as well as finding charts for interesting objects. The observer will often find many minor planets far outside the zodiac, owing to the steep orbital inclinations of some of these objects.

The continuously changing distance of a minor planet from the Earth and from the Sun requires some reductions for brightness determinations:

1. The light-travel time. Light traverses a distance of 1 AU in 8.317 minutes, and the distance from the minor planet to Earth in $8.317 \times \Delta/r$ minutes. This amount is subtracted from the observing time at which the minor planet showed the measured brightness.
2. Reduction to unit distance. In order to obtain comparable brightnesses, the observed values m_obs are converted into those which the planet would have at $\Delta = 1$ and $r = 1$.

$$m_r = m_\text{obs} - 5(\log r + \log \Delta). \tag{19.7}$$

The observed magnitudes can also be referred to mean opposition distance $r_0 - 1$:

$$m_0 = m_\text{obs} - 5(\log r + \log \Delta) - \log r_0(r_0 - 1), \tag{19.8}$$

where r_0 is the semimajor axis of the orbit of the minor planet in AU.

3. Allowance for phase angle. The phase angle α_p is the angle at the minor planet in the Sun/minor-planet/Earth triangle (see Fig. 19.18), calculated from Eq. (3.38) in Vol. 1 for $\tan(\alpha_\text{p}/2)$. The phase coefficient expresses the diminution in apparent brightness of a minor planet in magnitudes for a 1° increase in the phase angle. Observations of light curves of minor planets, however, have shown that the brightness depends not only on α_p but also, because of rotational light variations, on shape and surface properties. For specific observing data for minor planet ob-

Table 19.5. Bright variable minor planets.

Minor Planet	Amplitude
(5) Astraea	$0.^m3$
(6) Hebe	0.2
(7) Iris	0.3
(9) Metis	0.3
(15) Eunomia	0.5
(18) Melpomene	0.4

servers, the reader is directed to the Supplemental Reading List (at the end of this volume) and also Binzel [19.56].

19.5.4.2 Observational Tasks. Locating minor planets in the sky is not always simple, as the observer needs good star maps (see the list of maps in Appendix A7.1 in Vol. 1) in addition to the almanac. Minor planets change their celestial positions rapidly, and it is precisely by this motion that they can be distinguished from stars. Their average distances from the Sun are about 3 AU, which, according to Kepler's third law, yields periods of revolution of around 5 years. The mean daily heliocentric motion is hence about 0.2°. Changes in the angular distance from reference stars can be noted from one night to the next. A good pair of binoculars or a small refractor will suffice to view the brightest minor planets. Observers may find it helpful to graph the apparent orbit of the minor planet onto a photocopy of the star map. In order to search for minor planets, a map with stars down to at least magnitude +10 is needed. The *Photographische Sternatlas* by Hans Vehrenberg reaches a limiting magnitude around +13.0, and is thus an ideal map for observers of these small objects.

In view of the large number of objects known, it is perhaps not surprising that their investigation is less complete than would be desirable. Minor planets can thus serve as the focus for amateur activity. While those objects distinguished by an unusual orbit or by peculiar photometric features are as a rule covered by observing programs at large observatories, there remain innumerable other minor planets whose behavior has been deemed "normal." Such surveillance requires control observations (to check on the orbital elements), and this mandates many more observers and observations than are feasibly available in professional research centers. An amateur with suitable equipment can expect to photographically monitor minor planets down to magnitude +15 and reach coordinate accuracies of $1''$ to $2''$ (see Fig. 19.19).

Primary observational tasks include:

1. Position measurements;
2. Discovery and recovery of minor planets;
3. Photometry;
4. Observation of stellar occultations by minor planets.

It is not uncommon for the orbit of a numbered minor planet to be somewhat uncertain, as it is usually based on only a few oppositions. Even well-studied objects

Fig. 19.19. The minor planet Vehrenberg (3030), captured on film by R. Bendel, Traunstein, with a 210-mm Newtonian, $f = 1680$ mm, indirect guiding. Exposed 60 min on Kodak 103a, developed 4 min in D-19. The minor planet had an apparent magnitude of about $+16$ at this time (1986 October 01, 22^h43^m–23^h43^m UT).

may deviate by surprisingly large amounts from their predicted paths. For example, when the positions of (85) Io, which was discovered in 1865, were measured during the 1977 opposition, they were found to deviate from the ephemerides values by 14′ [19.57].

In principle, visual position measurements by micrometer should be possible, but only for the very bright and well known objects. Discovery and recovery of minor planets is likewise beyond the reach of most visual observers.

Since data on magnitudes and light variations are sometimes unreliable even for the brighter minor planets, the observer may wish to employ visual estimates by way of the Argelander step method. To achieve scientifically significant results on rotational variations in brightness, however, a precision of the order of 0.01 mag is required. Such high precision demands photoelectric measurements, as even photography cannot reach this accuracy.

The primary goal of photometry is to determine the magnitudes and their variations. The irregular shape of a minor planet often causes a distinct rotational variation. Another photometric task is to determine the color indices. With sufficient data on colors, the albedo of a minor planet can be determined.

Color measurements can be performed visually as differential brightness measurements with suitable filters (see Sect. 8.3). Comparison stars should have well established magnitudes and colors, as may be found in a good star catalogue (see Appendix A7.1 in Vol. 1). A star, preferably of the solar type (spectral type G), is selected from the SAO Catalogue. If that chosen star lies in the direction of motion of the planet, it can be used as the comparison star for several nights. The observations may be made by comparing selected stars in blue and yellow light with that of the minor planet using a *wedge photometer* [19.58].

The observation of occultations of stars by minor planets supplies data which can be used to determine minor planet diameters. The objective is to determine the duration of the occultation as accurately as possible. In principle, instruments and tools are the same as for observing occultations by the Moon and major planets (see Chap. 17). In this connection, there is the remote possibility of discovering a satellite of a minor planet [19.59]. The asteroid Hebe (6), for instance, is suspected of having a satellite with a diameter of about 20 km. The monitoring of the occultations of stars for the possible occurrence of double occultations is therefore a worthwhile project for the bright minor planets. For more details on these tasks and for further reading, the reader should consult [19.3,54,56].

19.5.4.3 Photography. A substantial improvement in the previously mentioned observing programs (excluding occultations) on minor planets can be realized if they are performed photographically instead of visually. In particular, the limiting magnitude of the program is extended; amateur observers have photographically reached down to magnitude $+13$ or even $+14$. For position measurements, a focal length of at least 500 mm is advised.

Although photographs have occasionally been obtained with versatile but powerful 35-mm cameras, systematic work of this kind is more feasibly undertaken using an astrocamera with an aperture of 15 cm. Besides astrocameras, there are Newtonian, Schmidt, and Maksutov systems with which amateurs have obtained satisfactory results [19.60].

The simplest procedure is to photograph a celestial area which is presumed to contain a minor planet. With normal tracking, the exposure may last up to one hour or longer. While the fixed stars yield point images (if the guiding was done carefully), the minor planet, because of its orbital motion, reveals itself as a more or less pronounced streak. Consequently, the limiting magnitude of the optical system cannot be fully reached, as the motion of the minor planet smears its intensity over the emulsion. It has been estimated that a 15-cm aperture will lose about 4 magnitudes in this way. The alternative is to compensate for the planetary motion with appropriate guiding, as had already been suggested at the beginning of this century by J.H. Metcalf. The minor planet should then show a point image while the stars are trailed on the photographic emulsion.

As an example, if a camera or a telescope with $f = 1000$ mm is guided on a star, then a 1-hour exposure of a minor planet with a daily motion of $0°\!.4$ will produce a trail 0.3 mm long. The resolution of the photographic emulsion is around 0.03 mm. Thus, to obtain a point image with this focal length, exposures should not exceed 6 minutes, but exposures up to 60 minutes long are allowable for a focal length $f = 100$ mm. Modern emulsions can capture minor planets as faint as $+13.0$ with exposure times of just 13 minutes (see Chap. 6). Of course, the geocentric motions of minor planets differ, and the observer should consult published ephemerides.[1] At the 1977 opposition of the minor planet (85) Io, the ephemerides gave a maximum hourly geocentric motion of $29''\!.7$, so that that Io could be exposed for 25 minutes at $f = 500$ mm while keeping the trail length under 0.03 mm [19.61].

Photographic positional observations have been reported by several amateurs recently. These publications (cf. [19.3], pp 162) are mentioned here as they exemplify the method. R. Hempel summarizes the study of position observations, and finds that a 150/1200-mm Newtonian reaches a positional precision of about $1''$ [19.60]. H. Vehrenberg [19.62] cites previous work on the photographic positions of minor planets, and also points out some difficulties:

1. The planet is photographed at two different times. This presents no problems; on the contrary, it may lead to underestimating the worth of the photograph.
2. Finding the planet and determining its approximate position is not easy for a faint object.
3. The accurate measurement of the minor planet's position against reference stars requires a measuring machine and can only be performed in this manner.

The essential materials for this study are a photographic star atlas, star catalogues for the reference star positions, and minor-planet ephemerides. In recent years, some amateurs have acquired adequate measuring facilities and have reported on the determination of positions [19.3]. The formulae to reduce photographic positions measurements have been given in Sect. 3.4.

To identify the minor planet, the use of an enlarger or a blink comparator will be found helpful [19.63]. Determination of magnitudes from photographs, for instance to detect light variations, are within reach of the amateur astronomer; see Sect. 8.1.3 and [19.62]. One should be aware of the fact that amplitudes of > 0.2 mag can also be detected by direct inspection of photographs. In fact, for light variations of this amplitude, the experienced observer may also use visual estimates.

The longest rotation period found for a minor planet thus far is that of (654) Zelinda with 31^h9. Most minor planets have rotation cycles between 4 and 10 hours. To obtain a scientifically useful light curve of a minor planet requires intensive photometric observations. During one night, estimates or photographs should be made over several hours at 5- to 10-minute intervals; these intensive observations should be continued, if at all possible, over several nights. Such concentrated series are more rewarding than single observations scattered over a much longer time period.

19.5.4.4 Photoelectric Measurements. The use of photoelectric techniques is particularly suited for minor planet work, for photometry as well as for the observations of occultations of stars. In photometry, these measurements allow one to find light amplitudes of 0.05 mag and less. A photoelectric photometer is of unmatched value for reliable color measurements (see Sect. 8.3). In principle, the equipment can also

---

[1] In addition to the data for bright minor planets readily accessible in almanacs, the ambitious observer may also obtain ephemerides from the Minor Planet Center, 60 Garden Street, Cambridge, MA 02138, USA.

be used in conjunction with visual estimates [19.64]. The photometric method of observing minor planets has been described in Sect. 8.6.1.1. It should also be noted that reduction techniques for variable stars (Chap. 25) can also be applied, in principle, for the minor planets.

Apparent diameters of minor planets range from $0''\!.6$ (Ceres at mean opposition) down to $0''\!.001$. The diameters are best derived by observing occultations of minor planets by the Moon and of stars by the planet. A highly sensitive photoelectric photometer is used here as well as for the attempts to find possible satellites of minor planets [19.58,59].

19.5.5 Jupiter

19.5.5.1 Visibility. By contrast with Mars, which becomes a rewarding object to view for only a few brief weeks every 25 to 26 months, the planet Jupiter is, after the Earth's moon, the easiest target for an amateur. Its apparent diameter exceeds $40''$ for several months each year. Jupiter, like the other superior planets, experiences opposition, at which time it can be observed all night. The pattern of visibility for Jupiter, beginning with the conjunction with the Sun, is as follows:

Day 0: Conjunction with the Sun;
Day 13: Jupiter appears in the morning sky;
Day 140: Jupiter begins retrogression;
Day 200: Jupiter in opposition with the Sun;
Day 260: Retrogression ends;
Day 386: Jupiter disappears from evening sky;
Day 399: Next conjunction.

A good almanac will provide the following data:

1. Right ascension and declination;
2. Apparent equatorial diameter (between $31''$ and $48''$);
3. Apparent polar diameter;
4. Apparent magnitude (ranging between -1.3 and -2.7);
5. Position angle of the axis of rotation;
6. Jovigraphic latitudes and longitudes of Earth and Sun;
7. Distance from Earth in AU;
8. Phase angle and the defect of illumination;
9. Light travel time from Jupiter to Earth;
10. Jovigraphic longitude of central meridian in rotation systems I, II, and III;
11. Times of transit;
12. Satellite phenomena.

In addition to the geocentric coordinates, some calendars also list the heliocentric coordinates of the planets.

Jupiter's impressive size makes it a favorite object for viewing in even a small telescope. Moreover, the observer always sees essentially the same projection of coordinates since Jupiter's rotation axis is only slightly tilted ($2°\!.1$) against the ecliptic axis. The maximum of tilt is $3°\!.5$. Only then and with high power will any curvature

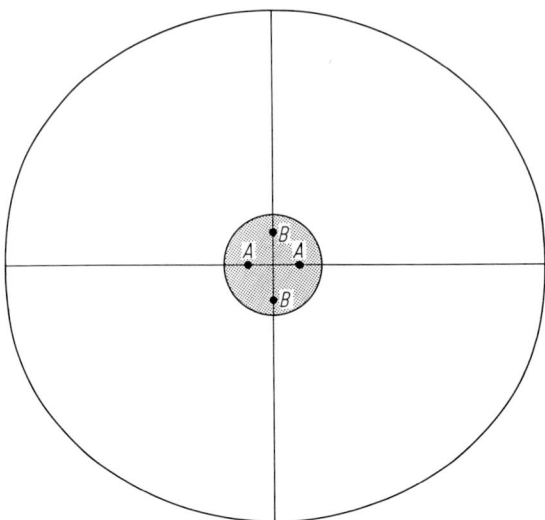

Fig. 19.20. Basket-shaped contour of the planet Jupiter.

in the parallels be noted. When observing with smaller instruments, the bright zones and dark bands parallel to the latitude circles appear as straight lines.

The oblateness of Jupiter, however, is a respectable 1/16, which can by no means be neglected when making drawings. Indeed, the elliptical nature of the Jovian disk can be seen even under low power. Denoting the semimajor and semiminor axis of the rotation ellipsoid by a and b respectively, then

$$\frac{a-b}{a} = \frac{1}{16}. \tag{19.9}$$

Drawings are made preferably on commercially available printed stencils.

With the aid of a compass, one can draw an approximate "basket-shaped" contour as shown in Fig. 19.20. The difference of semimajor and semiminor axes

$$a - b = \frac{1}{16} a \tag{19.10}$$

is graphed twice on the minor axis, 1.5 times on the major axis off the center. Circular arcs are drawn to connect the points A and B.

The phase angle is at most $12°$, which means that the illumination defect reaches at most 1% of the diameter. The phase angle may thus be neglected. The dark segment will be noted when viewing occultations of the moons of Jupiter, and especially if a large telescope is used.

19.5.5.2 Atmospheric Phenomena and Events. All of the features which the observer sees or photographs on Jupiter (see Fig. 19.21) are clouds! Measurements of Jupiter's size and density reveal that it consists predominantly of light elements (hydrogen, helium), which are in the gaseous state in the upper atmosphere, but liquid in the deeper-lying layers. Essential information on phenomena and events in the atmosphere were provided by the U.S. space missions *Pioneer 10* (flyby date 1973 December

Fig. 19.21. Photograph of Jupiter taken by the *Voyager 1* spacecraft on 1979 February 01 at a distance of 32.7 million km. NASA photograph.

04), *Pioneer 11* (1974 December 03), *Voyager 1* (1979 March 05), and *Voyager 2* (1979 July 09). The features created by atmospheric currents were found to be quite diverse: waves, eddies, cyclones, lightning storms, short-lived (a few days) and long-lived cloud formations. The meteorological events on Jupiter are triggered by its rapid rotation and internal heat sources; solar radiation does not play the same role as on Earth. The space missions confirmed a strong magnetic field which had been previously discovered in 1955 by radio astronomical observations. These missions also detected a system of Jovian rings, which can be observed from the Earth only in the infrared and with large telescopes.

The visible details are characterized by bright and dark stripes called zones and bands, respectively; the nomenclature is shown in Fig. 19.22.

Attention is to be paid to the rapid rotation of Jupiter, which takes place in two visible systems, I and II. The EZ and all of the bright and dark clouds which occur within it rotate as System I, while the NAB and SEB, usually the most prominent dark bands on both sides of the EZ, represent System II. The shortest (fastest) rotational period of $9^h 50^m 30^s$ is observed at the equator (I), the slowest in jovigraphic latitudes between $0°$ and $\pm 20°$ with a period of $9^h 55^m 40^s$ (II). The angular speed increases again toward the poles. The diminution of the rotational speed between latitudes $0°$ and $\pm 20°$ is very conspicuous, and must be allowed for when observing and charting other details such as bright and dark clouds.

Radio astronomical observations have defined a third rotation system (III) with a rotational period $9^h 55^m 30^s$ which some researchers expect to be that of the solid planetary body, in contrast to Systems I and II which are linked solely with the atmosphere. Astronomical almanacs publish the central meridians referring to the two systems. They can be interpolated with the angular rotations:

System I: rotation $877°\!.9$ in 24^h or $36°\!.6$ per hour;
System II: rotation $870°\!.3$ in 24^h or $36°\!.3$ per hour.

See also Appendix Tables B.20 and B.21 in Vol. 3.

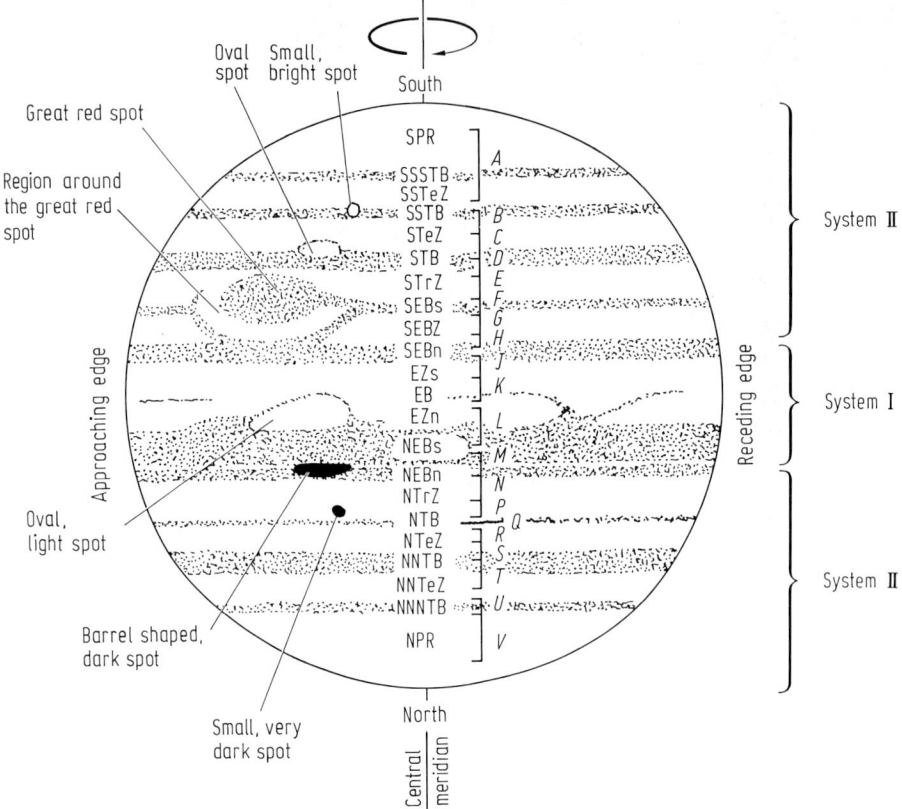

Fig. 19.22. Names of bands and zones on Jupiter. The abbreviations have the following meanings: P = preceding, F = following side of an object in the Jovian atmosphere, for instance the Great Red Spot (GRS); other designations include north (N), south (S), equatorial (E), tropical (Tr), temperate (T or Te), arctic (A), polar (P), band (B), zone (Z), region (R), barge = barrel-shaped region, dark cloud, plume = oval, bright spot, spot = small, bright, round marking. From G.D. Roth *Taschenbuch für Planetenbeobachter*, 3rd edn., Sterne und Weltraum, Munich 1987, p 183.

19.5.5.3 Visual Observations. Visual observations of the planet Jupiter are still superior to those made via photography, particularly concerning fine detail. According to R. Sopper [19.65],

"In any event it has been seen with respect to certain details (as occurred in 1971 with the strongly retrograde SEB_s-spots) that the visual observer can perceive and record fine detail (about $0\rlap{.}''5$) more unambiguously than photography could do with whatever great effort."

For the variety of observed atmospheric features, there is not only the previously mentioned, internationally unified terminology on bands and zones (Fig. 19.22),

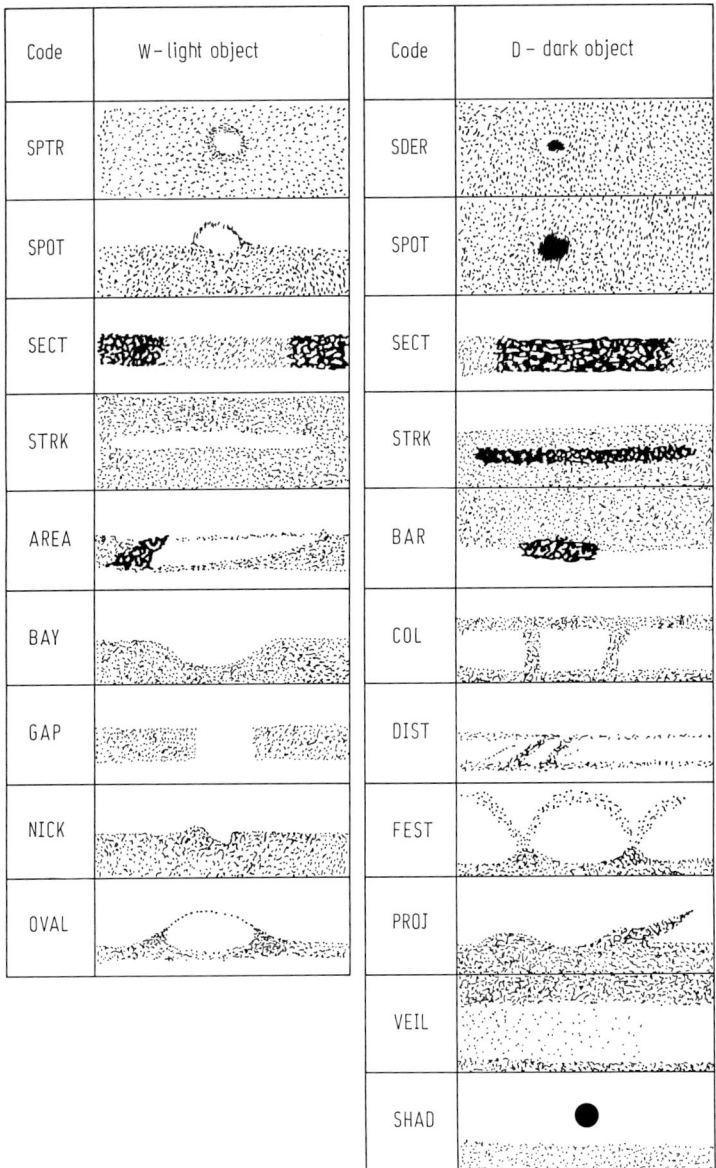

Fig. 19.23. Schematic representation of the designated codes for observed features in the atmospheres of Jupiter and Saturn. From G.D. Roth [19.3], p 179.

but also a computer-adapted code of observable object types to be described after [19.3, 66] as follows (see also Fig. 19.23).

The abbreviated name of a feature consists of six or seven letters or numbers:

1. The first letter divides the feature into two categories: "W" for bright or shining features and "D" for dark features.
2. The second letter refers to the center "C," the preceding limb in the sense of rotation, or eastern limb "P," the following or western limb "F," or—for micrometer measurements—the northern "N" or southern "S" limb.
3. The third digit gives a rough designation regarding the visibility of the object: from "1" (prominent) to "3" (difficult to see or very small).
4. The following three or four letters characterize the type of object. Their short code and description is (after Haug and Kowalec):

1. Bright (W) objects

Code: *Description*
SPOT: a small, bright round spot.
SPTR: a small, shiny spot surrounded by a dark ring.
OVAL: an oval feature of medium or large size, fairly bright, well bounded; often found in the equatorial zone.
BAY: a large indentation, usually half oval at the edge of a dark band.
NICK: a small, semicircular notch at the edge of a band, often somewhat brighter than the adjacent zone.
GAP: a rather wide, weakened, or missing part in a band.
STRK: a bright, very oblong spot; when appearing in a dark band, it may look like part of a tear.
AREA: an extended, bright, and irregularly bounded region.
SECT: a particularly bright section of a band or a zone.

2. Dark (D) Objects

Code: *Designation*
SDER: a small, very dark spot surrounded by a bright ring.
SPOT: any single spot not oblong.
SECT: a noticeably darker section of a band or zone.
BAR: dark oblong or barrel-shaped spot.
PROJ: something akin to a prominence at the edge of a band, which may also be darker than the main part of the band. Shapes vary, the dents may be rounded or peaked.
VEIL: an extended, homogeneously smooth, dark region, sometimes occurring in the zones or in polar regions.
FEST: a dark "fiber" or garland crossing a zone. One end of it or both may originate from a dark condensation in the band.
COL: a column-shaped dark area in a zone, either perpendicular to it or somewhat tilted. Such columns are occasionally seen in STrZ or SEBZ.
DIST: disturbance: a dark, extended area, more or less well defined containing often very fine detail in irregular distribution and unusual shapes.
STRK: a very oblong, dark, stripe-shaped object.
SHAD: the shadow of a moon.

On Jupiter there are a few other long-lived or frequently appearing features with special designations:

WOS–FA, –BC, –DE: white oval spots FA, BC, and DE.
GRS: Great Red Spot.
RSH: Red Spot Hollow.
STRD: Disturbance in the STrZ.
SEBD: Disturbance in the SEB.

Example: During a transit through the central meridian, the preceding (P) end of a prominent (1), dark (D), stripe (STRK) is measured. The code to be chosen according to the preceding list is: DP1 STRK.

There exist certain decidedly long-lived features on Jupiter, including the Great Red Spot, the Grey Veil, and also, according to recent observations [19.65], some white oval spots, as observed, for instance, in the STB. These form the category "WOS." In general, a series of observed features appears to be more stable than one would expect for purely atmospheric phenomena.

The first observation of the GRS dates from 1664 and is ascribed to the English physicist R. Hooke [19.43]. The central column of the Spot extends 8 km above the cloud level and receives its energy from some long-lived internal source. The age of the GRS is estimated to be over 10^5 years. It is not fixed with respect to the surface (Fig. 19.24), but moves westward in the atmosphere at a rate of $0°.5$ per day. The reddish color of the GRS sometimes changes dramatically [19.67].

The observer will find it worthwhile to investigate the following:

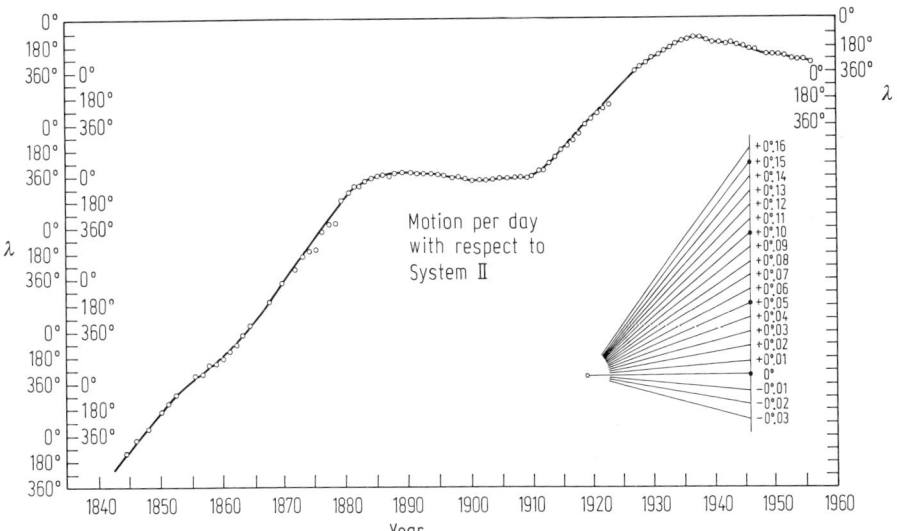

Fig. 19.24. Motion of the GRS on Jupiter during the period 1840–1950. After Löbering.

1. The speed of rotation of the Jovian atmosphere;
2. Motions and changes in the GRS, the Grey Veil, and the WOS;
3. Motions and changes in latitude and width of the bands and zones;
4. Motions, occultations, and transits of the bright moons and their shadows.

Within the frame of visual observations, there exist the following possibilities for tackling the above tasks:

(a) To represent the state of the visible surface by individual graphs (black-and-white or color) or overall maps, perhaps using color filters (see Sect. 19.2.2);
(b) Observation of central meridian transits of bright and dark spots and other well-bounded objects. This may be done by visual estimates or, for especially well-defined features, by micrometer measurements;
(c) Intensity estimates of bright and dark objects, in conjunction with color estimates using filters;
(d) Determination of the width of bands and zones by estimation or by micrometer.

A description of atmospheric features on Jupiter and Saturn and of their time variations necessarily includes accurate positions, since only by so doing can the motions and interactions of such features be identified. Haug and Kowalec mention the various possibilities:

Positions may be derived from sketches, provided the observer was careful in graphing and timing. Stencils of Jupiter and Saturn are readily measured with commercially available overlays which contain coordinate lines at 10° intervals. The resulting latitudes and longitude differences against the central meridian are usually uncertain by at least ±5°.

More precise, and well suited to studying motions of atmospheric features, is the timing of transits of an object (or a distinct point thereon) through the central meridian. This requires merely a good watch, checked against a time signal. For each transit, it is advisable to measure three readings (1/4 minute each):

t_1: transit possibly occurred
t_2: transit probably occurred
t_3: transit certainly passed.

The adopted average t_m is computed using $t_m = (t_1 + 2t_2 + t_3)/4$, giving double weight to the presumably closest estimate t_2. A crosswire eyepiece oriented so that one wire marks the central meridian will be of advantage. A trained observer can obtain a substantial number of positions during an observing season of Jupiter or Saturn, and with a precision (±1° or 2°) adequate for determining motions. Extended objects such as the Jovian GRS can be measured during one night at several points: the preceding edge (p), the center (c), and the following edge (f); cf. Fig. 19.22.

Estimating transits over the central meridian is for most planetary observers the only way to achieve reasonably reliable longitude data. Such time estimates can be inserted in between regular observations (graphing) and thus permit one to test the graph for correctness in longitude.

The precision in longitude, and even more so in latitude, is improved by using a filar micrometer (with at least one fixed wire and one parallel, movable wire).

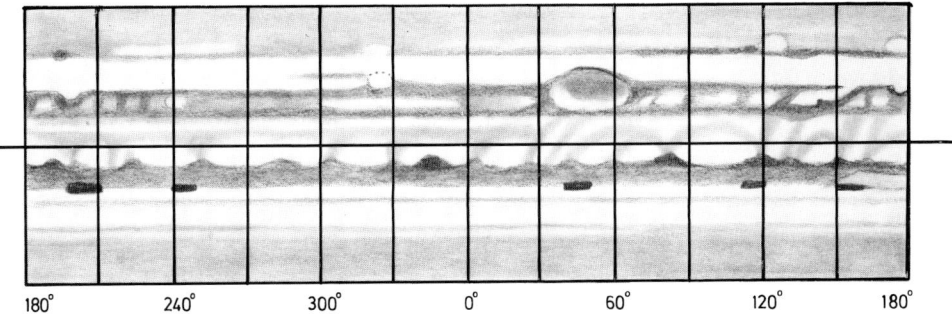

Fig. 19.25. Global map of the Jovian atmosphere. Date: 1979 March 24–25, 17^h10^m–01^h40^m UT. Instruments: 6- and 12-inch refractors of Wilhelm-Foerster Observatory, Berlin. Observer: Christian Kowalec. The labeling is for rotation system II.

Measuring atmospheric features on Jupiter and Saturn in this way requires more time and experience, but is not restricted to the central meridian, and trained observers may attain a precision better than $\pm 1°$.

Whenever seeing permits, the planetographic latitudes of the north and south edges of zones and belts should be regularly measured, as well as the longitudes of well-defined features (GRS, white oval spots, dark bars). The following procedures are schematized in Fig. 19.26:

The zero reading at the superposition of the two wires is determined first, and subtracted from all subsequent readings. The equatorial diameter d is measured with the wires parallel to the central meridian (at right angles to the bands on Jupiter and Saturn). Objects are then measured in longitude (x) with the fixed wire at the left (E) edge of the disk, and the moving wire approaching the point to measure alternately from left and from right. Rotating the micrometer by $90°$, so that the wires now parallel the equator, points are then measured in latitude (y) with the fixed wire at the lower (N) edge of the disk, and again readings are repeated. The polar diameter need not be determined as it is computed from the equatorial diameter. Record the time (to 1/10 minute) at the beginning and end of each measurement. Needless to say, readings without good timings are useless as the information on longitudes is lost.

Coordinate grids for processing drawings allow for the oblateness of the planet; see Fig. 19.27 and [19.68].

Visual estimates of the jovigraphic latitudes of bands and zones have been performed from systematic sets of estimates; motions and changes have been deduced. Again, a micrometer may be useful [19.4].

Intensity estimates may be made with the aid of a scale used by P. Fauth and which runs from $+5$ (or 5 B) = brilliant white through 0 = neutral gray to -5 (5 D) = deep black.

Observations and processing may proceed according to the following instructions, which are excerpted from an observing report by the author:

"It is advisable to choose a magnification which is not too low; in the present case, $161\times$ to $210\times$ was used with a 110-mm mirror. A slightly extrafocal setting has sometimes been found to be of advantage in more easily comparing contrasts. The method does not allow for a

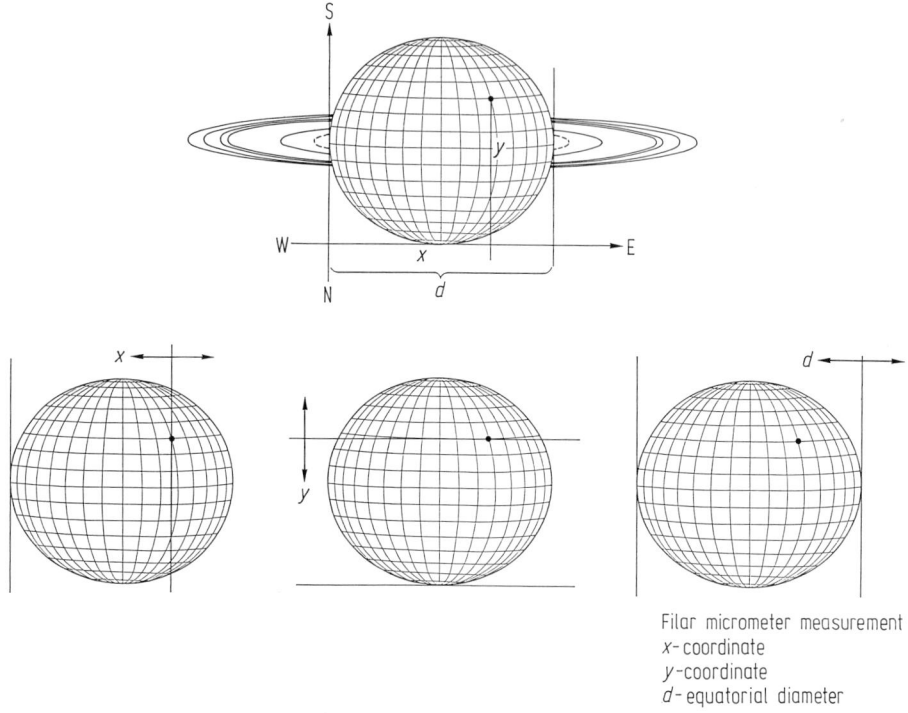

Fig. 19.26. Definition of quantities measured with a filar micrometer. After Kowalec.

really exact basis of comparison; the estimates are related to one another in an approximative fashion. For the opposition in 1951, the NEB was used as a standard of darkest intensity, and the intensity of other bands estimated relative to it. After completion of the observing series, the data were evaluated by forming averages of individual values and eliminating some large deviations, which should to some extent smooth out the errors. It is impossible, however, to really eliminate subjective errors, as is evident from the results of purely visual photometry. B.M. Peek has suggested a scale for intensity estimates specifically for the GRS." [19.3], pp 190.

Area photometry to measure intensities is described in Chap. 8.

These observations can be supplemented by recording colors. R. Sopper [19.65] observed on Jupiter the following tints: (1) red (magenta to bright orange), (2) brown, (3) blue (spots and garlands), (4) yellow (zone areas), (5) white, (6) gray (apparently at the highest level). Confirmation of color perceptions may be obtained by filter observations and by comparison with color and spectral photographs.

19.5.5.4 Photography. The substantial apparent diameter of Jupiter favors photographic work even with small- and medium-sized optics. Photographs showing the most prominent bands and zones can be obtained with just a 4-inch telescope. Also, the image of the Great Red Spot should not be too difficult, provided the Spot is dark enough. Besides focal photography, the eyepiece projection method is also well-suited for Jupiter; see Sect. 6.6, where hints on the choice of films are also given.

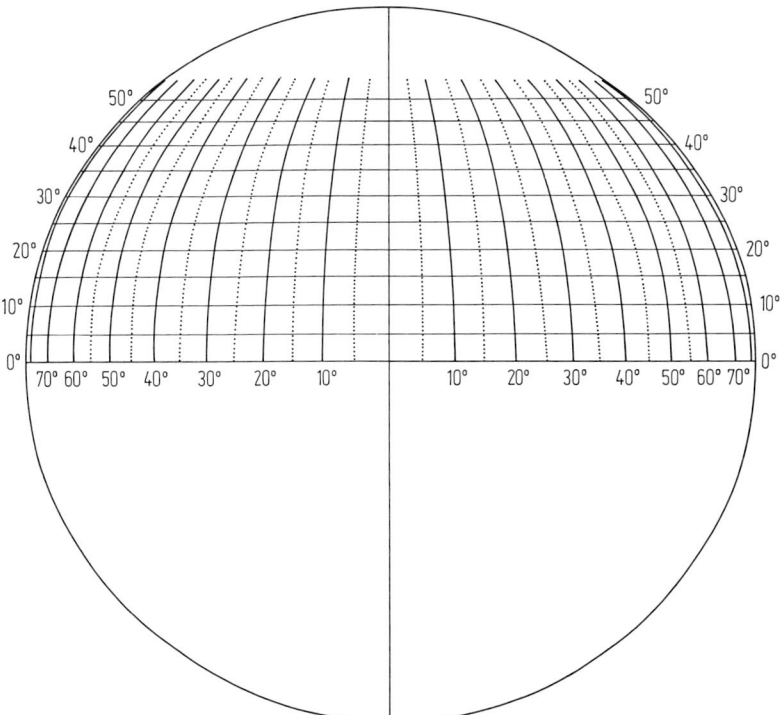

Fig. 19.27. Coordinate grid for measuring the positions and sizes of features in the Jovian atmosphere. Original from S. Cortesi, Locarno-Monti.

The tasks of visual observation as outlined in Sect. 19.5.5.3 can, in principle, be tackled photographically as well. Within a working group, it will usually be desirable to perform photographic and visual observations *simultaneously*. On the other hand, the solitary observer will find that taking photographs immediately after a visual drawing is of help in checking the graph and the positions determined. Similarly, filter photographs can supplement corresponding filter-visual observations. The principle is illustrated by pictures taken under the very best conditions by J. Dragesco (Fig. 19.4). But opportunities also exist here for observers with smaller instruments (from about 20 to 25 cm).

Photographic studies are well suited for measuring positions of distinct spots, for finding latitudes of bands and zones, and also for determining the positions of the Jovian moons (see Fig. 19.28). But position measurements from photographs are not without problems; in particular, the location of the true limb is made difficult owing to the strong limb darkening effect of the planetary atmosphere. In order to determine longitudes, C. Kowalec [19.68] advises:

"A series of photographs is made around the transit of an object through the central meridian, and the distance of the object from the central meridian then graphed against time so that the instant of central transit can be interpolated. This avoids the use of the planetary limb, since

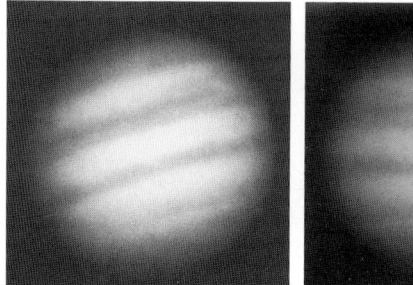
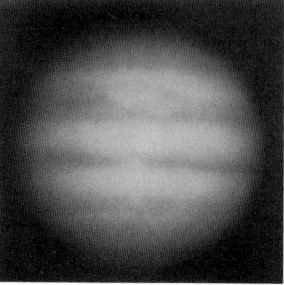

Fig. 19.28. Jupiter, as photographed by Gino Farroni, Saint Avertin. *Left*: 1987 September 09, 1^h38^m UT. *Right*: 1987 September 09, 3^h00^m UT. The GRS can be found in the right-hand picture at the zero meridian. Taken with a 406-mm Newtonian reflector, $f/5$. Magnified with two Barlow lenses to $f' = 12\,228$ mm ($f/30$). Exposure time 1/5 s.

Table 19.6. Errors in measuring jovigraphic latitude caused by limb darkening. After C. Kowalec.

True Latitude	Measured Latitude with Limb Darkening in %			
	−1	−5	−10	−20
90°	—	—	—	—
80	84°.0	—	—	—
70	72.0	82°.0	—	—
60	61.2	66.0	74°.8	—
50	50.9	54.0	58.4	73°.8
40	40.4	43.6	47.1	54.8
30	30.3	32.4	33.8	38.6
20	20.2	21.2	22.5	25.3
10	10.1	10.6	11.1	12.5
0	0.0	0.0	0.0	0.0

only the position of the central meridian, as found by orientation by the bright zones, needs to be known. Thus, a useful longitudal determination is obtained whose precision can be improved by increasing the number of exposures."

The photographic limb-darkening effect causes substantial errors when determining latitudes, as is shown in Table 19.6. It is preferable to use a computed polar diameter, reconstructed from the measured equatorial diameter (the bright Equatorial Zone) and the known oblateness.

The positions of Jovian moons can be suitably determined from photographs taken using the eyepiece projection method. For further processing, O. Zimmerman recommends projecting the negatives onto tracing paper using a slide projector [19.69].

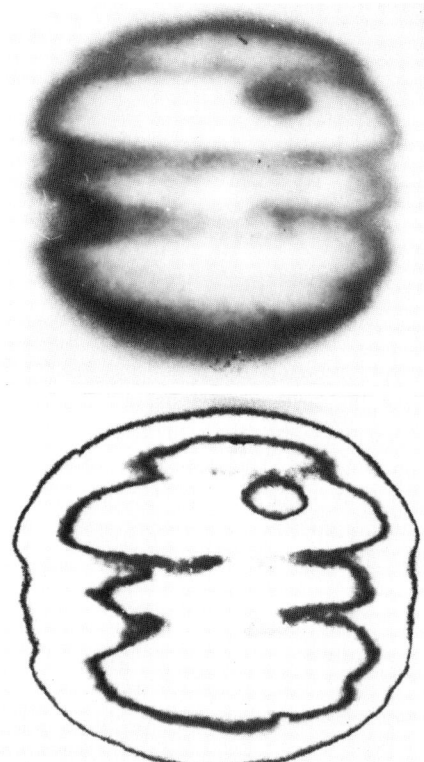

Fig. 19.29. Equidensities of the first and second orders from a photograph of Jupiter. Original photograph: 1974 August 15, 11^h43^m (CET), ZM II: 4°. Instrument: 1-m Cassegrain, Pic du Midi. Emulsion: Kodachrome II, 1-s exposure. Photograph by R. Sopper. In photographic equidensitometry, information is extracted by way of the density curve of the emulsion (see Sect. 6.7.3).

19.5.5.5 Photometry. Measurements of individual bright or dark objects can be done photometrically, either by estimates (see Sect. 19.5.5.3) or by area photometry (see Chap. 8). The apparent brightness of the moons I and II is somewhat variable. Observations also suggest that I (Io) has a slightly higher apparent brightness than usual immediately after reemergence from the planet's shadow and for about 15 minutes thereafter. The polar caps of Io exhibit a reddish tint which has been confirmed by recent color photographs. By making comparative magnitude estimates of moons which do not differ significantly in brightness and occasionally comparing with nearby stars, the observer will gain much valuable experience.

Photoelectric photometry is well suited for the above-mentioned tasks, as well as for following occultations and eclipses of the moons; see the report of an amateur group on the photoelectric monitoring of mutual eclipses of Jovian satellites in 1973 [19.70]. Those phenomena which occur close to the bright disk of the planet, however, must be excluded because of the interference of the scattered planetary light. Not to be forgotten is integrated photometry of the planet and its satellites, with the goal of, for instance, finding evidence for relations between solar activity and brightness changes in planets and satellites. Since 1950, measurements of this kind have been performed regularly at the Lowell Observatory using a 21-inch telescope, as described by Lockwood [19.71]. For details on photoelectric photometry, see Chap. 8.

Table 19.7. Apparent diameters and magnitudes of the four Galilean moons of Jupiter.

Moon	Diameter	Magnitude
I (Io)	$1.''05$	$5.^{m}43$
II (Europa)	0.87	5.57
III (Ganymede)	1.52	5.07
IV (Callisto)	1.43	6.12

19.5.5.6 Satellites of Jupiter. From the results of the *Voyager* missions, one can assume that Jupiter has at least 16 satellites (see Appendix Table B.17 in Vol. 3). Most of these moons are beyond the reach of amateur equipment. The more interesting moons from the amateur standpoint are I–IV, which were discovered by Galileo centuries ago. Together with the Earth's moon and Saturn's satellite Titan, they are the largest and most massive moons in the solar system. Their apparent diameters and apparent magnitudes are given in Table 19.7.

The shadows of the moons on the disk provide good test objects for the quality of 3-inch apertures. To see the moons as disks requires a 5- or 6-inch aperture. When transiting the planetary disk, moons I and II (Io and Europa) appear bright on a gray background, III (Ganymede) gray on a brighter background, and IV (Callisto) almost as dark as a shadow.

Observers equipped with telescopes of 30 cm aperture and over may try to find surface shadings or details on Ganymede. This satellite is somewhat larger than the planet Mercury; Io and Callisto are somewhat smaller, but still larger than Earth's moon. The Galilean satellites are thus comparable in size to the smaller terrestrial planets. The recent *Pioneer* and *Voyager* missions have made possible the construction of detailed surface maps of the four large moons. *Voyager* discovered eight active volcanoes on Io and a crust of frozen sulfur and sulfur dioxide. The surfaces of the other satellites are most likely covered with ice formations. Ganymede is assumed to consist of half water and half ice [19.23].

The motions of the moons and their frequent transits, occultations, and eclipses provide an interesting demonstration of orbital motion. Such phenomena occur regularly for moons I to III, and for moon IV when the Jovian axis has a low tilt, which occurs when Jupiter is around right ascensions 8^h or 20^h. At these points, the lines of sight and of shadow fall in the Jovian equatorial plane, which coincides with the orbital plane of the satellites. It is at these times that mutual occultations and eclipses of the moons can occur. During such an occurrence, an observer using a small telescope will notice brightness changes, while a large instrument will capture the actual event, for instance the eclipse of III by I. Also, the moons III and IV can then approach each other closely, resulting in a combined magnitude of +4.8, which at moon III's elongation can then be seen by keen-eyed observers without a telescope.

The occurrence of these phenomena make it possible to measure photometrically the limb darkening of the satellites. Brinkmann and Millis [19.72] describe the possibilities offered by the observation of occultations and eclipses. Needed are instruments with apertures larger than 12 inches, a one-channel photometer with online data recorder, and an accuracy of some tenths of a second in time recording.

The albedos of the Jovian moons are markedly different: highest for I and II at over 0.6, lowest for IV at 0.2. During transits across the Jovian disk, moons I and II appear bright against a darker background, while moon III is grayish and IV could almost be mistaken for the *shadow* of a moon.

The four large moons provide the astronomer with an opportunity to "rediscover" Kepler's third law by observations, taken either photographically or by micrometer, of the relative positions of the moons with respect to Jupiter on several consecutive nights [19.69].

Systematic observations of the satellite motions have helped to improve the accuracy of the orbital elements. P. Ahnert [19.73, 74] reports:

" Observations of shadow ingresses and egresses as obtainable with any amateur telescope and a good watch checked against a time signal will give the positions of the moons in their orbits to an accuracy of a few hundredths of a degree, or with an error of about $0\rlap{.}''1$ at the sphere."

The shadow ingresses and egresses last several minutes. Detailed predictions and graphs can be found in Section F of the *Astronomical Almanac*.

19.5.6 Saturn

19.5.6.1 Visibility. Saturn is one of the most interesting and beautiful objects for the observer, owing in particular to its famous ring system. The observable atmosphere shows similarities to that of Jupiter. The advice for observing Jupiter in general also holds for Saturn. As a superior planet, Saturn is observable all night during opposition. Its visibility, beginning with the conjunction with the Sun, is as follows:

Day 0: Conjunction with the Sun;
Day 18: Saturn appears in the morning sky;
Day 125: Saturn begins retrogression;
Day 189: Saturn in opposition with the Sun;
Day 253: Retrogression ends;
Day 360: Saturn disappears from evening sky;
Day 378: Next conjunction.

A good almanac will supply the observer with the following data:

1. Right ascension and declination;
2. Apparent equatorial diameter (ranging between $15''$ and $21''$);
3. Apparent polar diameter;
4. Apparent magnitude (between $+0.9$ and -0.6);
5. Position angle of axis of rotation;
6. Latitudes of Earth and Sun with respect to Saturn's equatorial plane;
7. Distance from Earth (in AU);
8. Phase angle (or deficit of illumination, from $0\rlap{.}''00$ = fully illuminated to a maximum of $0\rlap{.}''04$);

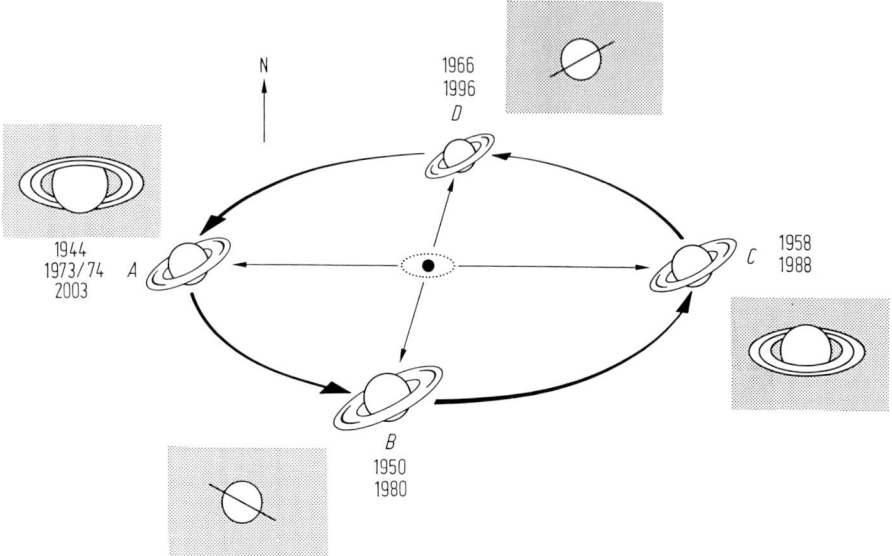

Fig. 19.30. The ring system of Saturn. After R.A. Naef, *Der Sternenhimmel*, 1976.

9. Light-travel time from Saturn to Earth;
10. Longitude of the central meridian in both rotation systems I and II;
11. Time of transit;
12. Orientation and opening of the rings;
13. Phenomena of the brighter satellites.

In addition to the geocentric planetary coordinates, some almanacs also contain heliocentric coordinates.

The apparent brightness depends upon not only the continuously varying distance from the Earth but also (and to a much larger extent) on the changing position of the rings. Saturn's equatorial plane, which contains the rings, is inclined against its orbit by 26° and against the ecliptic plane by 28°. This inclination must certainly be taken into consideration, as the observer will view only the northern side of the ring system for about 15 years, and the southern side for the next 15 years; see Fig. 19.30.

Also to be noted is the flattening of the planet, a full 1/10. For drawings, it is advisable to use prepared stencils, which also indicate the position of the ring system with respect to the line of sight during the observing season. The phase angle never exceeds 6° and can be neglected.

Bright zones and dark bands similar to those in the atmosphere of Jupiter are observed on Saturn (see nomenclature for Jupiter). However, the features, in particular the bright spots and clouds, are less well-defined than they are on Jupiter. Occasionally more prominent features appear. A recent example was a Great White Spot first noticed in September 1990 and which subsequently developed into a quite conspicuous object in the northern part of the EZ; see Fig. 19.31. D. Parker [19.75] described it as "the brightest thing ever seen on the planet, a quarter of the diameter of the disk, and as bright as a Martian polar cap." A report on the Spot has been given by O'Meara [19.76].

The first known Great White Spot was detected in December 1876 by American astronomer Asaph Hall in Washington D.C. The next one was found in June 1903 by E.E. Barnard with the 40-inch refractor at Yerkes Observatory, near Williams Bay, Wisconsin. The third and fourth were both found by eagle-eyed amateurs: in August 1933 by Will Hay in England, and in March 1960 by J.H. Botham in South Africa. All of these spots were seen in the northern hemisphere of Saturn, those in 1876 and 1933 at about the same latitude as the present one, while the two others were further north at $+40°$ (1903) and $+58°$ (1960).

Most planetary astronomers agree that the Great White Spots are upwellings from the lower atmosphere, whereby large clouds move upwards and become visible when they penetrate the uppermost, hazy layers. They resemble the towering cumulonimbus clouds often seen in the Earth's atmosphere. However, the lifting mechanism is not yet known; one possibility is that their upward motion is due to to the release of heat by condensation of water vapor, perhaps in combination with strong updrafts from sublimating ammonia grains.

The spots become longer as the clouds are carried along by strong winds in the upper atmosphere. Eddies and whirl patterns undoubtedly develop because of the different wind velocities at different latitudes, but because of their smaller size they are very difficult to observe from Earth. This may imply that the spots, perhaps in particular those which have emerged more recently, are actually gigantic storm centers, such as the Giant Red Spot on Jupiter, which has now been visible for almost 400 years.

Micrometer measurements and transit times of suitable objects across the central meridian to determine the rotation are therefore difficult [19.77]. Analysis of the observed data [19.78] indicated that there are two rotation systems; since 1969, the Association of Lunar and Planetary Observers (ALPO) has published tables of the central meridians:

System I (NEB, EZ, SEB) $10^h 14^m 13\overset{s}{.}08$ (= $844°$ per day);
System II (middle latitudes) $10^h 38^m 24\overset{s}{.}2$ (= $812°$ per day).

In 1980, the International Astronomical Union (IAU) slightly changed System I and added a System III, derived, as in the case of Jupiter, from radio observations:

System I (IAU) Period of rotation $10^h 14^m 0\overset{s}{.}00$;
System III (IAU) Period of rotation $10^h 29^m 59\overset{s}{.}42$ [19.3].

19.5.6.2 Visual Observations. For information relating to atmospheric phenomena on Saturn, the corresponding text on the planet Jupiter (Sect. 19.5.5) should be consulted, as in principle it also applies to Saturn. However, the number of the bright and dark objects perceivable with amateur instruments is significantly less. The intensities of the objects also are usually so low that micrometer measurements and central meridian transits are more difficult.

Figure 19.33 shows Saturn's rings and divisions before *Voyager*. The space missions discovered numerous new sections and divisions in the ring systems (see

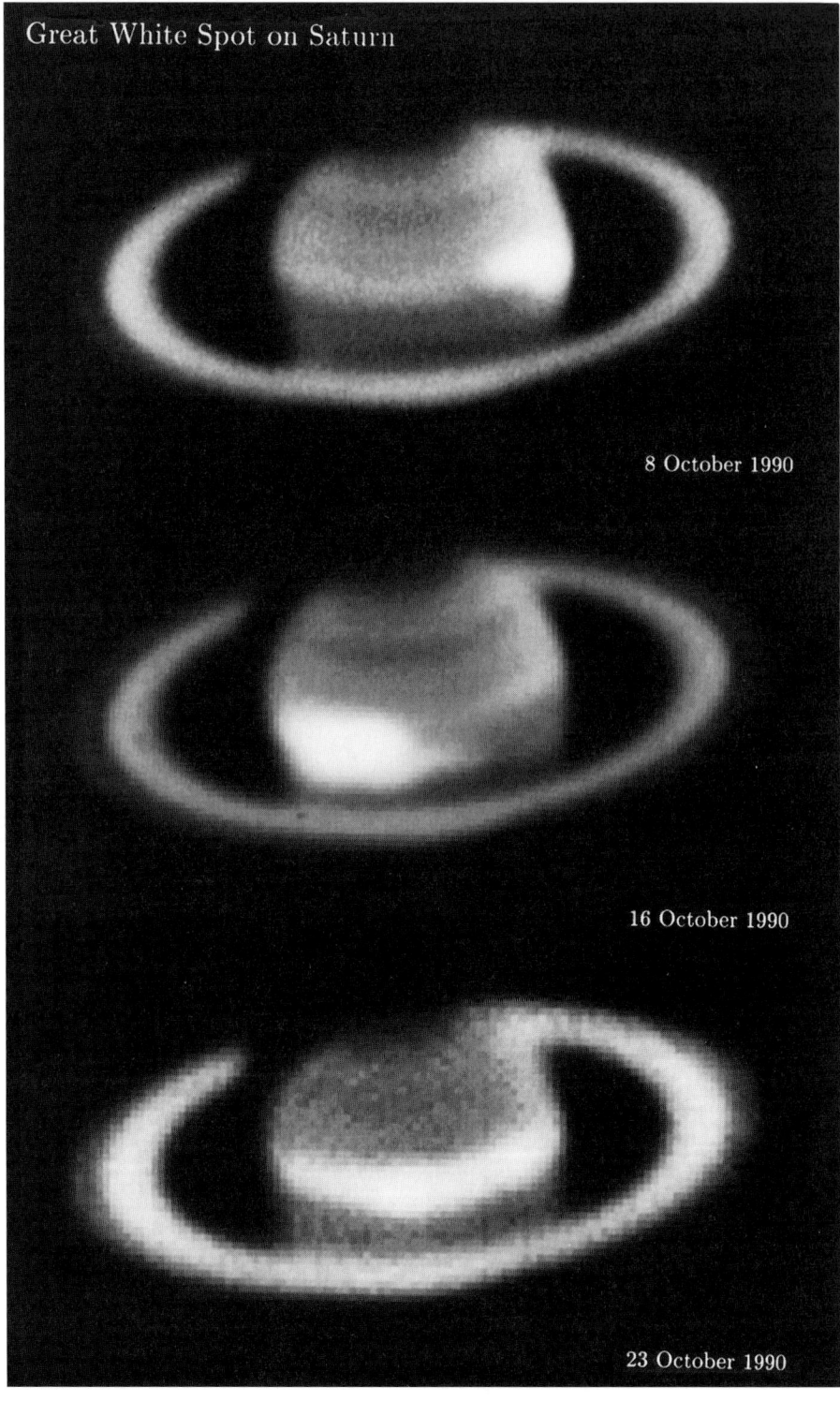

Observations of the Planets   249

Fig. 19.31. The Great White Spot on Saturn. This series of three exposures from the ESO La Silla Observatory shows the development of the Spot over a period of two weeks. Between October 8 and 16, the Spot grew significantly in size; it became longer and probably also brighter. After October 16, it rapidly expanded to reach all the way around the equator of the planet. The images were made in blue light in mediocre seeing conditions with three different instruments, all equipped with CCD detectors. Photograph courtesy of the European Southern Observatory.

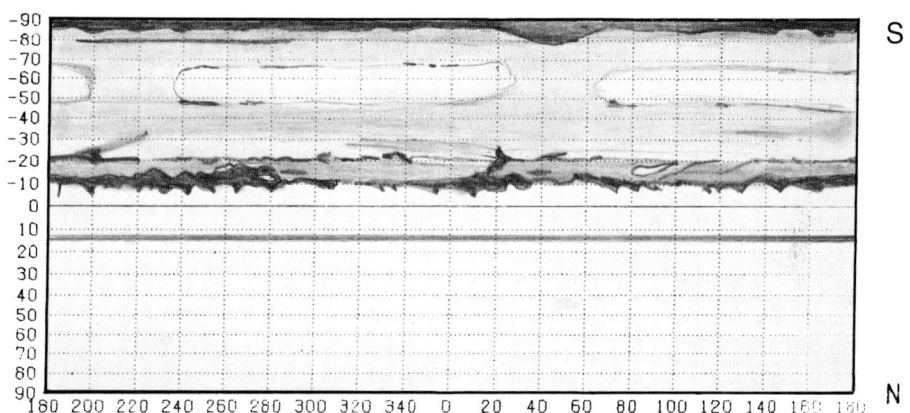

Fig. 19.32. Global map of the planet Saturn as obtained in 1974 April 1–8. Cylindrical projection distance true at the equator; flattening of 1:10 is allowed for. Instruments: 6- and 12-inch refractors of the Wilhelm-Foerster Observatory, Berlin. Observers: Wolfgang Anklam and others.

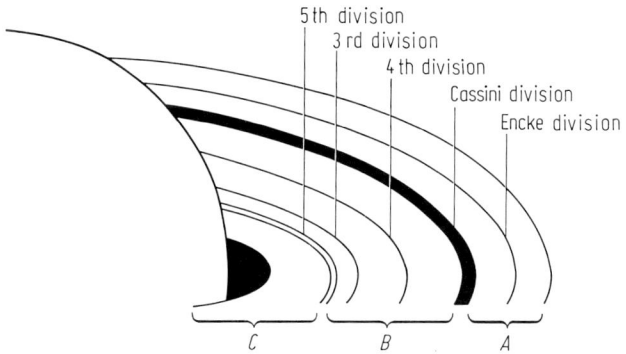

Fig. 19.33. Rings of Saturn (A, B, C) and the ring divisions known before the *Voyager* missions.

Fig. 19.34). When the rings are fully opened, Cassini's Division, the dark dividing line between the outer ring A and the middle ring B, can be seen even in a 3-inch telescope. The visibility of other rings and divisions also depends on the de-

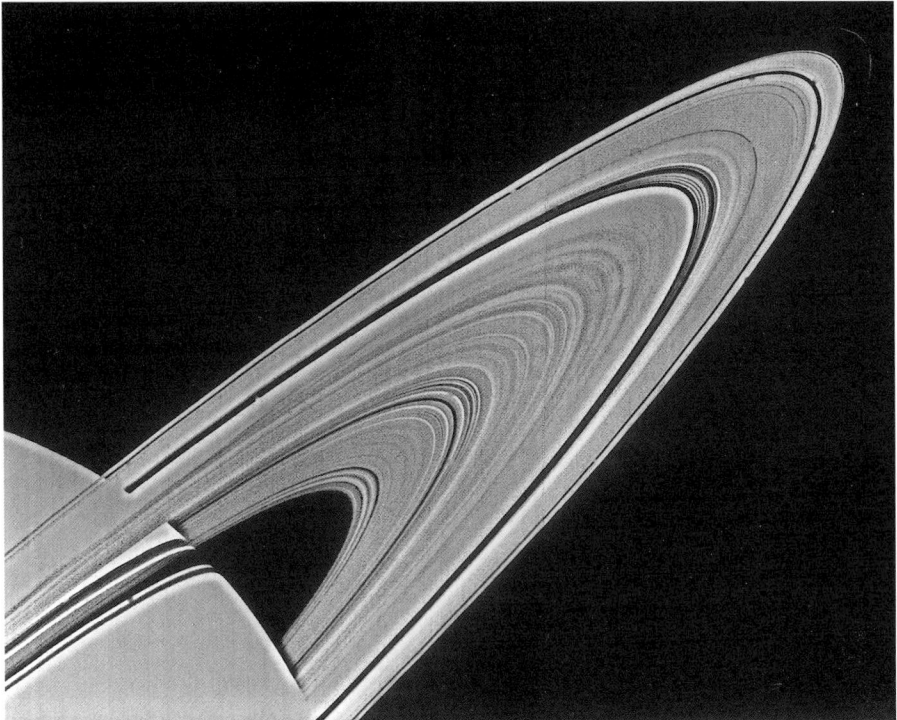

Fig. 19.34. *Voyager 1* photograph taken on 1980 November 06 at a distance of 8 million km from Saturn, showing copious detail in the ring system.

gree of opening. When open widely enough, the rings can be seen to differ distinctly in brightness. Intensity estimates may be tried. In all graphs it should be remembered that the shadows cast by the planet onto the rings and vice versa must be correctly represented. Observed deformations of these shadows appear to be optical illusions. However, thickenings of the ring system as seen with the rings edge-on coincide with zones of elevated brightness. They also indicate that the brightness of the ring structure is far from uniform. It is also quite interesting to observe the "transparency" of the ring system when Saturn occults a star [19.79].

19.5.6.3 Photography. Published reports [19.80, 81] have suggested that photography of Saturn be carried out using instruments with apertures of 20 cm and larger. Good results have indeed been obtained with instruments in this size range. Besides SEB, EZ, and NEB, the photographs show shadows on the rings and the Cassini Division. Compared with Jupiter, Saturn requires longer exposures: about 3 to 4 s at moderate aperture for focal photographs, and about 10 to 12 s for eyepiece-projection photographs. Photographic monitoring of the planet makes sense only under very good optical and climatic conditions. This also holds for photographic photometry of, for instance, the ring profiles.

Table 19.8. The brightest satellites of Saturn.

Moon		Magnitude
II	Enceladus	11.7
III	Tethys	10.3
IV	Dione	10.5
V	Rhea	9.7
VI	Titan	8.3
VIII	Iapetus	10–12, variable

19.5.6.4 Photometry. Suitable features for photometric studies include, in addition to the bright and dark surface detail (see Sect. 19.5.5.5 on Jupiter), the brighter moons and the ring system. In particular, photometry can be performed in conjunction with photographs and with color filters of known transmissivity (see Sect. 7.5). Reasonable results, however, require apertures over 25 cm and a total focal length over 20 m. J. Dragesco [19.82] reports on the photometric work of the ring system which was performed from 1974 to 1976 in Naples. Using a 60-cm aperture telescope and a total focal length of 40 m, photographs were taken in the violet (Wratten filter 34 and Tri-X film), in the yellow (Wratten 9 and Tri-X), and in the infrared (Wratten 89 B and High-Speed Infrared Kodak Film).

19.5.6.5 The Satellites of Saturn. As a result of the *Voyager* missions, 17 Saturnian moons are now known (see Appendix Table B.17). Several of them which of especial interest for the amateur observer are given in Table 19.8

The *Astronomical Almanac* also includes data on the fainter moons I (Mimas) and VI (Hyperion). With the ring system seen edge-on, the brighter Saturnian moons show similar phenomena (eclipses, etc.) as those of Jupiter (see Sect. 19.5.5.6). But even the transit of the brightest and largest moon Titan and the observation of its shadow on the planet require an aperture of at least 20 cm.

Of interest are the apparent magnitudes of the moons, which may vary; visual photometry either by direct estimates or by photometer is within the reach of amateurs. It is well known that Iapetus exhibits a light variation with an amplitude of about 2 magnitudes. The period of variation corresponds to the orbital period of $79^d 2^h$. The maximum brightness is observed near western elongation, the minimum near eastern elongation [19.83]. Evidence for a relation between solar activity and brightness changes of Titan and Rhea requires long-term and very precise photometry, obtainable only by photoelectric means. Seasonal changes of 0.01 or 0.02 magnitudes for Titan have been reported.

19.5.7 Uranus

19.5.7.1 Visibility. Uranus was discovered only two centuries ago, in 1781, by F.W. Herschel. With a diameter of 50 800 km, Uranus is the third largest planet in the

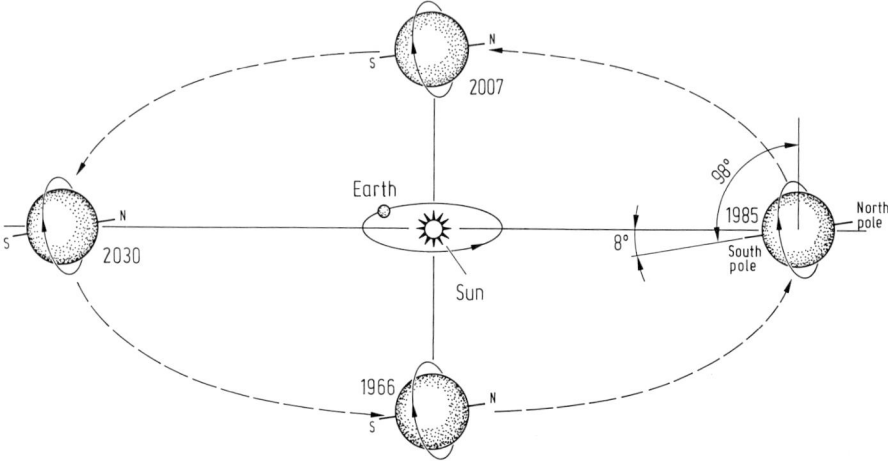

Fig. 19.35. Configurations of the planet Uranus relative to the Sun. The rotation axis is inclined 98° from the orbital axis. Taken from *Taschenbuch für Planetenbeobachter*, 3rd edn., Sterne und Weltraum, Munich 1987.

solar system, revolving about the Sun in 84 years. Owing to its extreme distance, the planet's apparent diameter remains between $3''$ and $4''$, limiting its interest for telescopic observation.

On 1986 January 24, *Voyager 2* passed Uranus at a distance of 93 000 km. The pictures obtained by the probe reveal an unstructured atmosphere, with a veil of haze surrounding the planet. Filter photographs showed clouds in the lower layers of the atmosphere moving with speeds of 100 m s^{-1}, about the same as the jetstreams on Earth 9 km above the surface, whereas the westerlies on Saturn attain speeds of up to 480 m s^{-1} near the equator. While the upper layers of Uranus' atmosphere consist of hydrogen and some helium, the lower parts are composed of methane and other hydrocarbons.

Stangely enough, the atmosphere over the dark pole of Uranus is found to be warmer than over the insolated pole. This may be explained by heat accumulated over the 42 years during which the now dark pole was exposed to the Sun. The rotation axis of the planet is almost exactly in the orbital plane (Fig. 19.35). Since the inclination of the planet against the ecliptic is low, Uranus' equator is almost perpendicular to the ecliptic. The passage of *Voyager 2* permitted direct measurements of the rotation period. Data from radio observations yield a period of $17^h239 \pm 0^h009$. Using measurements of the magnetic field, a rotation period of $17^h29 \pm 0^h20$ was derived. Uranus also displays a kind of polar aurora.

During the processing of observations from the 1977 occultation by Uranus of the star SAO 158687 (magnitude +8.8) in Libra, it was discovered that the planet is apparently surrounded by five narrow, thin rings; the total breadth of the ring system was found to be 7100 km. The ring system is in the equatorial plane of the planet, and its lower edge lies 17 700 km above the surface. Whether or not this system reaches the density of Saturn's rings still needs to be ascertained.

Four more rings had been detected even before the *Voyager* flyby. Again, it is assumed that all rings are in the equatorial plane. The following are the radii of the nine rings now known:

Ring (Code)	Radius (km) (±30 km)
6	41 980
5	42 360
4	42 663
α	44 844
β	45 799
η	47 323
γ	47 746
δ	48 432
ϵ	51 697

Between the ϵ and δ rings, *Voyager 2* found a tenth ring, which has been given the preliminary designation 1986 U1R. This ring lies 50 040 km from the center of the planet. There are indications that other rings may exist outside the epsilon ring. By contrast with Saturn, whose rings consist largely of golf-ball-sized particles, the particles in the rings of Uranus have diameters of mostly about 1 m. This may be explained by assuming that the particles which make up the Uranian rings are composed of a sturdier material than that of their their Saturnian counterparts, and are thus more resistant to the continual collisions [19.3].

19.5.7.2 Observations. Some observers had reported that telescopes with apertures larger than 25 cm are large enough to show dark shadings (e.g., the "dark equatorial band") [19.84]. No indications of such structure were found on photographs taken in 1970 by *Stratoscope II*, a 91-cm telescope flying 24 km above the Earth's surface. These photographs, however, show a very small oblateness and a distinct limb darkening which is even stronger than that of Jupiter.

Uranus shows a pronounced variation of brightness with total amplitude 0.9 mag. This change has geometric as well as physical causes [19.84] and may be monitored with amateur equipment, either by photometer (see Sect. 8.7) or by the well-known Argelander "step method" (Sect. 8.4.2). While viewing Uranus, the observer may also note the greenish coloring of the planet.

Photography of the two outer moons Titania and Oberon (magnitude +14) is within reach of the amateur [19.85]. These moons were discovered by W. Herschel in 1787. Ariel and Umbriel were discovered in 1851 by Lassell while observing with a 61-cm reflector on the island of Malta; their brightness is around magnitude +16. In 1948, G.P. Kuiper discovered Miranda at magnitude +17, and finally, the *Voyager 2* pictures showed ten new moons (see Appendix Table B.17, Vol. 3), so that Uranus now has a total of 15 known moons.

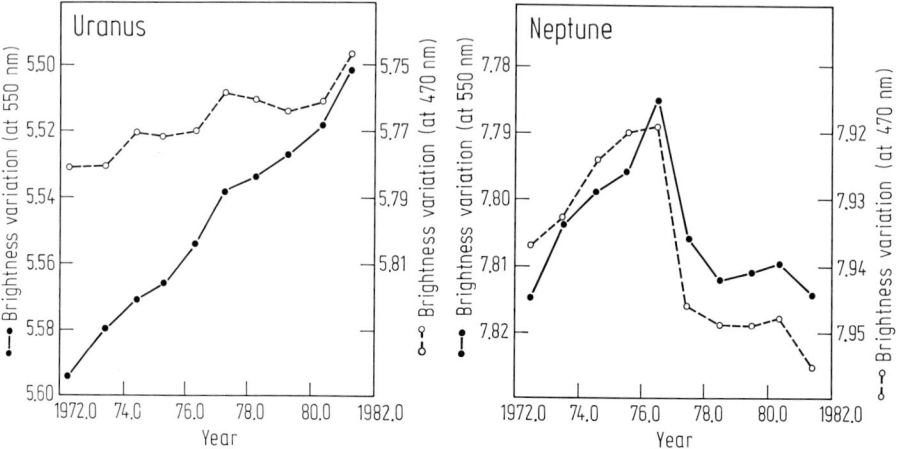

Fig. 19.36. Long-term change in apparent brightness in Uranus and Neptune during the period 1972–1982. Observations from the Lowell Observatory at 470 nm and 550 nm. From R.M. Genet (ed.), *Solar System Photometry Handbook*, Willmann-Bell, Richmond 1983.

19.5.8 Neptune

19.5.8.1 Visibility. Neptune was dicovered in 1846 by J.G. Galle, who had followed the theoretical predictions of U. Leverrier. Despite its linear diameter of 49 000 km, the apparent diameter of Neptune reaches only 2″.5, and is, with a mean opposition magnitude of +7.75, below the naked-eye limit of visibility. Neptune revolves in 164 years about the Sun; its oppositions recur every 367 days. It has an extremely low orbital eccentricity of 0.009.

In 1989, the space probe *Voyager 2* reached Neptune, approaching it on August 25 to within 29 240 km. The most important discovery in the atmosphere was the Great Dark Spot (GDS), a long-term stable cyclone with high white clouds ("cirrus") at the edges. It is comparable with the GRS on Jupiter [19.67]. *Voyager 2* found several other, less prominent bright and dark spots, as well as a band structure, the brightest zone being at latitude 20°S, and dark bands at 6° to 25°N and at 45° to 70°S. Compared with Uranus, Neptune shows a dynamical atmosphere; the existence of a yet unexplained internal heat source is conjectured. Fast westerly winds reaching up to 300 m s^{-1} were recorded.

A clear periodicity of Neptune's radio radiation over several rotations around *Voyager's* closest approach, $16^h 13^m \pm 4^m$, is interpreted as the rotation period of the core of the planet. Its magnetic field was measured as 0.13 gauss, which is weaker than that of other planets (Jupiter 4.2, Earth 0.3, Uranus 0.2 gauss).

Voyager 2 confirmed the ring system previously suspected from observations of occultations of stars [19.86]. The Earth-based measurements had suggested rather incomplete arcs, whereas *Voyager* reported closed rings like those around Jupiter, Saturn, and Uranus. The radii of the four rings know so far are given in the short Table 19.9.

Table 19.9. Radii of the rings around Neptune.

Ring	Radius (km)
1989 N3R	41 900
1989 N2R	53 200
1989 N4R	53 200–59 000
1989 N1R	62 900

As for the other Jovian planets, an extended disk of very fine dust surrounds Neptune in its equatorial plane, causing up to 300 impacts per second on the *Voyager* craft as it passed through it.

To the known moons Triton (discovered by Lassell in 1846) and Nereid (Kuiper 1949), *Voyager 2* added six more (see Appendix Table B.17 in Vol. 3). Triton's diameter was measured as 2760 km, and pictures of it were quite surprising as they revealed a wealth of surface detail. Resembling Mars and the moons Enceladus (Saturn) and Ariel (Uranus), no other solar-system body appears to show such a variety of surface formations. Triton's atmosphere was found to consist of a layer of nitrogen-methane haze.

19.5.8.2 Recent Observations. The amateur observer can see the planetary disk with magnifications over $300\times$, but atmospheric detail is ruled out. Similar to Uranus, the color of Neptune is observed to be bluish-green.

Photometric monitoring of the planet may have some scientific merit. The apparent brightness varies between $+7.5$ and $+8.0$; a rotational variation is suspected [19.3, 87]. The satellite Triton reaches $+13.6$ and is, at a diameter of 4000 km, one of the largest moons in the solar system. As with the brighter of Uranus's moons, attempts to photograph Triton may prove successful.

19.5.9 Pluto

19.5.9.1 Visibility. The outermost planet Pluto was found on photographs in 1930 by Clyde Tombaugh. The opposition brightness is about $+13.8$, the apparent diameter about $0\rlap{.}''3$. Even at high magnification, this planet does not look much different from a star.

19.5.9.2 Observations. Locating Pluto has recently become a popular "sport" among some amateur astronomers. This may be achieved visually or photographically using a telescope with an aperture of at least 20 cm [19.87 88].

"The visibility very strongly depends on transparency, extinction, and scattered light" (F. Frevert).

Amateurs have also reported to have successfully taken photographs of Pluto even with 15-cm optics.

Moon Charon was discovered in 1978.

References

19.1 O'Meara, S.J.: Observing Planets: A Lasting Legacy. *Sky & Telescope* **76**, 474f (1988).
19.2 Arbeitskreis Planetenbeobachter an der Wilhelm-Foerster-Sternwarte Berlin: *Mitteilungen für Planetenbeobachter* (Neue Folge).
19.3 Roth, G.D.: *Taschenbuch für Planetenbeobachter* (3rd edn.), Sterne und Weltraum, Munich 1987, p. 242.
19.4 O'Meara, St.J.: Jupiter's North Equatorial Belt Erupts. *Sky & Telescope* **79**, 94 (1990).
19.5 Koppmann, R.: Die Entdeckungen der Planetensonden. Alles alte Hüte?. *Sterne und Weltraum* **22**, 346 (1983).
19.6 Wichmann, H.: Die Verwendung von Binokularen insbesondere für das Celestron 5. *Sterne und Weltraum* **20**, pp. 27, 69, and 110 (1980).
19.7 Martheray, M. du: *Orion* **28**, 109 (1950).
19.8 Wichmann, H.: *Sterne und Weltraum* **4**, 175 (1965).
19.9 Parker, D.C., Dobbins, T.A.: The Art of Planetary Observing–II. *Sky & Telescope* **74**, 603 (1987). and Optische Glasfilter. Ein Katalog der Schott Glaswerke, Hattenbergstr. 10, D-55122 Mainz.
19.10 Haug, H.: Zum Thema Filterbeobachtungen. *Mitteilungen für Planetenbeobachter* (new series) **2**, No. 5, p. 2 (1978).
19.11 Minnaert, M.G.J.: *Practical Work in Elementary Astronomy*, Reidel, Dordrecht 1969, p. 106.
19.12 di Cicco, D.: S & T Test Report: A Versatile CCD for Amateurs. *Sky & Telescope* **80**, 250f (1990).
19.13 Martinez, P.: *Astrophotographie*, Darmstäder Blätter, Darmstadt 1985, p. 354.
19.14 Spangenberg, W.W.: Über einige Probleme der visuellen Planetenbeobachtung. *Die Sterne* **41**, 94 (1965).
19.15 Dragesco, J.: La Vision dans les instruments astronomiques et l'observation physique des surfaces planétaires. In: *L'Astronomie*, September 1969, p. 355, and October 1969, p. 399.
19.16 Spangenberg, W.W.: Über einige Probleme des alternden Beobachters. *Die Sterne* **50**, 109 (1974).
19.17 Kowalec, Ch.: Hilfsmittel zur Positionsbestimmung auf Riesenplaneten. *Die Sterne* **49**, 233 (1973).
19.18 Fischer, D.: Planeten-Erfolge aus England. *Sterne und Weltraum* **28**, 612f (1989).
19.19 Macfarlane, A.W.: A Primer for Video Astronomy. *Sky & Telescope* **79**, 226f (1990).
19.20 *Mitteilungen für Planetenbeobachter* (new series) **2**, part 3 (1978) and **3**, part 1 (1979).
19.21 Hartley, Ch.: Astronomical Computing: Drawing Grids on a Planet's Disk. *Sky & Telescope* **79**, 86f (1990).
19.22 Dorst, F.: Tagesbeobachtungen des Merkur. *Sterne und Weltraum* **10**, 336 (1971).
19.23 Moore, P., Hunt, G.: *Atlas des Sonnensystems*, Herder, Freiburg, Basel, Wien 1985, pp 78.
19.24 Strom, R.G.: Mercury: The Forgotten Planet. *Sky & Telescope* **80**, 256f (1990).
19.25 *Sky & Telescope* **55**, 543 (1978).
19.26 *Sky & Telescope* **50**, 196 (1975).
19.27 Sandner, W., Bernhard, H.: Beobachtungen des Merkur-Durchgangs vor der Sonne. *Orion* **32**, 28 (1974).
19.28 Ashbrook, J.: A Well-Observed Transit of Mercury. *Sky & Telescope* **47**, 4 (1974).
19.29 Spangenberg, W.W.: Über den "schwarzen Tropfen" und verwandte Erscheinungen. *Die Sterne* **47**, 22 (1971).

19.30 Dorst, F.: Venus während ihrer Konjunktionen mit der Sonne in der Zeit 1968–1970. *Sterne und Weltraum* **10**, 79 (1971).
19.31 Graff, K.: Die Planeten. *Astronomicsches Handbuch* (2nd ed.), Franckh'sche Verlagshandlung, Stuttgart 1925, p. 146.
19.32 Beatty, J.K.: Magellan at Venus: First Results. *Sky & Telescope* **80**, 603f (1990).
19.33 Benton, J.L.: Visual Observations of Venus: Theory and Methods. *ALPO Handbook*, 1988.
19.34 Ahnert, P.: Phasenbeobachtungen der Venus vor und nach der unteren Konjunktion 1974 Januar 23. *Die Sterne* **51**, 114 (1975).
19.35 Kunz, W.: Größte Elongation und Dichotomie der inneren Planeten. *Sterne und Weltraum* **16**, 334 (1977).
19.36 Richter, N.: Das Übergreifen der Sichelspitzen beim Planeten Venus. *Die Sterne* **17**, part 7 (1937).
19.37 Phillips, J.L.: The Ashen Light of Venus. *Sky & Telescope* **75**, 250 (1988).
19.38 *Sky & Telescope* **43**, 394 (1972).
19.39 Kimmel, B.: *Sky & Telescope* **44**, 197 (1972).
19.40 Benton J.L.: The 1983–84 and 1985–86 Western (Morning) Apparition of Venus: Visual and Photographic Observation. *JALPO* **33**, 93 (1987).
Baumm, J.L.: Unusual activity on the night side of Venus 1986 September. *JBAA* **97**, 1 (1986).
Benton J.L.: The 1984–85 Eastern (Evening) Apparition of Venus: Visual and Photographic Observation. *JALPO* **34**, 1 (1989).
19.41 Wepner, W.: Rechnen mit dem Taschenrechner. Ortsbestimmungen auf Sonne und Planeten. *Sterne und Weltraum* **16**, 374 (1977); cf. also: Montenbruck, O.: *Practical Ephemeris Calculations*, Springer, Berlin Heidelberg New York 1989, pp 88.
19.42 Sandner, W.: Die Beobachtung der Planeten. In: *Handbuch für Sternfreunde* (2nd edn.), Springer, Berlin Heidelberg New York 1967, pp 305.
19.43 Roth, G.D.: *Kosmos Astronomiegeschichte*, Franckh'sche Verlagshandlung, Stuttgart 1987, p. 130.
19.44 Gordon, R.W.: Mellish and Barnard—They did see Martian Craters! *The Journal of the Association of Lunar and Planetary Observers* **25**, 196 (1975).
19.45 Sheehan, W.: Mars 1909: Lessons Learned. *Sky & Telescope* **76**, 247f (1988).
19.46 Beish, J.D., Capen, C.F., Parker, D.C.: The Meteorology of Mars, Parts I, II, and III. *Journal of the Association of Lunar and Planetary Observers* **31**, 229 (1986), **32**, 12 and 101–114 (1987).
19.47 Freydank, E., Freydank, H.: *Marsbrevier*, Wilhelm-Foerster-Sternwarte, Berlin 1984, p. 20.
19.48 A Mars Observer's Guide. *Sky & Telescope* **75**, 516 (1988).
19.49 Gerber, P.: Bilder vom Mars. *Orion* **34**, 157 (1976).
19.50 Batson, R.M., Bridges, P.M., and Inge, J.L.: *Atlas of Mars*, NASA Spec. Pub. **438** (1979).
19.51 Carr, M.H.: *The Surface of Mars*, Yale University Press, New Haven/London 1981. Contains a very large index of relevant publications.
19.52 Sandner, W.: In: *Handbuch für Sternfreunde* (2nd edn.), Springer, Berlin Heidelberg New York 1967, p. 306.
19.53 Börngen, F.: Die Renaissance der Planetoidenastronomie. *Kalender für Sternfreunde 1988*, p. 151.
19.54 Schober, H.J.: Asteroids. In: Hearnshaw, J.B., and Cottrell, P.L. (eds.): *Instrumentation and Research Programmes for Small Telescopes*, published for the International Astronomical Union (IAU), 1986.
19.55 Schmadel, L.D.: Kleine Planeten II. Quartal 1978. *Sterne und Weltraum* **17**, 145 (1978).
19.56 Binzel, R.P.: Photometry of Asteroids. *Solar System Photometry Handbook*, Willmann-Bell, Richmond 1983, pp 1-1.
19.57 Schmadel, L.D.: Der Kleine Planet (85) Io in der Opposition 1977. *Sterne und Weltraum* **17**, 98 (1978).

19.58 Hodgson, R.G.: Minor Planet Work for Smaller Observatories. *The Minor Planet Bulletin* **4**, 1 (1976).
19.59 Dunham, D.W.: Satellite of Minor Planet 532 Herculina. *The Minor Planet Bulletin* **6**, 13 (1978).
19.60 Hempel, R.: Eine Methode zur Vermessung photographischer Planetoidenpositionen. *Sterne und Weltraum* **16**, 259 (1977). See also Bendel, R.: Verfahren zur Astrometrie von Kleinplaneten. *Sterne und Weltraum* **18**, 142 (1979) and Wepner, W.: Verfahren zur Astrometrie von Kleinplaneten. *Sterne und Weltraum* **18**, 182 (1979).
19.61 Greßmann, M.: Positionsbeobachtungen an Kleinen Planeten. In: *Kalender für Sternfreunde 1979*, p. 162.
19.62 Vehrenberg, H.: Photographische Kleinplaneten-Beobachtung durch Amateure. *Sterne und Weltraum* **11**, 36 (1972).
19.63 Schnitzer, A.: *Der Blink-Komparator der Sternfreunde*, Sterne und Weltraum-Verlag, Düsseldorf.
19.64 Wallentine, D.: Observations of 1580 Betulia in 1976. *The Minor Planet Bulletin* **4**, 36 (1977).
19.65 Sopper, R.: Jupiter 1971—Visuelle Beobachtungen und ihre Auswertung. *Die Sterne* **49**, 146 (1973).
19.66 Oivarez, J.: Jupiter's Best Show in Twelve Years. *Astronomy* **15**, 64 (1987).
19.67 Beebe, R.F.: Queen of the Giant Storms. *Sky & Telescope* **80**, 359f (1990).
19.68 Kowalec, Ch.: Hilfsmittel zur Positionsbestimmung auf Riesenplaneten. *Die Sterne* **49**, 230 (1973).
19.69 Zimmermann, O.: *Astronomisches Praktikum II* (3rd edn.), Verlag Sterne und Weltraum Dr. Vehrenberg, Munich 1983, p. 53.
19.70 Jasicek, H. et al.: Photoelektrische Verfolgung von gegenseitigen Jupitermondverfinsterungen 1973. *Sterne und Weltraum* **13**, 236 (1974).
19.71 Lockwood, G.W.: Photometry of Planets and Satellites. In: *Solar System Photometry Handbook*, Willmann-Bell, Richmond 1983, pp 21.
19.72 Brinkmann, R.T. et al.: Mutual Phenomena of Jupiter's Satellites in 1973–74. *Sky & Telescope* **45**, 93 (1973).
19.73 Ahnert, P.: Das System der Jupitermonde I–III. *Die Sterne* **39**, 180 (1963).
19.74 Ahnert, P.: Die Jupitermonde 1971 bis 1974. *Die Sterne* **52**, 39 (1976).
19.75 *Sky & Telescope*, **80**, 591 (1990).
19.76 O'Meara, S.J.: Saturn's Great White Spot Spectacular. *Sky & Telescope* **81**, 144ff (1991).
19.77 Haug, H.: Periodische Aktivitäten in der Saturnatmosphäre? *Sterne und Weltraum* **27**, 116 (1988).
19.78 Sandner, W.: Die Rotation des Saturn. *Orion* **30**, 58 (1972).
19.79 di Cicco, D., Robinson, L.J.: Saturn and 28 Sgr Highlights. *Sky & Telescope* **78**, 361f (1989).
19.80 Benton, J.L.: Visual and Photographic Observations of the Saturn: the 1974–75 Apparition. *The Journal of the Association of Lunar and Planetary Observers* **26**, 85 (1976).
19.81 Saturn: A Photographic Roundup. *Sky & Telescope* **46**, 124 (1973).
19.82 Dragesco, J.: La Planète Saturne de 1973 à 1976. *L'Astronomie*, 1977, p. 203.
19.83 Böhme, D.: Der Saturnmond Japetus und seine Beobachtung durch den Amateur. *Sterne und Weltraum* **15**, 410 (1976).
19.84 Alexander, A.F. O'D.: *The Planet Uranus. A History of Observations, Theory, and Discovery*, Faber and Faber, London 1965.
19.85 Haar, F., Schatzmann, B.: Photographische Aufnahme von zwei Uranus-Monden. *Sterne und Weltraum* **16**, 271 (1977).
19.86 Brahic, A., Hubbard, W.B.: The Baffling Ring Arcs of Neptune. *Sky & Telescope* **77**, 606 (1989).
Baum, R., Smith, R.W.: Neptune's Forgotten Ring. *Sky & Telescope* **77**, 610 (1989).
19.87 Frevert, F.: Beobachtung von Pluto. *Sterne und Weltraum* **15**, 300 (1976).
19.88 Zehnder, F.: Visuelle Beobachtungsmöglichkeiten von Pluto mit Amateurinstrumenten. *Orion* **29**, 150 (1971).

Current observational tasks and possibilities for cooperation are subjects of reports in the *Journal of the British Astronomical Association*, the *Journal of the Association of Lunar and Planetary Observers* (USA), *Astronomy*, and *Sky & Telescope*. Advanced observers will find more information in *Icarus*, an international journal of solar system studies.

20 Comets

R. Häfner

20.1 The Nature of Comets

For thousands of years, the appearances of comets in the sky have caused fear and panic for humans, as nearly everyone thought them to be harbingers of disaster. Aristotle had unintentionally, with his interpretation of the comet phenomenon, given the "scientific" basis for that belief, since he thought comets to be atmospheric phenomena, evaporations or outgassings from the ground, fire in the air, correlated with drought, earthquakes, and other catastrophic events. In 1577, Tycho Brahe demonstrated from parallax measurements that comets lie beyond the orbit of the Moon and therefore have no connection with the Earth's atmosphere. Subsequently, comets were recognized as members of the solar system, and their orbits were investigated. E. Halley used the law of gravity which had just been established by Newton, and postulated that the appearances of comets in the years 1682, 1607, and 1531 were due to the same object, which had a recurrence time of about 76 years. He therefore ventured the prediction that "his" comet would reappear in the year 1758, which it did, but he did not live to see it. By then, comets had to some degree become fairly predictable phenomena.

Some centuries have elapsed since Halley's time, and astronomers have made great progress in understanding the nature of comets. Development of better observing instruments and techniques has revealed a wealth of detail. The current peak of interest was generated by the results gained at the rendezvous of an armada of spacecraft to Halley's comet in 1986. This chapter will give a brief survey of our present knowledge of comets; for detailed information, the technical literature should be consulted.

Particularly during the 19th century, the orbits of a great many comets were calculated. Many different orbital shapes were found with almost randomly distributed inclinations to the plane of the planetary system, and with both direct and retrograde motion. Comets were seen to change their orbits and periods under the influence of the major planets, and were even observed to fragment and decompose. On the other hand, their appearance rate of three or four new and about three periodic comets per year was about constant. Based on work by E. Öpik and A. van Woerkom, J. Oort concluded in 1950 from this result that a comet reservoir exists, and he suggested the existence of a spherical shell-shaped comet cloud which now bears his name and which surrounds the Sun at a distance of about one light year (i.e., about one thousand times further out than the orbit of Pluto). It is at this point, according to this idea, where interstellar space begins. The approximately 10^{11} members of the cloud with a total mass of about 300 Earth masses move with speeds of about 0.1 km s^{-1} around the Sun with periods of millions of years. Although direct observation does not support the existence of such a cloud (such studies may become feasible with the Hubble Space Telescope), it is accepted as a working hypothesis by the majority of professional astronomers. Recently, *IRAS* (the Infrared Astronomical Satellite) found shells—partly flattened—around several main-sequence stars which may be linked with Oort-type clouds surrounding these stars.

It is now also agreed upon that this cloud is the region of comet formation, which occurred at the same time as the birth of the solar system, about 4.6×10^9 years ago. Gas molecules of the pre-solar nebula accreted onto dust particles and, under the influence of turbulent flows which occurred during the formation of Sun and planets, formed larger bodies with diameters of some km, the nuclei of comets. This concept is supported by the results of the Halley missions: the isotope ratios found for certain elements exclude an origin near the Sun or the planets but yield a time of origin approximately coinciding with that of the solar system. Furthermore, some of the molecules found could have formed only under temperatures near absolute zero, and such conditions existed in the outer parts of the pre-solar nebulae. Comets can therefore provide information on the original composition of the nebula since they were preserved in the Oort cloud, which acted like a cosmic refrigerator. The compositions of all other members of the solar system, on the other hand, have been altered more or less strongly since formation owing to the influence of the Sun or to the heat generation as the planets were accreted.

The passage of stars or of huge interstellar molecular clouds, and also the influence of the galactic disk itself, disturbs the orbits of the cometary nuclei in the Oort cloud, and occasionally swarms of these nuclei drift into interstellar space or into the inner solar system. When comets approach the Sun the gravitational interactions with the large planets can deflect some of them into short-period orbits; the result is the creation of the so-called *comet families*. The family of Jovian comets is distinctly the largest: it contains about 70 members.

Also in 1950, F. Whipple presented his "dirty snowball" model of cometary nuclei, interpreting them as solid structures with a mass of about 10^{15} to 10^{18} g, composed of dust and water ice, and mixed with some frozen gases (NH_3, CH_4, O_2). This model was largely confirmed and refined by recent space missions to comets. Also, for the first time the nucleus of a comet could be studied in great detail; in the case of Halley's comet, the nucleus had an irregular shape, very dark (albedo $\approx$ 4%) and porous crust at the surface, spherical structures similar to impact craters, and hills and valleys.

When a cometary nucleus in its orbit approaches the Sun at about the same distance as Jupiter, the ice on the side facing the Sun gradually heats and sublimates; because of zero pressure in space, it is immediately transmuted into the gaseous state. Thus, with the coma and perhaps also the tail, the well known comet shape develops. The nucleus and coma collectively are called the *head* of the comet. The gases evaporate off the nucleus with a speed of about 1 km s^{-1} and carry dust particles along with them. Quite often this occurs in eruptions which form huge jets thousands of km long. This outflow has a rocket-like recoil on the nucleus, and, depending on the direction of the impulse, such non-gravitational forces cause an acceleration or deceleration of the comet in its orbit. Clues as to the composition of the dust, on which little was previously known, were also obtained by the Halley missions: carbonaceous chondrites do not, as was once suspected, form the main constituent of the dust; rather, silicates (compounds of Si, Mg, O, and Fe) and combinations of light and heavy elements, particularly the so-called CHON particles enriched with C, H, O, and N, predominate. The mass spectrum ranges from about 10^{-4} to 10^{-17} g, and the diameters range from the size of sand grains down to macro-molecular dimensions. Concerning the gases, the very high contribution by water of about 80% was a surprise. This seems to be the dominating parent molecule in the coma. Knowledge of various molecules and radicals, such as CN, C_2, C_3, CH, NH, NH_2, and OH, had been obtained earlier. Even complex molecules like hydrogen cyanide (HCN) or methylcyanide (CH_3CN) have been found. Of course, this does not reflect the composition of the nucleus, as chemical processes, insolation, and the solar wind effect a continuous dissociation, recombination, and ionization of the gases after they leave the nucleus. In addition, in the close proximity of the Sun various atoms are observed as having been freed by evaporation from dust particles. The gas density decreases from about 10^{12} to 10^{14} cm^{-3} immediately near the nucleus to 10^2 to 10^4 cm^{-3} at the boundary of the coma, which may have a diameter of several 10^5 km. The mass ratio between gas and dust varies from one comet to the next, ranging between about 1:1 to about 10:1.

Many comets develop a longer (up to 10^8 km) straight tail and a shorter curved one. The straight tail, which has a density of about 10^2 cm^{-3}, consists essentially of singly ionized carbon monoxide (CO^+), whose emissions are partially responsible for the bluish hue. In addition, emissions of CO_2^+, CH^+, CN^+, C^+, N_2^+, OH^+, and H_2O^+ normally occur. For a long time it was unknown what forces accelerate the gases flowing off the coma with speeds of about 10 km s^{-1} into the tail to more than 100 km s^{-1} and why the tail is always averted to the side away from the Sun—a phenomenon already noted 1500 years ago by Chinese astronomers and first documented in the Middle Ages by Apianus and independently by G. Fracastoro for the comet of 1531. The riddle was solved in 1951 by L. Biermann: as the solar wind, a plasma of electrons and protons continuously flowing away from the Sun at several hundreds of km s^{-1}, ionizes the gases in the coma, it sweeps them along and carries the gas tail always away from the Sun like a smoke trail, since the random speeds of gas molecules are small compared with the wind speed. The variable solar activity, in conjunction with the interplanetary magnetic field, then leads to the sometimes-observed peculiar shapes of tails with knots, bends, or waves. Apart from these features, the curved and often strongly fanned tails are usually without structure; they consist of dust particles, which merely reflect sunlight and thus appear yellowish. These particles are accelerated by the radiation pressure of the Sun, but since the speeds reached are only of the order of the orbital speed of the comet, the dust tail is usually distinctly curved in the direction of the comet's motion. Figure 20.1 is a photograph of the comet Mrkos 1957 V, showing both types of tails. The sometimes-observed *anti-tail* is actually part of the dust tail, which—for certain relative positions of Sun, comet, and Earth—is then seen in projection toward the Sun.

The existence of a halo of neutral hydrogen with dimensions up to 10^7 km had been predicted by L. Biermann in 1968, but this optically invisible halo has subsequently been found around all comets investigated. Thus, comets, which actually are among

Fig. 20.1. Comet Mrkos 1957 V with a straight, narrow gas tail and broadly fanned dust tail. Photographed on 1957 August 13 at $12^h 10^m$ UT by A. McClure, Frazier Mountain, California using a 10-cm refractor ($f = 50$ cm) on Kodak 103a-E emulsion. Exposure time was 13 minutes.

the smallest bodies in the solar system, "inflate" as they approach the Sun to become among the largest objects of that system.

How long will a comet visible from Earth in the inner part of the solar system survive? It undoubtedly loses some fraction of its mass with each successive approach to the Sun. Estimates put this amount possibly far in excess of 10^{10} kg. The lost material is distributed along the orbit of the comet and will, when the Earth crosses that orbit, burn up (as meteor showers) in the Earth's atmosphere. The lifetime of a comet thus amounts to only a few hundred revolutions, unless the core, as was noted earlier, fragments owing to the tidal action of the Sun, or evaporates in the solar corona, or possibly ends as an outgassed asteroid.

20.2 Searching for Comets

To discover a new comet is the dream of many amateurs; statistics show that even today this dream can come true with some persistence. Only very few observatories operate systematic searches for comets, and even then such comets are found serendipitously on photographs taken for entirely different purposes. About 20% to 30% of all discoveries and recoveries are still made visually by amateurs active in that area. The following will provide some advice which—combined with some perseverence—will increase the chances of success.

An important factor for success is the site of observation, which should be characterized by frequently clear skies, the absence of artificial illumination, and an unimpeded view down to the horizon. Nights with moonlight or zodiacal light are unfavorable as the sky brightening impairs the contrast threshhold and thus the perception of extended objects. During morning and evening twilight, the horizon near the Sun can be examined with the unaided eye. With exceptionally good luck, one might, perhaps, find a bright comet which, though near the Sun, has escaped the notice of other observers for various reasons. Otherwise, one should proceed methodically and use binoculars or a telescope with a large field of view and good light gathering power. Instruments with apertures of about 10–20 cm and a focal ratio of about $f/6$, and used with low power ($\approx$ 20–30×), are best suited for the task. Such an instrument will detect comets with limiting magnitudes of around +9.5 to +10.5, the range at which most amateur discoveries are made. Apertures being equal, refractors are of better optical suitability than reflectors. Double telescopes (for viewing with both eyes) or large-aperture binoculars (over 8 cm) have the advantage that they reduce eye fatigue and therefore are well suited for longer observations. They also permit one to better recognize objects which are faint or diffuse.

An important requirement for successful work is to be well acquainted with the celestial region intended for comet searches. This may be done with atlases such as that of Bečvář [20.1], which was designed specifically for comet hunters. It shows stars to about magnitude +8 and also galaxies, star clusters, and nebulae, as these may easily be confused with comets. The nighttime handling of maps, even under low light, can for some time impair the dark adaptation of the eye and thus the attainable limiting magnitude. The size of the selected sky region depends on how much time is available per night and on how fast the scanning procedure is done. Where is the most promising observing region and when is the best time to observe? Discoveries have been made in all regions of the sky, but the best place to look for a comet is near the Sun, i.e., in the western evening and the eastern morning skies. Since the evening observer normally has stiffer competition, the opportunities are better for morning comet watchers. Also, investigations show that nearly three-fourths of all comets first become "discoverable" in the morning sky. The less experienced observer should first select star-poor regions, but not those which, such as Virgo or Coma Berenices, are known to contain numerous galaxies, as these could be mistaken for comets and thus impede the search. Having selected the area, it should be scanned, beginning at an edge, horizontally or vertically with a speed of $0°.5$ to $1°$ per second; the instrument can then be shifted by about one-half the field diameter, and the scanning run repeated. An alt-azimuthal mount is most convenient in this case. Statistics show that the probability of a discovery is good after an average of 200 or 300 observing hours. The history of comet hunters is rich in anecdotes, for example when success occurred after only 3 nights in one instance, but after 6 years in another. In yet another case the observer quit the search in exasperation after 15 years without success. It thus would make sense to consolidate the search for comets with the examination of the sky for novae or the recording of meteors.

If on one night a diffuse patch of light is seen in a particular region of the sky where no such object has been recorded previously, the observer should perform several clarifying tests. One should first check that he/she has not fallen victim to a simple optical illusion, such as the reflection of bright starlight at the lens. The observer should then use a higher power in order to verify that the object is not a faint double star or perhaps a star cluster. The most important test, however, is for motion, as a comet should have changed its position relative to neighboring stars within half an hour to one hour by some recognizable amount. In the meantime, the position of the object is graphed onto a star map, the universal time (UT) recorded, and an attempt made to estimate its magnitude. To this end, the telescope is defocused so that the stellar disks have about the same diameter as the object in question. From the known magnitudes of at least two of these stars bracketing the object, its magnitude is easily assessed by estimating the magnitude difference between the stars in equal steps and interpolating for the suspected comet. A precision of about 0.5 mag, which should suffice for the initial estimate, can easily be reached. The brightnesses of the stars which have to be clearly identified can be obtained later from a suitable star catalogue. If, after the time in question, the object has moved then one can be fairly certain that a comet has been found. An attempt should then be made to estimate the amount and direction of motion and the diameter of the coma, and it should be examined for any trace of a tail. The meteorological conditions at the observing site should also be recorded. There is still the possibility that the comet seen has already been discovered or is a known periodic comet. If the necessary information from, for instance, the *IAU Circulars* is not available, then the nearest large observatory should be informed of the observation before it is forwarded by telegram to the proper data center. The address is: IAU Central Bureau for Astronomical Telegrams, Smithsonian Astrophysical Observatory, Cambridge, MA 02138, USA. The brief message should contain all observing data and of course the name and address of the discoverer. This message can also be forwarded by the observatory which has been contacted.

A newly discovered comet carries the name of the observer, or up to three names if it was independently and nearly simultaneously discovered by several observers. In addition, comets are coded by a preliminary designation consisting of the year and a Latin letter depending on the sequence in the discoveries of the year (e.g. 1987d). In recent years this procedure has occasionally caused slight difficulties since nearly two dozen reports per year were received. After the orbital elements have been computed, the comets are ordered after their times of perihelion passage, and characterized by a Roman numeral after the year (e.g., 1987 II)- A complete designation might then be "Comet Meyer-Miller 1987 II." Finally, periodic comets carry the prefix "P/".

The coordinates α and δ of newly discovered or returning comets and data on the distances from the Sun and Earth and on their expected brightness are published in particular in the *IAU Circulars*. These data usually cover an interval of several months and are normally given at 10-day intervals. Corresponding references, particularly for brighter comets, are often found, for instance, in *Sky & Telescope*, in *The Handbook of the British Astronomical Association*, or in the almanac *Der Sternenhimmel*. To find a brighter comet then presents no difficulties. For fainter objects, the orbit is graphed onto maps of the region copied from, for instance, the *Photographische Sternatlas* of Vehrenberg [20.2]. Comet positions, of course, have to have been previously reduced to the equinox of the map. The actual position is then easily found by interpolation. If a telescope with equatorial mount and setting circles is not used, the useful orientations provided by neighboring stars can be of help. It may happen that the observed position as well as the brightness differs considerably from the predicted values. The construction of rising and setting diagrams [20.3] will be helpful when planning the observations.

20.3 Determining the Positions

An important task when observing comets consists in the repeated determination of its position so that the orbital elements can be determined or improved upon if they were known before only approximately. In principle, this requires visual and photographic methods, some of which are presented in the following subsections.

20.3.1 Visual Observations

Many observers find estimating positions an engaging activity. Positions can be estimated with a telescope similar to the one used for comet searches. A graph with typical star configurations in the field (with known angular diameter) is made. Finally, the position of the comet is marked and the date and time recorded. Later, the observed configurations can be identified on a star map, the cometary position graphed, and the coordinates for this specific observation found. The error will be of the order of several arcminutes. Estimates will be more accurate if an eyepiece with crosswire is used, and especially when even one reference star of known coordinates α_*, δ_* is in the field. The eyepiece is oriented such that by an east–west motion in a parallactic mount, the reference star moves parallel to one of the wires. It will then be advantageous to bring it into the center of the crosswire or at least onto one wire. When the orientation in the field is determined, the distances from star to comet (a, d) are estimated parallel to the directions of right ascension and declination in units of the field diameter. Then,

$$\alpha - \alpha_* = \pm \frac{a}{15 \cos \delta_*}, \qquad (20.1)$$

$$\delta - \delta_* = \pm d. \qquad (20.2)$$

The differences are in minutes of time or, when the field diameter is counted in minutes of arc, in minutes of arc. The signs depend on the relative positions of comet and reference star. Of course, the data thus obtained cannot serve to determine the orbital elements, but they can be used to follow the geocentric path of the comet on a map.

In order to reach the precision of about one arcsecond needed for orbit calculations, careful measurements must take the place of estimates. A position filar micrometer consisting of a crosswire rotatable around the optical axis and a movable wire will be found best suited. Relative positions can then be found by a divided circle and by the reading drum of the micrometer screw. Since such instruments or variations thereof normally are not accessible to the amateur and can be made only with much technical skill and machining tools [20.4], a brief description of the crossbar micrometer, which does not present too many problems for do-it-yourself construction, will be presented here. This device has been successfully used, in comparison with the ring micrometer in particular, especially for cometary observations. It consists of wires forming a right angle in the focal plane of the eyepiece. The cross is then oriented in the equatorially mounted telescope so that the wires form an angle of $45°$ with the direction of diurnal motion. Illumination can be dispensed with since the disappearance and reappearance of objects behind the wires can easily be noted, and therefore faint and especially extended objects are more accessible for measurement.

The procedure uses only one reference star whose coordinates α_*, δ_*, corrected for proper motion, must be known. With the telescope drive turned off, star and comet trail through the field (Fig. 20.2) and the sidereal times of their transits behind both wires are recorded by the eye–ear method. The simple geometry of passage gives the following relations for the coordinate differences:

$$\alpha - \alpha_* = \frac{(t_1 + t_2) - (t_{1*} + t_{2*})}{2}, \qquad (20.3)$$

$$\delta - \delta_* = \pm 7.5 \, (\Delta t - \Delta t_*) \cos \delta_0. \qquad (20.4)$$

Here $t_{1,2}$ and $t_{1,2*}$ are, respectively, the transit times of comet and star, Δt and Δt_* the corresponding differences, and $\delta_0 = (\delta + \delta_*)/2$. An estimate of the comet's declination δ suffices to determine $\cos \delta_0$. For the declination difference, the positive sign holds if transits are observed to the north of the crosswire, negative for those south of it.

Potential errors caused perhaps by incorrect angles of the wires may be compensated by a second observation with the cross rotated by $90°$. Near the celestial pole, the objects do not move straight through the field, and a correction to the $\Delta \delta$ would be needed. But even at $\delta = 80°$, the correction amounts to, for instance, only $0.''08$ for a difference of $10'$, and thus can normally be neglected, as can the influence of

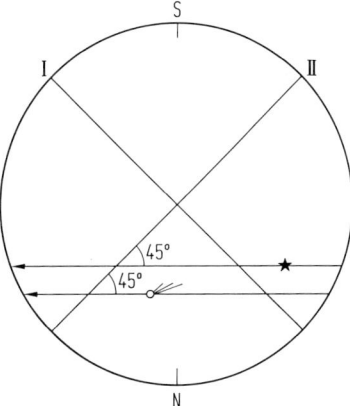

Fig. 20.2. Field of view of a telescope with a crosswire micrometer. See text for details.

refraction. If in special cases the reduction is desired, the appropriate formulae can be found in books on spherical astronomy [20.5]. Depending on the unit of time in which the transits are measured, e.g., the minute, the results from Eqs. (20.3) and (20.4) come out also in this case in minutes of time and minutes of arc.

So far the position of the comet has been assumed to be constant relative to the reference star. However, it has a more or less appreciable motion which can influence the measurement, and this has to be allowed for. Let α, δ be the coordinates computed from Eqs. (20.3) and (20.4) and $\Delta\alpha$, $\Delta\delta$ the motion expressed, for instance, in units of minutes of time or minutes of arc per minute of time. Then the final values α_C, δ_C of the comet are given by

$$\alpha_C = \alpha \pm \frac{\Delta\delta \Delta t}{30 \cos \delta_0}, \qquad (20.5)$$

$$\delta_C = \delta \pm 7.5 \, \Delta\alpha \Delta t \, \cos \delta_0. \qquad (20.6)$$

Here the upper sign denotes southern and the lower sign northern transits. $\Delta\alpha$, $\Delta\delta$ can be easily found from an ephemeris, and thus the position is valid for the time $T = \Delta t/2$ determined. This position, however, is normally reduced to the equinox 1950.0 before possible forwarding. If, on the other hand, no ephemeris is available, then two observations sufficiently separated in time may yield the comet's motion by an iterative application of Eqs. (20.3–6).

20.3.2 Photographic Observations

Most frequently, positions are determined by photography, i.e., when the relative position of the comet with respect to a reference star is measured on a photographic plate. To obtain the required precision of one arcsecond, several rules must be observed starting with the exposure. Owing to their relative motion, the comet and stars cannot simultaneously be recorded as point-like images, as would be the ideal for measurement. In this connection, since it is only the *head* of the comet (or, more precisely, the starlike nucleus which marks the position of its center of mass) that is of interest, the exposure time should generally be kept as short (a few minutes) as possible and high-speed film used (see Eq. (20.7)). It will usually suffice to guide the exposure on a star or to use automatic tracking and to tolerate the potential slight elongation of the comet head. Only in cases of rapidly moving comets is it advisable to follow its motion with the telescope (see Sect. 20.4) or with the plateholder. In this event, the stellar images will be somewhat elongated or streaked. Since at least three reference stars should be distinctly identifiable, a distortion-free field diameter of several degrees will usually be required. This requirement limits the choice of instruments to, ideally, an astrograph or a Schmidt camera. The beginning and end times of the exposure should be recorded to the second; their average defines the instant of observation. It has been found advantageous to take several exposures using stepped exposure times on the same plate or film, moving the camera or telescope slightly after each successive exposure (always in the same direction, e.g., in δ). This procedure will provide security against possible flaws in the photographic emulsion and permits the evaluation of several exposures.

Measuring the photographs should be done with a coordinate measuring machine which permits the determination of the position of the comet and reference stars with high precision (better than 1 μm). Financial constraints will normally preclude the purchase of such an instrument for amateurs. The observer may turn to a professional observatory or might, depending on his technical skill, consider constructing a measuring device him/herself. Regarding the latter, suggestions have been given, for instance, by Hempel [20.6], Bendel [20.7], and Everhart [20.8]. Simple methods such as projection onto a wall or enlarged photottgraphic reproduction and then using a ruler are not advisable, as uncontrolled errors may creep in all too easily in such cases. If the photograph has been made on film, it must during measuring be placed between two glass plates to avoid warping. The reference stars selected should be well distributed around the position of the comet; they should also, if possible, not be too bright. They are then identified in a star atlas and the coordinates and their correction for proper motion (from the equinox of the catalogue to the date of observation) taken from a positional catalogue (e.g., the *SAOC* [20.9]). It is advisable not to mix data from several catalogues.

The rectangular coordinates of the reference stars are determined quite simply by positioning the crosswire of the microscope at the measuring machine onto the point of strongest photographic density. If the star images are elongated, the beginning and end of the trails may be measured and an average taken. It is a bit more difficult to correctly position the crosswire on the comet head, which always appears extended and somewhat diffuse, and the geometric mean does not necessarily coincide with the point of highest density. The measuring procedure then becomes somewhat subjective. It has therefore been proven helpful to repeat the measurement of the photograph after a certain time, and perhaps more often, and to take the mean. The rectangular coordinates of the comet found are then converted into spherical coordinates by the method of Turner (see Chap. 3 in Vol. 1) and usually also refer to the equinox 1950.0. The Turner method requires a minimum of three reference stars; a somewhat larger number to be solved by least squares improves the accuracy of the result.

From at least three measurements of the position obtained at intervals of several days, it is possible to determine the orbital elements of the comet. The calculation time, given the power and speed of modern pocket calculators and personal computers, is minimal. Experiences in astrometric reduction of photographic positions and/or the calculation of orbits are given, for example, by Mandel and Klare [20.10], Kneissl [20.11], and Kasten [20.12].

20.4 Studies of Structure

20.4.1 Visual Studies

The coma in particular of a comet can present a wealth of detail which quite often cannot be shown photographically owing to the large brightness contrasts that occur. Thus, even today visual observation and drawings are considered useful procedures. To this end, a magnification between 50× and 150× is used, and a drawing made "positive" with white pencil on rough, black drawing paper, as this matches the visual impression best. Gradual changes and shades can, with some practice, be obtained by simply wiping with a cloth or similar method. This technique also has the advantage that the graphing surface needs to be only weakly illuminated, and thus the dark adaptation of the eye is not significantly interfered with. For fine detail, however, the "negative" method (black on white paper) is preferable. Apart from the correct orientation, all structures should be graphed true to scale. This is not difficult with some experience and with the field diameter known.

Regarding the overall appearance of the coma, its outer shape (round, elliptical, fanned, etc.) and its structure (homogeneous, irregular, etc.) are of particular interest. Often, a bright, starlike condensation is observed around the nucleus where the gas and dust densities are highest. For this concentration, a scale running from values 0 (no condensation seen) and 9 (starlike condensation) is used. It can happen particularly very close to the Sun that matter streams out from this concentration toward the Sun and produces jets which then, at some distance from the origin, bend toward the tail and thus give the impression of huge fountains. As a result of such activity, shell-shaped layers are created which taper to a sharp point toward the tail. Sometimes a dark region is also recognized extending from the condensation far into the tail. In particular, phenomena near the condensation may suffer substantial changes over timescales of just a few hours owing to the rotation of the nucleus or its variable activity. This is illustrated by Fig. 20.3 (from Innes [20.13]) and Fig. 20.4 (from Schmidt-Kaler [20.14]), showing as negative graphs some of the detail described in the inner corona of the comets P/Halley and Bennett 1970 II.

The use of color filters when observing bright comets may give interesting indications as to the distribution of gas and dust. Gas flows are best seen in blue light, dust in yellow or reddish light. Drawings of the same phenomena made using different filters thus may separate the constituents and reveal their relative dimensions. Also, the coma diameter will differ depending on the particular filter used. It normally appears smaller in yellow or red light than in blue.

When the comet has one or more tails, various other observations can be made. The often fanned dust tail is normally structureless, but the gas tail may under certain

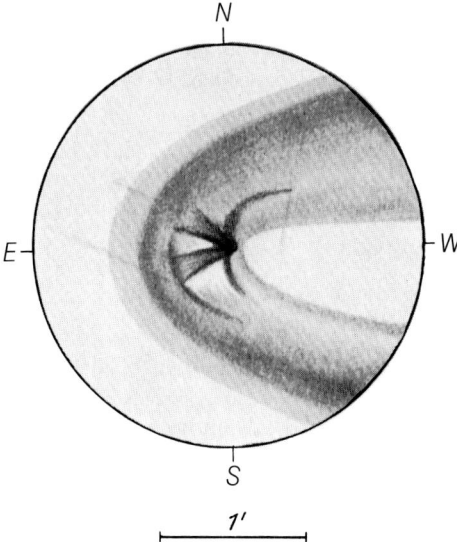

Fig. 20.3. Head of Halley's Comet on 1910 May 09 from a negative drawing by R.T.A. Innes [20.13]. Visible are the central condensation with two rays and two fanned streams, one with bright cloud stripes at the end, as well as envelopes and a dark area.

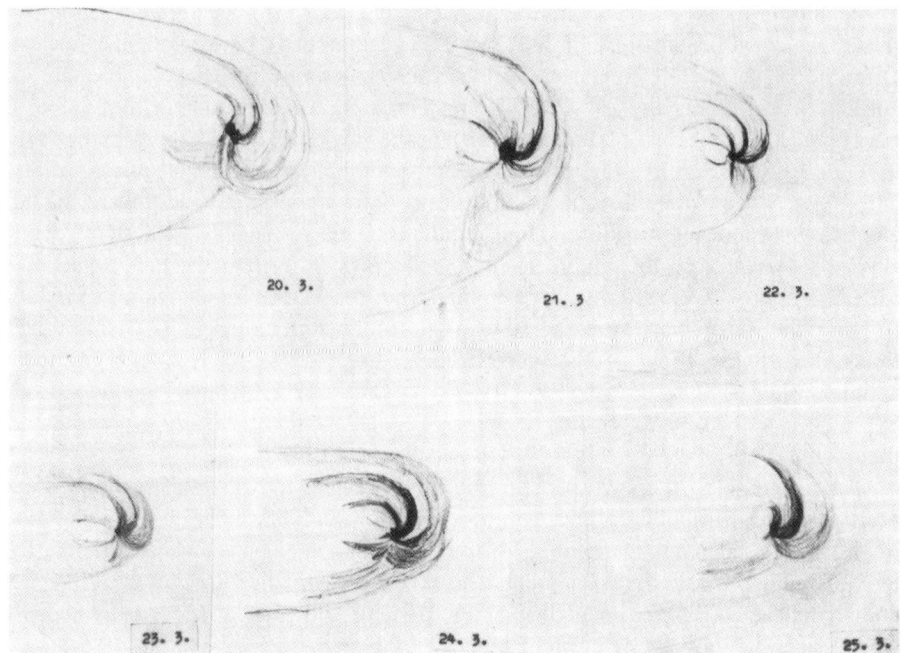

Fig. 20.4. Changes in the head of Comet Bennett 1970 II, seen in late March of 1970. From negative drawings by T. Schmidt-Kaler [20.14] (not to scale).

circumstances display, as has been mentioned, pronounced structures. Inhomogeneities like knots and ripples, and even the detachment of parts of the tail, are frequently observed. Owing to the low surface brightness of the tail and the rather low magnification used to view it, binoculars can even be used, and the tail phenomenon entered into a prepared star map. Changes may also be seen here within a short time. Corresponding filters may, for bright comets, provide a distinction between dust and gas constituents near the coma. Besides structures in the tail, the position angle (i.e., the direction of the tail counted from north toward east), its angular width, and its dimensions are also of interest.

Suggestions and hints for visual comet observations and the subsequent processing can be found in, for instance, Bortle [20.15] or Kammerer and Linder [20.16].

20.4.2 Photographic Studies

It is, however, advisable to perform investigations of structure, in particular of the outer coma and of the tail, photographically. Aside from their aesthetic appeal, photographs may supply additional information in conjunction with certain techniques. Also, dimensions, orientations, velocities of various phenomena, and similar quantities are more conveniently determined.

Most of the tasks mentioned require photographs with long exposure times. It will therefore often be necessary to guide on the comet itself. If it is bright enough and has a central condensation in the coma, then this may be achieved using the guiding scope as for a star. If this is not the case, then the guiding must be done indirectly. Ideally, the relative motion of the comet is computed from the ephemeris, and the corresponding correction in α and δ applied to the automatic tracking, allowing also for refraction when observing near the horizon. Another option is to use an eyepiece with special scales which permit guiding on a suitable star with a speed corresponding to that of the comet but opposite to the direction of its motion. The construction and handling of such an eyepiece is described by Stättmeyer [20.17]. An interesting alternative is to adjust the polar axis of the telescope by just the right amount so as to make the automatic tracking follow the comet as suggested by Liller [20.18]. The exposure times needed depend on the purpose of the photogragh and on the brightness of the comet, as well as on the properties of telescope, camera, and photographic emulsion. A first guide is the following relation (Babu [20.19]):

$$t = \frac{a^2 \ 10^{0.4(m+13)}}{sf^2}. \qquad (20.7)$$

Here, a is the f-ratio, f the focal length of the telescope or camera, s the speed of the photographic emulsion in ASA, and m the apparent magnitude of the comet. The exposure time t is in seconds. It can sometimes substantially deviate depending on the brightness of the sky background, the air mass, and atmospheric conditions.

To photograph the inner coma region, the use of long-focus telescopes is advised. Nevertheless, visually perceivable details will often not be found on the photograph. Methods of contrast exaggeration in copying can then be used in order to exhaust the contents of the latent image and to reveal finer details (Eccles et al. [20.20], Högner et al. [20.21]). To photograph the overall appearance of the comet, often a standard single-lens reflex camera with a normal or telephoto lens will amply suffice. It is attached with an adaptor to the telescope tube, and the telescope is used for guiding. In order to avoid imaging errors at the edge of the photograph, the aperture should be stopped down to at least 1:2.5, even at the expense of some light loss.

In addition to integral photographs, it advisable to use a special color filter, which is attached in front of the camera objective. Since this makes the exposure time substantially longer, very fast black-and-white emulsions must be used. The optimal interference filters are those with a bandwidth of about 5 to 10 nm, which corresponds to the characteristic cometary emissions (see Sect. 20.5.3) or emission-free regions in the spectrum. The resulting pictures then show the distribution of the cometary matter typical for the chosen wavelength over coma and tail, and they permit the study of

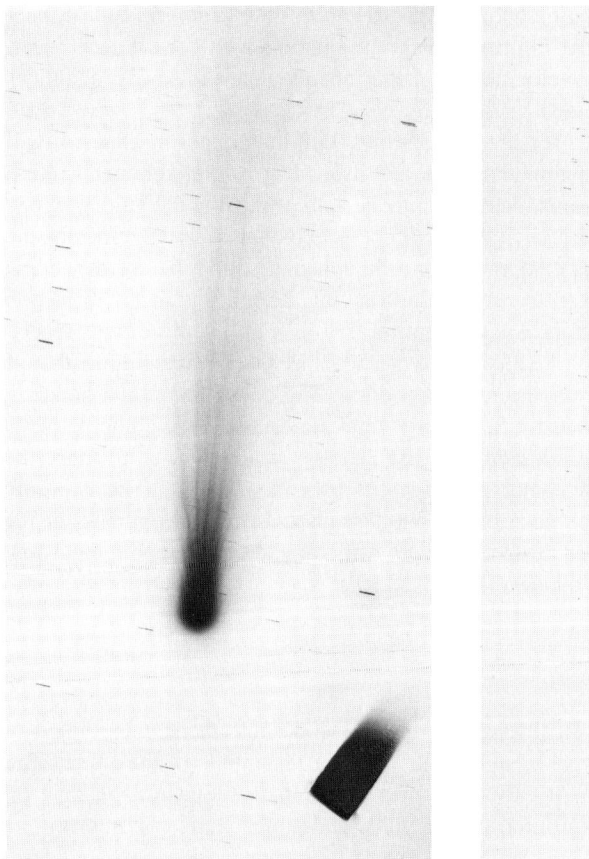
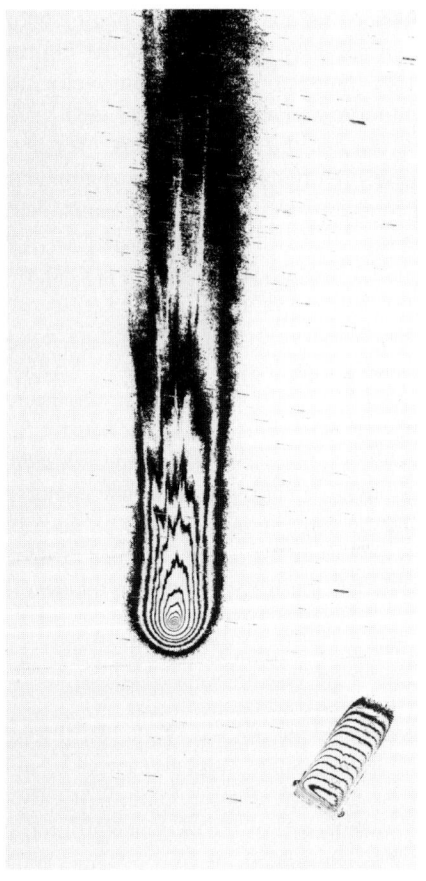

Fig. 20.5a, b. Equidensity diagram of Comet Bennett 1970 II, after N. Richter. **a** Original negative; **b** Equidensity representation; calibrating wedge in the lower right, respectively.

the production rate as a function of heliocentric distance of the comet. To study the gas component, filters at the following wavelengths (in nm) are suggested: 388 (CN), 426 (CO^+), 405 (C_3), and 513 (C_2). The dust component is reached with filters at, for instance, 365 or 485 nm, where the dust-reflected sunlight is not affected by molecular emissions. Broad-band negative filters which, for instance, truncate wavelengths shorter than 570 nm and thus the strongest emissions, can also be used.

The method of *equidensitometry* is well suited to visually displaying the density structure in such integral or filter photographs (Richter [20.22]). Then, a positive copy onto "hard" film material is made from the original negative, and, after development, is not fixed. Instead, it is post-exposed with diffused light and developed further. The central part of the comet is thus blackened on the positive copy except for a narrow band along the comet's structure. This band then corresponds to the positions of equal density on the original negative, hence the term "equidensity." If, now, a whole series of such copies is made with different exposure times and then treated simultaneously by the method described, a series of *equidensites* is obtained ranging from the outer edge of the comet to its center. With the aid of actual stars or artificial markings, these can then be exactly superposed and assembled into an equidensitogram. Figure 20.5 shows the result of such a procedure. If, in addition, the photometric wedge has been copied in for calibration (as in this case), then the relative intensities of the individual equidensites can be determined. Geyer [20.23] reports on his experiences with a special film (Agfacontour) which simplifies the preparation of equidensides.

For structural studies of this kind, the *charge-coupled device*, or CCD, will in the near future certainly become widely used, even by many amateurs. This electronic area detector, with its very wide dynamical range and high quantum efficiency, is far superior to the photographic plate. The cost of such a receiver, including the electronics needed for its operation, and for a computer system offering the image processing, is now approaching reasonable figures. First-hand reports on the use of such equipment can be found in Buil [20.24], Bickel [20.25], and Harris [20.28].

20.5 Special Techniques

20.5.1 Photometry

The brightness with which a comet appears to the observer is a function of its distances from the Earth and from the Sun, and also from its production rate of gas and dust. Normally, the brightness is given in magnitudes and described by the following formula:

$$m = m_0 + 5\log \Delta + 2.5n \log r, \tag{20.8}$$

where Δ and r are the distances from Earth and Sun respectively, m is the apparent magnitude, and m_0 the "absolute" magnitude, which holds, by definition, for distances $r = \Delta = 1$ AU. The parameter n is an indicator of the physics of the luminescence.

Pure reflection of sunlight gives $n = 2$, but with the resonance light of the gases added, n can assume values over 7. Young comets usually have a value $n = 2$, while older ones average between 4 and 6. The n value for a particular comet may easily differ before and after perihelion passage. In principle, two magnitude observations, sufficiently spaced in time, would suffice to determine the unknowns m_0 and n, and to predict the development of brightness. In practice, however, comets exhibit strong changes of brightness (a factor of 3 is not uncommon) and even eruptions. It is thus preferred to graph the brightness over longer ranges of time and of distance r and to graph $m - 5 \log \Delta$ versus $2.5 \log r$ in a diagram. The shape of the resulting curve is approximated ideally by a straight line or by segments thereof. The slope of the line gives the prevailing n, and its intercept with the ordinate axis m_0. The parameters thus determined permit more dependable brightness predictions. Ideally, one could use a computer to determine a best-fit straight line, but this is not very instructive; a diagram shows the general trend in brightness and its periodic and erratic changes directly.

The apparent magnitude of the comet is determined by comparison with known magnitudes of one or more stars which—in order to avoid extinction errors—should be near the comet (but outside the bright coma and tail) and, if possible, of spectral type similar to that of the Sun. The amateur observer will usually work visually. If the head of the comet can be seen with the naked eye and appears almost starlike, then its brightness may be determined without a telescope using suitable comparison stars (cf. Chap. 8). If a telescope is needed, then it is advisable to use an eyepiece whose exit pupil is suited to the eye. This is achieved when the numerical value of magnification equals about twice the objective diameter in centimeters. For objective apertures less than 10 cm, however, a magnification of 20× should always be chosen. Using the method of defocusing (Sect. 20.2), the brightness of such comets can then be fixed. Since fairly precise values are desired, the procedure should be repeated a few times, perhaps with different comparison stars. With some practice, accuracies of about 0.1 or 0.2 mag can be attained. In case the comet is too faint for the telescope used and its defocused image is not discernable, its size and brightness may be compared with the defocused star images *from memory*. This may sound difficult, but trained observers using this method can obtain comparable results (Kasten [20.27]). When a central condensation can be seen, its brightness should be estimated by direct comparison with neighboring stars. A stronger magnification is then used to diminish the influence of the superposed coma. In any event, the interfering brightness of the coma limits the attainable precision. Since, as experience shows, brightness estimates depend on the type and aperture of the telescope used, astronomers have agreed to reduce such observations to a standard aperture of 6.8 cm, which presents the possibility of obtaining a comparative evaluation of results of various observers at different telescopes. The correction amounts to 0.066 mag per centimeter of aperture for refractors, 0.019 mag per centimeter for reflectors, and is to be subtracted (added) for larger (smaller) apertures than 6.8 cm. Since most amateurs now use larger telescopes, standardization on the basis of a larger aperture would be desirable (Festou [20.28]).

The precision of magnitude determinations is improved by at least a factor of 10 when methods of photoelectric photometry are employed. Since the number of amateur observers who own photoelectric equipment is continually growing, a few peculiarities which apply specifically for comets will be mentioned here; the basics are to be found in Chaps. 4 and 8 of Vol. 1 and in Chaps. 24 and 25 of Vol. 3. The use of interference filters (see Sect. 20.4) permits the study of the brightness distribution of the coma and tail in various emission lines and in the continuum. When performing photometry of the coma, successively larger diaphragms are employed, each time centered on the center of light or on the nucleus. Thus, various filters provide information on the radial distribution of the individual spectral contributions. Besides this classical method, one can also, using always the same diaphragm, study the coma piece by piece in the radial direction in discrete steps, for instance, along and perpendicular to the direction to the Sun. Constant diaphragms are also used when searching for short-term changes of a particular region. To this end, measurements are done, alternating in quick succession, of the sky background and a comparison star, both sufficiently far away from the comet and, as a precaution, off at right angles to the direction of the tail. Apart from this, it will suffice to measure the comparison star and sky at intervals depending on the variation of air mass and of sky brightness. Magnitudes of the comet are then given relative to the comparison star. The possibility of photoelectric recording of the occultation of a star by the coma, a fairly frequent event, should also be considered.

Photographs serving to determine positions can, of course, also be utilized to obtain the (photographic) magnitude if they contain suitable comparison stars (or, in addition, a calibration field) and if the comet head is reasonably starlike. The reduction methods are essentially those of the photographic photometry of variable stars (see Chap. 8).

Having ascertained the quality of the magnitude data, they should be published in a suitable serial and/or submitted to the Data Center of International Comet Quarterly, so that they will be available to interested astronomers and amateurs. The data should always be accompanied by a detailed observing record which contains technical details, a description of the reduction method, and also meteorological data. The contact address is: Daniel Green, Smithsonian Astrophysical Observatory, 60 Garden Street, Cambridge, MA 02138, USA.

20.5.2 Polarimetry

Light, with its randomly distributed vibrational planes and phases of its constituent electromagnetic waves, is termed *unpolarized*. If, however, it contains a predominant direction of vibration, it is called *partially linearly polarized*. The degree of polarization specifies the relative fraction of polarized to total light. Cometary light shows a very high linear polarization caused by sunlight scatter on dust particles and by fluorescent emissions of molecules. The contributions of these two mechanisms vary from one comet to another, as do the contributions from various regions within the coma and tail. While the degree of polarization caused by fluorescence is normally less than 10% and is nearly independent of the phase angle (i.e., the angle Sun–comet–Earth), the part generated by scattering depends very strongly on that angle. The scattering component is maximum for $90°$ and then reaches about 20% to 30% (and even up to 50% in exceptional cases). The vibrational direction of the electric vector is always perpendicular to the plane of scattering.

The high degree of polarization offers the possibility for polarimetric studies of bright comets even for amateurs with only simple means. By placing a polarization foil (e.g., Käsemann PW-64) before the eyepiece and rotating it slowly, a periodic change in the brightness of the comet will be distinctly visible. Brightness maxima will occur when the direction of transmission of the foil coincides with the plane of vibration of the linearly polarized light, and minima when perpendicular to this position. For photographic polarimetry, the foil is placed in a frame into the parallel beam, for instance in front of the telescope or telephoto lens. A device which can read angles is needed so that the foil can be rotated into defined orientations. In principle, then, three exposures with settings different by 45° each will suffice to determine the degree of polarization P and the polarization angle φ relative to the first arbitrary position of the foil. In the judging of the proper exposure time, one should allow for a light loss of about 1 to 2 magnitudes created by the foil. The intensities needed in the reduction are obtained by a calibration device, for instance by exposing a neutral wedge on the same photographic material. In order to reduce the Schwarzschild effect, the exposure times for comet and wedge should about match. The resulting density curve then permits the conversion of densities of the various parts of the comet image into intensities I (see Chap. 7). The number and size of the regions to evaluate will depend upon the image scale. Then, for each area where the measures have been corrected for sky background brightness, the following relations hold:

$$P = \frac{[(I_1 - I_3)^2 + (2I_2 - I_1 - I_3)^2]^{1/2}}{(I_1 + I_3)}, \tag{20.9}$$

$$\tan 2\varphi = \frac{2I_2 - I_1 - I_3}{I_1 - I_3}, \tag{20.10}$$

where the indices refer to the respective foil positions and exposures; one should not confuse their sequence. The quadrant of φ can be found by the different signs in Eq. (20.10). By observing standard stars with known polarization angles using the same technique, the observer's system can be related to the standard system. Regarding the degree of polarization, the result still contains the instrumental contribution. This effect, however, can be determined with the aid of unpolarized standard stars. A list of such standards is given, for instance, by Serkowski [20.29].

The polarization can be made "visible" through a double exposure: one image taken with the foil in the position of maximum light and the other with the foil perpendicular to it, and then photographic subtraction applied. For that purpose, one of the photographs is contact-copied to a positive, and this is arranged exactly on top of the other original negative. If all densities of the copy are within the linear part of the photographic density curve, the residual image must then be caused by polarization differences (Fountain [20.30]). Thus, qualitative changes of polarization across the image of the comet can be found and correlated to structural properties.

20.5.3 Spectroscopy

Normally the instrumentation available to amateurs will be sufficient to secure spectra of only very bright comets. In order to carry out successful spectroscopic studies, a large-aperture telescope is indispensible (see Chap. 7). Trials with a combination of objective prism/telephoto lens/single-lens reflex camera have not proven satisfactory, owing in particular to the long exposure times required, which cannot, for a variety of reasons, be adhered to (Pollman et al. [20.31]).

By the time a comet is within reach of amateur observations, the coma has long formed, and the CN (cyanogen) emissions appearing usually on approach to within 3 AU from the Sun have been joined by those of C_3 and NH_2 at about 2 AU, followed by those of C_2, NH, OH, and CH at distances within about 1.5 AU. Details regarding spectroscopy of the tails will not be discussed here, as cometary tails are, because of their low surface brightnesses, unsuitable for spectroscopy with smaller instruments.

In any event, spectroscopy of the coma should be made with the dispersion perpendicular to the direction of the tail, that is to say, the refracting edge of the prism or spectrograph slit are placed parallel to the tail. This will, for very bright comets, forestall a superposition of the various spectral contributions. The blue spectral range will be dominant, as it contains the strongest emissions. Even with modest spectral resolution, the complexes or aggregations at wavelengths (in nm) 388 (CN), 405 (C_3), 468 (C_2), and 516 (C_2) will be noticed and their intensity changes related to the distance from the Sun can be followed. Figure 20.6 shows a spectrum of Comet P/Halley over the normal spectral range; it may serve also for orientation when identifying the separate emissions. Superposed on the coma emission spectrum is the dust-reflected absorption spectrum of the Sun. As references, the wavelength tables in the *Atlas of Cometary Spectra* by Swings and Haser [20.32] and the identifications in the red

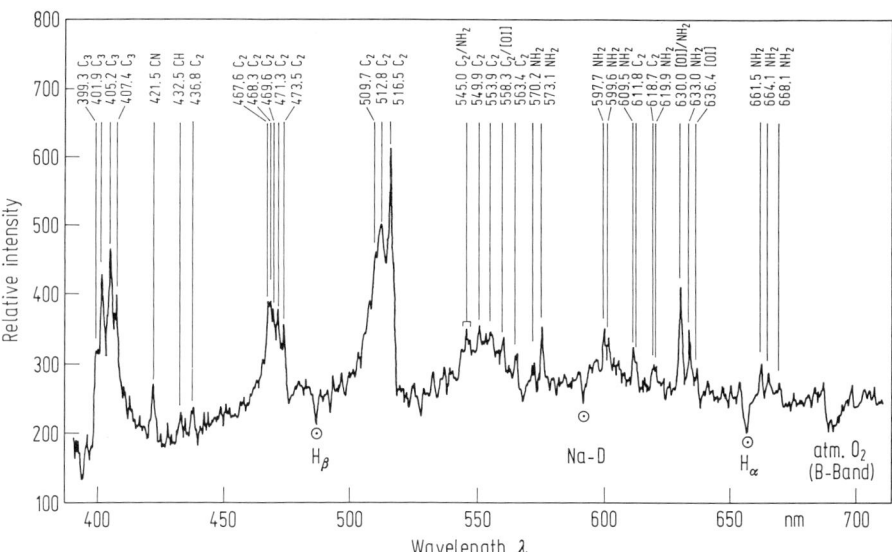

Fig. 20.6. Spectrum of the coma of Halley's Comet, obtained by the author on 1985 December 06 $1^h 35^m$ UT at the European Southern Observatory in La Silla, Chile. A 1.5-m telescope and Boller and Chivens spectrograph with image-dissector scanner was used with an integration time of 5 min. The spectral resolution is about 1 nm, $m = +5.8$, $r = 1.41$ AU, $\Delta = 0.68$ AU. The very strong CN emission at 388.3 nm is just outside the range represented. Those features labeled with a ⊙ are absorption lines from the scattered solar spectrum.

spectral range by Gary et al. [20.33] may be mentioned. Besides the molecular emissions, atomic ones appear as the comet approaches the Sun (within about 0.7 AU). Among a very large number of metallic lines, in particular the well-known Na doublet dominates. The forbidden lines of neutral oxygen [O I] (Fig. 20.6) deserve special attention. It was only recently shown that they actually do originate in the coma, as these lines are contaminated by the [O I] emissions of the Earth's atmosphere and also by the cometary emissions of NH_2 and C_2. The influence of contributions by the Earth's atmosphere has been eliminated in the spectrum of Fig. 20.6. These lines are not caused by resonance or fluorescence as are the molecular emissions, that is, by the absorption of sunlight followed by re-emission. Rather, they originate from the dissociation of H_2O or CO_2 with an excited, metastable level of the oxygen thus formed. For more information on the physical background of the processes in comets, the book by Krishna Swamy [20.34], can be recommended. Although requiring an elementary knowledge of atomic physics, it presents molecular physics as applied to comets in readable form.

References

20.1 Bečvář, A.: *Atlas of the Heavens (Atlas Coeli 1950.0)*, Czechoslovak Academy of Sciences, Praha 1958.
20.2 Vehrenberg, H.: *Photographischer Sternatlas*, Treugesell Verlag, Düsseldorf 1977.
20.3 Güssow: Komet Halley 1985/86. *Sterne und Weltraum* **23**, 209 (1984).
20.4 Polman, J.: A Homemade Filar Micrometer. *Sky & Telescope* **53**, 391 (1977).
20.5 Becker, E.: Mikrometer und Mikrometermessungen. In: Valentiner, W. (ed.): *Handwörterbuch der Astronomie, Vol. 3, Part 1*, E. Trewent Verlag, Breslau 1899. (W. Chauvenet: *A Manual of Practical Astronomy*, 1891, reprinted by Dover Publ., New York 1960, describes the similar ring micrometer.
20.6 Hempel, R.: Eine Methode zur Vermessung Photographischer Planetoidenpositionen. *Sterne und Weltraum* **16**, 259 (1977).
20.7 Bendel, R.: Verfahren zur Astrometrie von Kleinplaneten. *Sterne und Weltraum* **18**, 142 (1979).
20.8 Everhart, E.: Constructing a Measuring Engine. *Sky & Telescope* **64**, 279 (1982).
20.9 *Smithsonian Astrophysical Observatory Star Catalogue*, Smithsonian Institution, Washington D.C. 1966. Also: *PPM Star Catalogue*, Heidelberg 1991.
20.10 Mandel, H., Klare, G.: Astrometrische Auswertung von Kometenaufnehmen. *Sterne und Weltraum* **24**, 546 (1985).
20.11 Kneissl, R.: Zur Bahnbestimmung in der Amateurastronomie. *Sterne und Weltraum* **11**, 684 (1988).
20.12 Kasten, V.: Kometenbahn-Bestimmung mit dem HP 67. *Sterne und Weltraum* **16**, 299 (1977).
20.13 Innes, R.T.A.: Observations of Halley's Comet. Circular No. 4, Transvaal Observatory, Pretoria (1910).
20.14 Schmidt-Kaler, T.: Visuelle Beobachtungen des Kometen Bennett 1969 I. *Sterne und Weltraum* **9**, 200 (1970).
20.15 Bortle, J.E.: How to Observe Comets. *Sky & Telescope* **61**, 210 (1981).
20.16 Kammerer, A., Linder, J.: Die visuelle Kometenbeobachtung und ihre Auswertung. *Sterne und Weltraum* **23**, 466 (1984).
20.17 Stättmayer, P.: Eine einfache Methode zur indirekten Kometennachführung. *Sterne und Weltraum* **13**, 132 (1974).
20.18 Liller, W.: Tracking a Comet. *Sky & Telescope* **71**, 514 (1986).
20.19 Babu, G.S.D.: Techniques for Comet Photography. Preprint No. 154, Indian Institute of Astrophysics, Bangalore, (1986).

20.20 Eccles, M.J., Sim, M.E., Tritton, K.P.: *Low Light Level Detectors in Astronomy*, Cambridge University Press, Cambridge 1983).
20.21 Högner, W., Löchel, H., Richter, N.: Optimale Ausschöpfung des Informationsgeheltes von Kometenaufnahmen durch entwicklungstechnische Kontraststeuerung. *Die Sterne* **49**, 72 (1973).
20.22 Richter, N.: Photographische Methoden zur Informationsausschöpfung wissenschaftlicher Photogramme. *Leopoldina* **20**, 157 (1976).
20.23 Geyer, E.H.: Astronomical Applications of the Agfacontourfilm Technique. In: West, R.M. and Heudier, J.L. (eds.): *Modern Techniques in Astronomical Photography*, European Southern Observatory, Geneva 1978.
20.24 Buil, C.: A Charge-Coupled Device for Amateurs. *Sky & Telescope* **69**, 71 (1985).
20.25 Bickel, W.: Ein CCD-Versuch. *Sterne und Weltraum* **25**, 40 (1986).
20.26 Harris, C.: Silicon Eye: A CCD Imaging System. *Sky & Telescope* **71**, 407 (1986).
20.27 Kasten, V.: Zur Genauigkeit visueller Helligkeitsschätzungen von Kometen. *Sterne und Weltraum* **11**, 24 (1972).
20.28 Festou, M.: On Predicting and Analyzing Comet Light Curves. *International Halley Watch Newsletter No. 6*, 21 (1985).
20.29 Serkowski, K.: Polarimeters for Optical Astronomy. In: Gehrels, T. (ed.): *Planets, Stars, and Nebulae Studied with Photopolarimetry*, University of Arizona Press, Tucson 1974.
20.30 Fountain, J.W.: Spatial Distribution of Polarization Over the Disks of Venus, Jupiter, Saturn, and the Moon. In: [20.29].
20.31 Pollmann, E., Timm-Arnold, K.P., Alteweier, D.: Das Spektrum des Kometen Halley. *Sterne und Weltraum* **26**, 40 (1987).
20.32 Swings, P., Haser, L.: *Atlas of Representative Cometary Spectra*, Astrophysical Institute, University of Liège 1956.
20.33 Gary, G.A., Fountain, W.F., O'Dell, C.R.: Spectrographic Observations of Comet West (1975n). *Publ. Astron. Soc. Pacific* **89**, 97 (1977).
20.34 Krishna Swamy, K.S.: *Physics of Comets*, World Scientific Publishing Co. Pte. Ltd., Singapore 1986.

21 Meteors and Bolides

F. Schmeidler

21.1 General Information on Meteors

The term *meteor* is used to describe the brilliant phenomenon which occurs when a material body from space, called a *meteoroid*, penetrates into the Earth's atmosphere; the resulting fragment reaching the surface is called a *meteorite*. Observation of meteors is an area of astronomy which, at least in principle, continues today to provide opportunities for scientifically useful work by amateur astronomers.

One must also distinguish between the small "shooting stars," or meteors, and the large fireballs, also called *bolides*; the latter fall with substantial development of light and heat, and are often accompanied by thunderous noise. There is no sharply defined boundary between meteors and fireballs; as an approximate guideline, a meteor is considered a fireball when it reaches at least the apparent brightness of Venus (about -4^m). The German word "Sternschnuppe" is reminiscent of the snuffs falling off candle wicks. Not until about 1800 was it clearly recognized that all meteors are the result of cosmic bodies entering the Earth's atmosphere and glowing in it. The explanation of the problems associated with these bodies has been handled primarily by professional astronomers, as amateur observations in most cases were not performed in a sufficiently systematic manner. With well planned observing, however, amateur astronomers may supply useful contributions to scientific progress in this area. A good, detailed presentation on the history of meteor research has been given by Hughes [21.1].

The scientific interest in meteors pertains to data on their origin and physical constitution. The first question is tackled primarily by determining the spatial path of the meteor, the second by studying the luminous phenomenon of the fall; only for the large fireballs is there occasionally the opportunity to investigate a piece of the meteorite afterwards in the laboratory.

Only part of the desired results for meteor observations can be obtained from the records of a single observer. Many questions require observations by at least two observers at different locations. Suggestions for the planning and evaluation of observations are the subject of this chapter.

21.2 Methods of Meteor Observation

Meteors may be observed visually (with or without optical aid), photographically, or using radio-astronomical, spectroscopic, and video instruments. The spectroscopic aspects are separately detailed in Chap. 7.

21.2.1 Visual Observations

Observations by single observers as well as simultaneous observations from several different stations require the following important data:

1. Three points of the path are desired, including the beginning and end, if possible. It is best to state these points relative to fixed stars, for instance: "midway between α and β Ursa Majoris," or "1/4 of the line from α Cygni to α Lyrae," or a similar description.
   In case the points cannot be placed relative to stars (i.e., for a bright fireball during the daytime), estimates relative to the Sun or Moon may be made. Comparison with prominent terrestrial markings or estimates of apparent altitude and azimuth can also be used, but are less dependable. If the beginning point was not seen, the point where the meteor was first caught should be stated, of course noting this fact. All data are recorded immediately in order to minimize the risk of memory errors.
2. Time and place of observation. Of course, the error of the clock or watch used needs to be precisely determined, a task best done by comparison with a broadcast time signal. The place of observation is most suitably described by geographic longitude and latitude.
3. The duration of the phenomena. It is here that the largest errors occur. Only by substantial practice can good estimates be reached! Poor data on the duration are scientifically just as useless and confusing as good results are in demand. Observers lacking a stopwatch may test their memory scale of time by performing an experiment in which a watch is repeatedly checked, for instance, for the time needed to slowly count: "1001, 1002, 1003, ..." and so on. After practicing this for some time, one becomes accustomed to a certain counting rhythm, which then can be repeated for each individual case with closely matching speed.
4. The luminous phenomenon. Estimating the maximum brightness of a meteor requires considerable experience. It is suggested that the brightness be stated not in magnitudes but rather as compared with stars at about the same altitude (because of differential extinction). Typical descriptions could be: "as bright as β Aquarii," or "somewhat fainter than γ Aquilae but distinctly brighter than β Aquilae."
   A luminous trail often remains visible for some time along the path: in this case data on intensity and duration of afterglow are very useful. Statements on the color of the phenomena are occasionally possible if the event is bright.
5. Acoustic data. Large meteors are sometimes accompanied by the sound of thunder. The time between the light and sound phenomena should be recorded as accurately as possible.
6. Weather data, in particular the state of cloudiness during the observation, are needed in every case.

Of course, observations will frequently aim at a particular program item, and the other features listed here will then be omitted or attended to with lesser effort. It all depends upon the specific target of observation.

In any event, the observation of meteors requires much patience. A certain sector in the sky is selected to be monitored for a long time, and all meteors appearing there are recorded. It makes little sense to try to survey the whole sky in the hope that a meteor falling anywhere would be detected owing to its brightness and its motion. Using optical aids has the advantage of reaching fainter meteors, but the disadvantage that it limits the field of view; which aspect weighs heavier depends upon the purpose of observing. Under all circumstances, of course, the optical instrument used, such as binoculars, should have as large a field of view as possible.

Practice has shown that experienced observers can fix the end point of a meteor track in the sky to within few degrees. Often, instead of a statement giving the beginning and end of the apparent path of a meteor, the pertinent information is plotted onto a star map. In principle, there is no objection to this but in practice it seems to be less accurate.

21.2.2 Photographic Observations

The target of photographic observations of meteors is essentially the same as for visual ones, though the circumstances will cause some modifications in detail. First, few amateur astronomers own instruments of high light-gathering power and a large field of view, as will be needed for the observation of meteors. Also, a good tracking mechanism must be provided. Commercial photographic cameras can be used with success only in exceptional cases, since the small linear scale on the negative will reveal only a few meteor trails, if any. Two principal objections connected with the rapidity of a meteoric event can be raised against the photographic method *per se*:

1. The photograph shows only the brightest meteors.
2. It is biased toward those meteors with the slowest apparent motions.

The first difficulty has been largely overcome by the so-called Super-Schmidt camera, constructed by Baker; details may be found in Whipple [21.2]. There is principally no remedy against the second difficulty.

An advantage of photographic observations is the facility to measure the speed of meteors with suitable mechanical devices. If, by some means, perhaps using a rotating sector in front of the objective or the plate, the light is chopped at regular intervals (e.g., every 1/10 second), then the meteor will appear as a light trail which is broken at discrete intervals. Counting the interruptions and measuring their separations permits the determination of the duration and speed of the meteor. Of course, construction of such a device requires a reasonably well-equipped mechanical workshop.

When arranging photographic monitoring simultaneously from two different positions, both cameras are pointed to regions of the sky so that their lines of view intersect in the average height of meteors (about 80 to 100 km). One may consider 40 to 50 km the most favorable distance between two photographic observing posts.

In summary, it may be said that photographic observation of meteors is advised only when suitable optical and mechanical tools are available.

21.2.3 Radio Astronomical Observations

The detection of meteors at radio wavelengths is possible because the air along the path of the meteor becomes more strongly ionized than elsewhere, and a radar beam can then be reflected by the trail of residual ionization. This fact was noted already before and during World War II, and thereafter utilized for the systematic exploration of meteors. The radar instruments which comprise the primary technical components will not be described here. Details are to be found in *Meteor Astronomy* by Lovell [21.3], and important suggestions for useful radio equipment are contained in an article by Schippke [21.4].

Besides the recording of meteors, radar observations permit the determination of two quantities which are principally inaccessible to a single visual or photographic observation. These are the direction of the path of the meteor, or *radiant*, and its speed. The direction can be measured by the fact that a radar echo is received only when the beam meets the meteor orbit more or less perpendicularly. The meteor's speed is measured by small changes in the amplitude of the radar vibrations, which depend on the speed of the target. Details on this subject may be found in specialized literature, in particular the above-mentioned book by Lovell [21.3].

21.2.4 Observation by Television

In recent years, meteors have also been observed using video cameras and television monitors. The process is in principle the same as in photography, except that the photographic camera is replaced by a television camera. According to results obtained thus far, the limiting magnitude reached is somewhat better than the corresponding photographic limit. On the other hand, the mentioned disadvantage of photographic data, i.e., that slow meteors are recorded preferentially, occurs to the same or even greater extent for televison cameras.

Good results with this technique have been obtained by Hapgood, Rothwell, and Royrwik [21.5] for observations of the Perseid meteors, and further experiences can be anticipated [21.6].

21.3 Special Aspects of Observations

The following section addresses the specific questions which are expected to be answered by meteor observations. A summary of the present scientific knowledge is given first for meteors and later for bolides.

21.3.1 Meteors

There is a major distinction between *sporadic meteors* and those belonging to a *meteor shower* (see following page). The physical processes of ionization and incandescence, which are, of course, the same for both cases, are not yet fully explored. It is now assumed that air friction plays a relatively minor role and that the primary process is the collisional ionization of air molecules. Most meteors appear at 100 to 150 km above the ground and expire at about 50 to 100 km. The mass of most meteorites is under 1 gramm, or comparable to a grain of sand.

One of the most directly accessible data is the *meteor frequency*, defined as the number of events noted per hour. It is much larger than would be expected; on a clear night an observer without optical aid can see an average of 8 to 10 meteors. In addition, there are distinct periodicities with time of day and time of year. Usually the best time to see meteors is when the apex of the Earth's motion (i.e., the point toward which the Earth is traveling in its orbit about the Sun), is high above the horizon. All meteors appearing on the sphere near the apex move through space in the direction opposite that of the Earth, and therefore their velocities relative to Earth will tend to be quite high. The apex always lies in the ecliptic 90° west of the Sun, and so most meteors appear in the early morning hours; moreover, the fact that the apex is highly elevated above the horizon during autumn results in appreciably more meteors seen at that time than in spring. These rules are modified depending on what actual speed the meteors have and whether or not their space motions exhibit a preferential direction. Answers to both questions are not yet scientifically clear, and thus amateur observers can still contribute usefully by simply counting meteors, so long as they are not averse to working until dawn!

The determination of the spatial path of meteors requires at least two observers, located at different sites. If both observers see the same meteors and have noted the requisite data, then the parallactic displacement gives the altitude above Earth's surface, and the speed gives the path in space. The necessary formulae are compiled in Sect. 21.4.

The question of meteoroid velocities in particular has in recent years given rise to noticeable differences of opinion. The orbit determination (see Sect. 21.4) gives the heliocentric velocity at the instant of penetration into Earth's atmosphere. If it is larger than 42 km s^{-1}, then the meteoroid had a hyperbolic orbit; otherwise it was an elliptical one. It has sometimes been claimed that a large number of meteoroids have hyperbolic orbits; these meteors thus should have come from outside the solar system. More recent radio astronomical measures, however, have always indicated elliptical orbits. It is thus far unclear whether the discrepancy results from systematic errors of either visual or radio astronomical observing methods. Thus, the reality of "interstellar" meteoroids following hyperbolic orbits has been neither proven nor refuted. A decision awaits additional observations.

At specific times during the year, there appear unusually large numbers of meteors whose paths, when extrapolated backward, meet approximately in one point of the celestial sphere, the so-called *radiant*; the rule of perspective shows that these meteoroid follow parallel orbits in space and are members of a shower. The best-known meteor shower is that of the Perseids, which appears in early to mid-August. For some of these showers it has been proven that their paths coincide with the orbit of a specific comet. Thus, the particles which become meteors are in reality remnants of cometary matter. Table 21.1 provides data on the more prominent meteor showers.

Among the showers listed, the Draconids and the Leonids are *periodic showers*, which means that rich meteor displays occur only when the Earth crosses that particular part of the parent comet's orbit where matter is strongly concentrated. For the Draconids (Comet Giacobini-Zinner) this occurs every $6\frac{1}{2}$ or 13 years, but the rich shower that was expected from this rule in 1972 October did not materialize. Regarding the Leonids, particularly rich showers occurred in the last century every 32 or 33 years; the shower was subsequently absent for some time, but in November of 1966 it produced one of the most brilliant displays ever seen.

Table 21.1. Some prominent meteor showers.

Stream	Dates	Radiant		Richness	Associated Comet
		α	δ		
Qudrantids	Jan. 01–Jan. 04	230°	+50°	High	No comet known
Lyrids	Apr. 19–Apr. 23	273	+31	Low	Comet 1861 I
η Aquarids	Apr. 28–May. 16	340	0	Medium	Halley's Comet?
δ Aquarids	July 22–Aug. 10	344	−15	Medium	No comet known
Perseids	July 27–Aug. 18	40	+55	High	Comet 1862 III
Draconids	Oct. 10 only	267	+56	High	Comet 1900 III
Orionids	Oct. 15–Oct. 25	94	+14	Low	Halley's Comet?
Taurids	Oct. 26–Nov. 22	54	+15	Low	Encke's Comet?
Leonids	Nov. 15–Nov. 17	151	+22	Low	Comet Temple 1866 I
Geminids	Dec. 06–Dec. 16	113	+32	High	No comet known
Ursids	Dec. 21–Dec. 23	217	+76	Medium	Comet Tuttle 1939k?

The most impressive of meteor showers are the Perseids, which may produce nearly 100 meteors per hour. The Geminids are also rich, but the frequently inclement weather in December impedes their observation. The tabulated dates of beginning and end are averages and may vary by a few days either way for individual appearances.

To best observe these phenomena, the following points should be carefully attended to:

1. It is useful to make new observations of sporadic meteors as well as of the known showers. The results (e.g., regarding richness) may change within a few years in an unexpected fashion.
2. The known showers should, if possible, be observed anew each year.
3. The data from observations on numbers of meteors per hour increase in value, when also the brightnesses of these meteors can be stated.
4. Observations at zenith distances of over 70° and in moonlight are best avoided.
5. In the southern hemisphere of the Earth, the data collected so far are very scant. Observers who have the occasion to travel to southern latitudes may serve science considerably by observing meteors even without using any special instruments.

Radio-astronomical observations have also shown that daytime showers exist; they appear only in the daytime sky, and therefore are inaccessible to most other observing methods. Most of these showers occur during the months May and June. Reports by observers or groups thereof, having made systematic meteor observations and now state their results, are often found in various astronomy magazines.

21.3.2 Bolides

A falling body of large dimensions will penetrate deep enough into the Earth's atmosphere to be heated on the outside by friction, and thus originates the brilliant event known as a *fireball* or *bolide*. In many cases, the meteorite, as it is suddenly heated, bursts into several pieces. After the speed has been sufficiently braked by friction, the

Fig. 21.1. Iron meteorite from Cabin Creek, Arkansas, USA. After Berwerth [21.8].

visible meteor phenomenon terminates and the fragments fall to the ground. Among the meteorites found, the two groups of *stony* and *iron meteorites* are distinguished. The latter are composed almost entirely of iron which, on the other hand, is not completely absent in the stonies.

The methods and problems of observing fireballs are in principle the same as for the fainter meteors. In practice, however, the rarity and unpredictability of fireballs makes still greater demands on the observer's patience, and chance plays a much greater role. Some observers have successfully employed photographic methods, as a report [21.7] on photographic registration of a fireball shows.

Of particular interest are the acoustic phenomena which sometimes accompany the fireball. Depending on the height and the location of the observer relative to the path of the fireball, the sound may be quite different. For details on these matters, see Knöfel [21.9], who states that audible perceptions of fireballs should be reported to the Arbeitskreis Meteore (Working Group on Meteors), Neuer Garten 5, D-14469 Potsdam, Germany, for central scientific processing.

The meteorite fragments, which can have masses of several kilograms, may possess very darkened and crusty surfaces. Be warned, however, as not every stone with such characteristics is a meteorite! In each specific case, the decision as to whether or not a particular chunk found is actually a meteorite can be made only by a specialist. A characteristic feature of metallic meteors are the interlocking crystal patterns, the so-called *Widmannstätten patterns*, which become prominent after the stone has been cut, polished, and etched with acid. But there also exist iron meteorites without such patterns. Details can be found in the technical literature, in particular in the book *Kleine Meteoritenkunde* by Heide [21.10]. Other useful information, including new results of recent research, can be found in Hoppe [21.11].

21.4 Orbit Determinations of Meteoroids

The spatial path of a meteoroid is always a conic section with a focus at the Sun. As the meteoroid approaches the Earth, this path is converted into a hyperbola with a focus at the Earth's center. The direction of motion naturally approaches the zenith, that is to say, the meteor falls more steeply than it would in the absence of Earth's gravity. The direction from which the meteor comes, i.e., its radiant, thus suffers a *zenith attraction*.

Only a small portion of the hyperbolic orbit, with a focus at the Earth's center, is actually seen. This portion can in practice be approximated by a straight segment and thus a circular segment on the celestial sphere. The task of orbit determination is to reconstruct from observations of these orbit segments the original orbit in space.

21.4.1 The Path Within the Atmosphere

As a first step in calculating the orbit of a meteoroid, the geometric elements of motion within the atmosphere need to be found. The necessary formulae are presented here without proof; they have been taken from Bauschinger [21.12], who provides the rigorous mathematical derivations. The following mathematical quantities are introduced:

ρ = radius of Earth,
λ, φ = geographic longitude and latitude of the observer,
h, A = apparent height and azimuth of the endpoint of the orbit,
α, δ = right ascension and declination of the endpoint,
α', δ' = right ascension and declination of any other point of the orbit,
z = linear height of the endpoint above the ground,
A_0, D_0 = right ascension and declination of the radiant.

For the measurements from every observing site, the auxiliary quantities J, N, K, and i are computed from the relations

$$\begin{aligned} \sin J \sin(\lambda - N) &= \sin A \sin \varphi, \\ \sin J \cos(\lambda - N) &= \cos A, \\ \cos J &= \sin A \cos \varphi, \\ \tan i \sin(\alpha' - K) &= \tan \delta', \\ \tan i \cos(\alpha' - K) &= \frac{\tan \delta - \tan \delta' \cos(\alpha - \alpha')}{\sin(\alpha - \alpha')}. \end{aligned} \quad (21.1)$$

Then, the unknowns x, y, X, and Y are determined using

$$\begin{aligned} x \sin N \sin J - y \cos N \sin J + \cos J &= 0, \\ X \sin K \sin i - Y \cos K \sin i + \cos i &= 0. \end{aligned} \quad (21.2)$$

Data from the two sites give the unknowns from two pairs each of Eqs. (21.2); if data are available from more than two observing sites, the most probable values of the unknowns can be computed by the method of least squares.

The following formulae are then evaluated:

$$\cot \chi \cos \eta = x, \qquad \cot D_0 \cos A_0 = X,$$
$$\cot \chi \sin \eta = y, \qquad \cot D_0 \sin A_0 = Y,$$

$$\sin^2 \frac{s}{2} = \sin^2 \frac{\varphi - \chi}{2} + \cos \varphi \cos \chi \sin^2 \frac{\lambda - \eta}{2}, \qquad (21.3)$$

$$z = 2\rho \sin \frac{s}{2} \frac{\sin(h+s/2)}{\cos(h+s)},$$

and the coordinates A_0, D_0 of the radiant and the height z above ground for the termination point of the orbit are found. The linear length of the orbit segment for the two points observed is arrived at using three auxilliary quantities σ, τ, and d and the relations

$$\cos \sigma = \sin \delta \sin \delta' + \cos \delta \cos \delta' \cos(\alpha - \alpha'),$$
$$\cos \tau = -\sin D_0 \sin \delta - \cos D_0 \cos \delta \cos(A_0 - \alpha),$$
$$(\rho + z)^2 = \rho^2 + d^2 + 2\rho d \sin h, \qquad (21.4)$$
$$l = \frac{d \sin \sigma}{\sin(\sigma + \tau)}.$$

Finally, the speed of the meteor is found as the quotient of the length l divided by the duration in which that segment was passed.

In many cases, the radiant can be adequately found by graphical analysis. The apparent paths as seen respectively from the two observing sites are traced out on a star map. The traces are extrapolated backwards until they intersect, the point of intersection being the radiant.

21.4.2 The Orbit in Space

The directly observed quantities relating to the motion of the meteor in the atmosphere refer to the path described under the additional influence of Earth's gravity. From it, the spatial orbit which the meteor followed while still at considerable distance from Earth can be deduced.

There are essentially two influences to be considered, namely the elevation of the apparent radiant by zenith attraction, and the increase of the linear speed. There is also a shift of the radiant by diurnal aberration, but this effect is almost always negligible. Again, the relevant formulae are reported without proof. A detailed derivation is found in the book by Porter [21.13]. The following quantities (all in units of km s^{-1}) are introduced:

V = speed of meteor in its original orbit,
V_1 = original speed of meteor relative to Earth,
V_E = speed of Earth in its orbit,
V_2 = observed speed of meteor in the atmosphere.

The laws of celestial mechanics show that

$$V_1^2 = V_2^2 - 125. \tag{21.5}$$

Thus, V_1 can be found. This serves to compute the amount of zenith attraction of the radiant:

$$\tan \frac{1}{2} \Delta \zeta = \frac{V_2 - V_1}{V_2 + V_1} \tan \frac{1}{2} \zeta \tag{21.6}$$

This gives the amount by which the observed zenith distance of the radiant ζ must be increased. The radiant thus corrected is still only an apparent one as it corresponds to the direction of motion of the meteor relative to the Earth, and not of absolute motion of the meteor. The book by Porter [21.13] should be consulted for instructions on finding the true radiant. To determine the original speed V, the simple relation

$$V^2 = V_1^2 + V_E^2 - 2V_1 V_E \cos\beta \cos(\lambda - \lambda_A) \tag{21.7}$$

can be used, where λ, β are the ecliptic coordinates of the apparent radiant corrected for zenith attraction and λ_A is the ecliptic longitude of the apex. It is linked with the ecliptic longitude $\lambda_\odot$ of the Sun by the relation

$$\lambda_A = \lambda_\odot - 90° + 57\overset{m}{.}6 \sin(\lambda_\odot - 102°\!\!.2). \tag{21.8}$$

From the heliocentric velocity of the meteor thus found, the semimajor axis of the orbit can be derived using the well-known relation,

$$V^2 = k^2 \left(\frac{2}{r} - \frac{1}{a} \right), \tag{21.9}$$

so that a can be found from V. r is the Earth–Sun distance, which, with adequate precision, can be taken as constant. If r and a are expressed in astronomical units (AU), and k^2 is set equal to 1, then V is found in units of the Earth's mean orbital speed, which is 29.8 km s^{-1}. In the parabolic limit ($a \to \infty$) this equation also gives the speed of $\sqrt{2}\, V_E = 42$ km s^{-1} as was mentioned in Sect. 21.3.1.

A numerical example of the determination of an orbit has been published by Schmitz [21.14]. Suggestions on the visualization of orbits of meteor showers and the comets which spawn them are given by Zeuner [21.15].

References

21.1 Hughes, D.W.: *Vistas in Astronomy* **26**, 325 (1982).
21.2 Whipple, F.: *Sky & Telescope* **8** (1949).
21.3 Lovell, A.C.B.: *Meteor Astronomy*, Oxford U. Press, Oxford 1954.
21.4 Schippke, W.: *Sterne und Weltraum* **20**, 287 (1981).
21.5 Hapgood, M., Rothwell, F., and Royrwik, O: *Monthly Notices of the Royal Astronomical Society* **201**, 569 (1982).
21.6 *Sterne und Weltraum* **14**, 247 (1975).
21.7 *Sky and Telescope* **52**, 391 (1976).
21.8 Berwerth, *Ann. Naturhist, Hofmuseum Wien* 1913.
21.9 Knöfel, A.: *Die Sterne* **61**, 356 (1985).

21.10 Heide, F.: *Kleine Meteoritenkunde*, Springer, Berlin 1934.
21.11 Hoppe, G.: *Die Sterne* **58**, 352 (1982).
21.12 Bauschinger, J.: *Die Bahnbestimmung der Himmelskörper* (2nd edn.), Leipzig 1928, p. 587.
21.13 Porter, J.G.: *Comets and Meteor Streams*, London 1952, p. 81.
21.14 Schmitz, B.: *Sterne und Weltraum* **18**, 9 (1979).
21.15 Zeuner, H.: *Sterne und Weltraum* **23**, 256 (1984).

Translators' Note: For general reading on meteors and bolides, a list of monographs in English is provided in the Supplemental Reading List in this volume.

22 Noctilucent Clouds, Polar Aurorae, and the Zodiacal Light

Ch. Leinert

The extended celestial phenomena which are the subject of this chapter assume a somewhat ambivalent status in the context of the present book. On the one hand, they are all observable with the naked eye, and their beauty makes a lasting impression; it is supremely rewarding to have successfully observed or photographed them. On the other hand, at mid-latitudes they are observed either with difficulty or only seldom. Moreover, their investigation often requires elaborate tools of space research which are not as a rule accessible to the amateur observer. Thus in the following, the emphasis will be upon the description and—if possible—on the explanation of the phenomena. The sequence followed in the title has, as is easily seen, a direction: from an atmospheric phenomenon to one anchored in the ionosphere and magnetosphere and then into the interplanetary domain. The events and their physical conditions, however, are so different that their presentations belong, in principle, under different headings. Each section can be read and the ideas utilized independently of the others.

22.1 Noctilucent Clouds

On summer evenings at high geographic latitudes, bright silvery-white clouds with a cirrus-like appearance are occasionally seen above the northern horizon (Figs. 22.1–22.3) between the end of civil and astronomical twilights. Their occurrence at this late hour gave them the name "noctilucent," which is somewhat misleading, as they are not visible during the late evening and night. These tenuous formations, which diminish the light of stars shining through them by scarcely 1%, represent by far the highest clouds known to occur in the Earth's atmosphere.

22.1.1 Early Observations

The first recorded sighting of noctilucent clouds took place on 1885 June 08 by T.W. Backhouse in Kissingen, Germany. It was not mere coincidence that this discovery occurred just after the gigantic eruption of the Krakatoa volcano in 1883, in which enormous quantities of dust were ejected into the stratosphere and distributed around the globe within just a few weeks. This resulted in spectacularly bright and colorful twilight phenomena on Europe and other continents, which led to an increase in observations of the sky. Systematic investigations began as early as the year of discovery by O. Jesse (Berlin) and by Tserasky (Russia), who determined, among other things, the heights of the noctilucent clouds from simultaneous observations at different sites. The parallactic displacements gave the surprisingly high value of 75 km, at which altitude the atmospheric pressure is only 10^{-5} that of the surface, and therefore no meteorological phenomena would have been expected. It was soon found that noctilucent clouds are seen in both hemispheres of the Earth and occur only in the respective summer seasons, that their light is due to reflected sunlight, and also that many more observations would be needed to fully understand the phenomenon. Thus, in 1886 Jesse [22.1] made an appeal for all astronomers to participate observationally in the study of these "brilliant celestial features." Indeed, research on noctilucent clouds has continued to the present [22.2–4].

22.1.2 Characteristic Features

Noctilucent clouds occur at an average altitude of about 82 km. Compared with their horizontal extent of 100 to 1000 km, their effective thickness between 0.5 and 2 km is surprisingly small. As the determination of height shows, they originate in the so-called *mesopause*, the coldest layer of the Earth's atmosphere (see Fig. 22.8). They are rather transient, lasting usually just a few hours, but occasionally a few days. They consist of particles with diameters of about 0.2 μm, and which scatter light of short wavelengths much more strongly than that of long wavelengths. This explains the bluish-white color of these clouds. Their density of 0.1–10 cm^{-3} is much too low to significantly weaken the light which passes through them; the optical density for vertically passing light is only 10^{-3}–10^{-5} in the visual range. The temperature of the clouds is around 135 K ($-140°$C). It is undisputed that the clouds consist predominantly of ice particles which form via condensation at the position of the cloud itself. A direct determination of their composition, however, is still lacking. Experiments performed from rockets have not thus far totally resolved the question, but they do suggest that particles of the expected size of volatile substances must be present [22.4]. Thus the nature of the condensation nuclei which initiate the growth of the ice particles is still unclear. These clouds are not stable; the ice particles sink by a few cm s^{-1} while still growing, and reach, after about 1 day, warmer layers where they evaporate. This in part explains their changing appearance, which is contributed to by winds aloft propeling the clouds by up to 100 m s^{-1}. The separate patterns and waves in noctilucent clouds have sizes of from 1 to 10 km. Most of these events are not particularly prominent (cf. Figs. 22.1–22.3), but they can be bright enough to cast shadows and even permit the reading of large print outdoors. An excellent and detailed review on observational results is given by Fogle and Haurwitz [22.3].

22.1.3 Classification of Types

On the basis of appearance, noctilucent clouds are classed into four main types: *veils* (Type I), *bands* (Type II), *billows* (Type III), and *whirls* (Type IV). These terms are explanatory enough not to require systematic comment. A cloud usually shows several of the physical shapes simultaneously; e.g., waves resembling fish bones often appear across bands. In the photographs shown here, the bands dominate in Fig. 22.1, the

Fig. 22.1. Noctilucent clouds over Stockholm on 1966 July 20 at 0^h55^m CET. Photographed with 1-s exposure time and aperture ratio $f/1.8$ on AGFA CT 18. Printed by kind permission of Nathan Wilhelm, Stockholm.

Fig. 22.2. Noctilucent clouds in the form of a veil. Photographed on 1986 July 14 in southern Sweden. By kind permission of Nathan Wilhelm, Stockholm.

Fig. 22.3. Noctilucent clouds with wave structure over Edinburgh, photographed on 1986 July 23/24. By kind permission of David Gavine, Edinburgh.

veils in Fig. 22.2, and waves in Fig. 22.3. Veils are the simplest type of noctilucent cloud, usually showing distinct contour shapes about one-half hour after they first appear.

22.1.4 Visibility

To see noctilucent clouds one must be in the right place at the right time, and this depends upon both the formation of the clouds and the geometry of observation. The latter is represented in Fig. 22.4. The noctilucent clouds are far too weak to be visible in the daytime. Not until twilight is far advanced, with the Sun at least 6° below the horizon, will they stand out distinctly against the now darkened sky. Thus, these clouds can in principle appear along the arc HNW, which at the beginning may still extend past the zenith of the observer. Below point L the sky is still too bright, which leaves the arc LNW. This range slowly follows the sinking Sun toward the horizon until, with the Sun 16° below the horizon, sunlight has been cut off from the layer at which the noctilucent clouds occur. The most favorable conditions (i.e., sky already dark and visibility range still large) occur for positions of the Sun 9° to 14° below the horizon. The clouds thus seen are a few hundred kilometers distant.

The visibility of noctilucent clouds depends strongly on the geographic latitude. They have never been seen north of 77° (Greenland), nor south of 42° (New York state). As Fig. 22.5 shows, the chances of seeing such clouds increase for latitudes

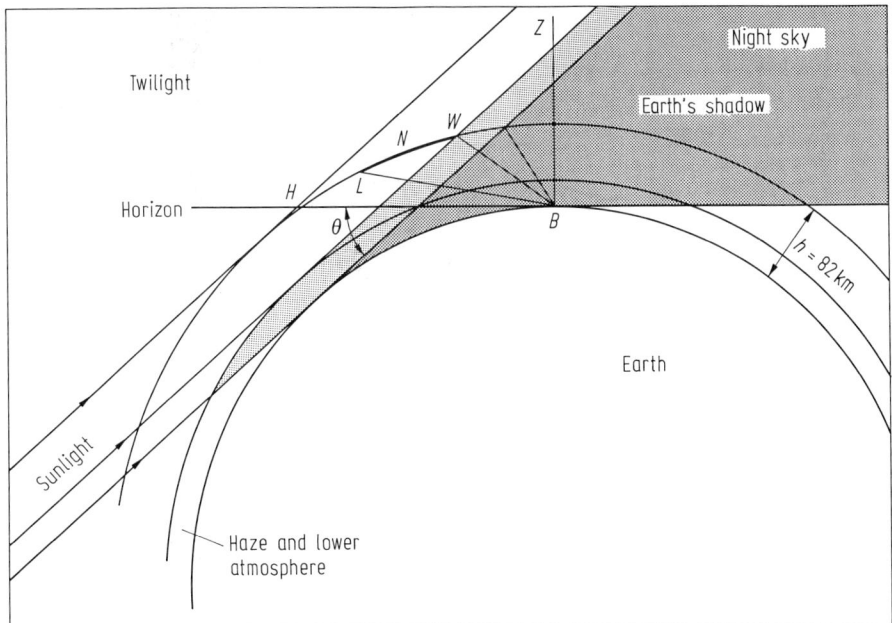

Fig. 22.4. Geometry of the observation of noctilucent clouds. The letters refer to the following: *B* observer, *Z* zenith, *W* beginning of obscuration of cloud layer by haze and the lower atmosphere, θ depth of Sun below the horizon, and *LNW* range in which noctilucent clouds can be seen.

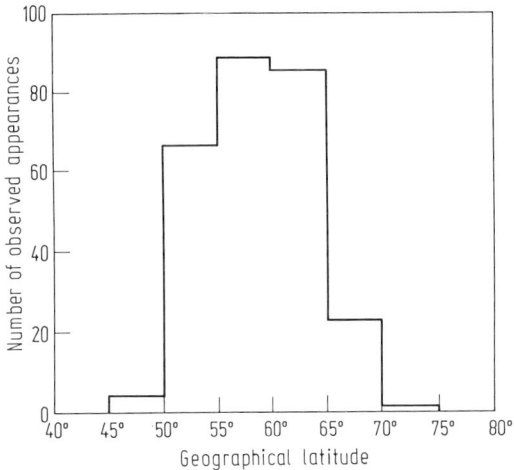

Fig. 22.5. Frequency of noctilucent clouds observed up to 1965 over North America versus geographic latitude.

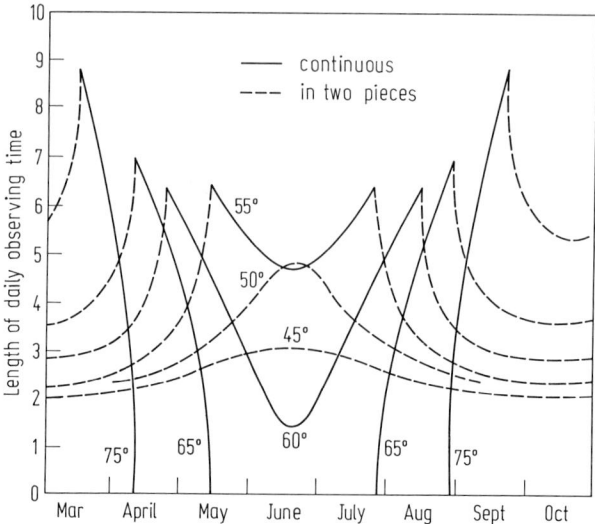

Fig. 22.6. Number of hours per night where the position of the Sun permits the observation of noctilucent clouds. When the Sun at night remains less than 16° below the horizon, the times of evening and morning visibility merge into one interval.

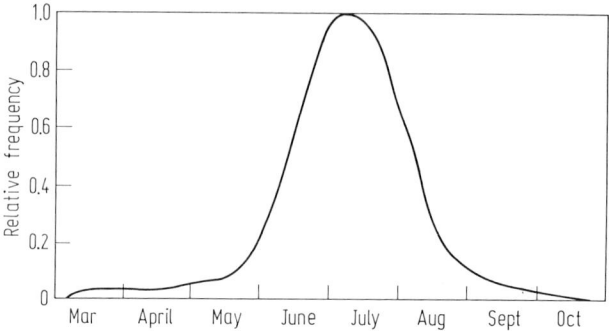

Fig. 22.7. Frequency of observation of noctilucent clouds during various seasons in the northern hemisphere. The maximum in early July is normalized to unity.

over 50°. The number of actual observations then distinctly increases northward, indicating the connection of the phenomenon with the polar region. The strong decline of observation frequency north of 65° is caused by the here less favorable observing conditions. The duration of visibility also strongly depends on the latitude. Figure 22.6 shows that at about 55° north latitude (e.g., southern Sweden or northern Scotland), up to 6 hours of twilight can occur during which noctilucent clouds may be seen, while at higher latitudes, summer observations are impossible because of the "midnight sun." In general, however, the conditions for seeing noctilucent clouds are best during the summer months. This is most fortuitous, as summer is precisely the time at which such clouds occur (Fig. 22.7). During the first half of July, they can be expected on more than half of clear nights at northern latitudes 55° to 60°. They do not occur during the winter. These relations hold for the corresponding latitudes in the southern hemisphere.

Noctilucent clouds are not to be confused with the similarly rare *nacreous clouds*, which also occur at high latitudes, but during winter and at distinctly lower heights of about 20 to 30 km. Their shape is more compact and lenticular.

22.1.5 Their Origin

The key to understanding the nature of noctilucent clouds lies in the temperature distribution of the atmosphere (Fig. 22.8). The lowest temperature occurs at a height of around 80 km. This layer is the mesopause, and is exactly the level at which noctilucent clouds are formed. From 1956 to 1958, during the International Geophysical Year (IGY), the temperature of the mesopause was more precisely fixed by rocket experiments. It was found, for instance, that over the launch site, Fort Churchill (Canada, 59°N), the temperature of the mesopause varied from an average of 230 K (−43°C) during the winter to 170 K (−103°C) during summer and frequently dropped to even below 140 K [22.3]. At this last-mentioned temperature, water vapor is supersaturated one-hundredfold and condenses quickly with about one dozen condensation nuclei (each at least 1 nm in size) per cubic cm. It was first believed that micrometeorites served as condensation nuclei, but G. Witt showed with rocket experiments that their frequency at the mesopause level is far too low. He suggested instead in 1969 that large molecule ions of the type $H^+ (H_2O)_n$ ($n \geq 10$) are the condensation nuclei. D.M. Hunten, R.P. Turco, and O.B. Toon then also suggested in 1980 that the very finest dust particles, figuratively called "smoke," may be responsible as they originate when the outer layers of meteorites evaporate during the high-speed penetration of the atmosphere. Condensation onto such "smoke" would occur so fast that ice particles would grow to some tenths of a μm before they could sink from the cold zone and evaporate again. The amount of water vapor, although at mesopause level comprising only a few ppm (parts per million) of the total atmosphere, is just adequate to produce noctilucent clouds at the observed density. A slight upwind speed of 1 cm s^{-1} or turbulence with similar strength would increase the lifetime of the clouds to over one day [22.5]. Thus, a fairly complete picture of their origin emerges, although the detailed processes are still under discussion. Of interest, for instance, is the relationship of the clouds with other events in the atmosphere. For instance, the "smoke" particles would become negatively charged and could, if so abundant as is assumed, play a role in various chemical reactions of the atmosphere, including the ozone budget. If, however, the mechanism of origin of these clouds is so widespread, why then are they such isolated and transient events? The answer lies probably in the limited optical conditions for visibility.

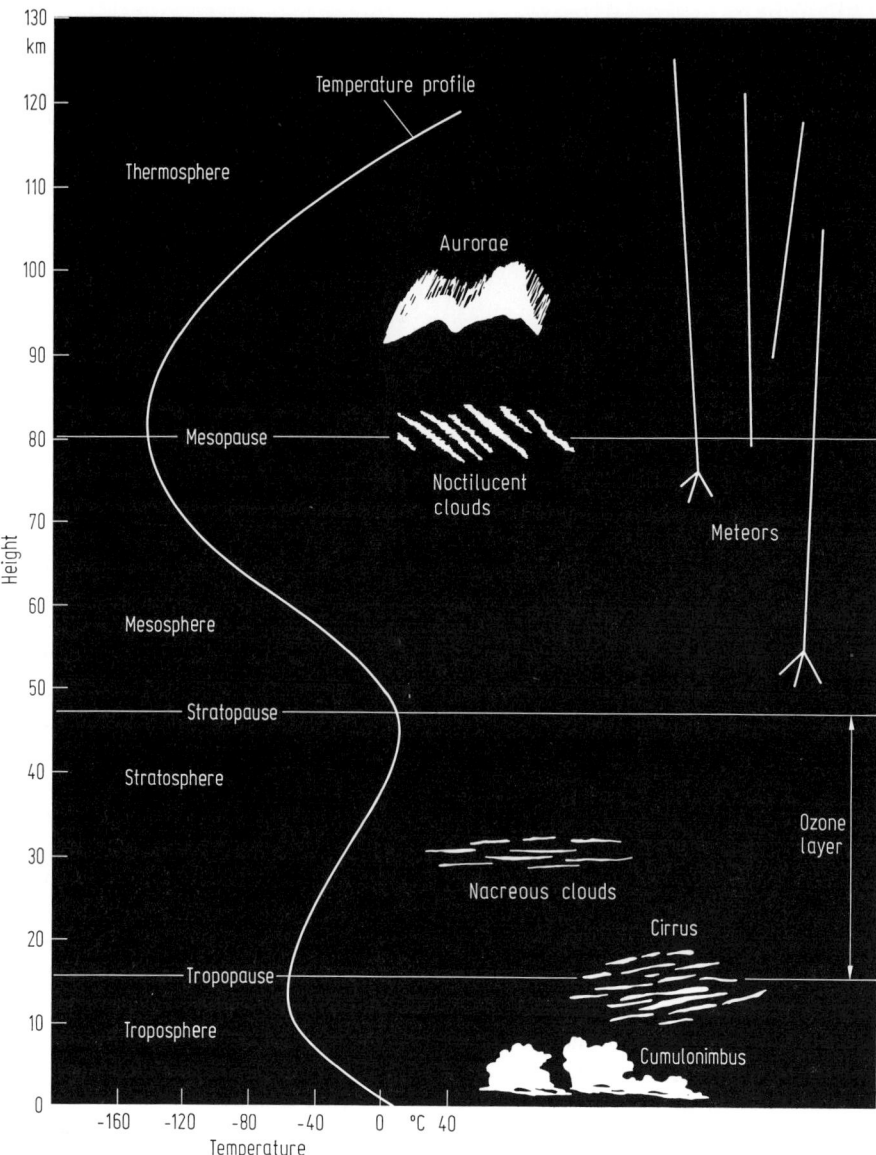

Fig. 22.8. The locations of the noctilucent clouds and aurorae in the atmosphere. The atmosphere is divided into various zones according to temperature characteristics: in the troposphere and the mesosphere, the temperature decreases with altitude, while in the stratosphere and thermosphere, temperature increases. The respective boundaries between layers are referred to as "pauses." The lowest temperature occurs at the mesopause.

22.1.6 The Larger Picture

Satellite measurements in the 1970s and 1980s (Orbiting Geophysical Observatory *OGO-6* [22.4] and *Solar Mesosphere Explorer* [22.6]) revealed that the polar caps of the Earth are often overlaid by a fine cloud layer at high altitude during their respective summers. The frequency of occurrence of this layer was found to increase dramatically with the geographic latitude, from 20% at 55° to over 40% at 65°, and even to 80% at 75°, so that in the polar regions the layer may be a permanent feature. The height of the cloud layer of 82 km, its thickness of less than 3.5 km, the particle size of maximally 0.14 μm, and the density of about 100 particles per cm^3 [22.6] agree so well with the values obtained for noctilucent clouds that it is probably one and the same phenomenon. The noctilucent clouds are therefore nothing more than the visible "frayed-out" edge of the summer mesospheric polar caps. The greater part of the phenomenon thus remains hidden to the ground-based observer in the brightness of white nights and of the midnight sun.

22.1.7 Future Observations

In order to understand these relations and the underlying atmospheric processes, a large-scale program is planned for the 1990s under the name Super-CAMP (*C*old *A*rctic *M*esopause *P*roject), which will include measurements taken from airplanes, rockets, and satellites. Can the amateur still make worthwhile contributions under these conditions? The author thinks so, since every picture of a noctilucent cloud described with time, place, and direction of the photograph documents the presence of the particular circumstances required to form ice particles in the mesopause. Such observations, when performed systematically over a long time base, may provide a meaningful supplement to professional research, which is geared to perform targeted deeper studies at specific times and places.

The British Astronomical Association has recently taken charge of the collecting of such photographs. Prospective participants may contact Dr. David Gavine, 20 Coillesdene Crescent, Joppa, Edinburgh EH15 2JJ, Scotland, for information on the observations required and for collection of observing reports. The primary motivation for conducting one's own observations, however, is the personal satisfaction gained in studying these fascinating phenomena.

22.1.8 Comment on the Literature

A very readable account of the subject is given by McDonnell [22.7]. An understandable compilation of the facts is in Fogle and Haurwitz [22.3], and the link to the technical literature in Gadsden [22.4], the latter of which does not yet include the papers of Turco et al. [22.5] and Thomas [22.6].

22.2 Aurorae

According to reports by numerous observers, the sporadic event known as an *aurora* is the most magnificent phenomenon to be witnessed in the night sky with the unaided eye. The coloring and richness of shapes it presents are indicated in Figs. 22.9 to 22.11, which of course cannot reproduce the distinct variability and the often rapid motion effects. The phenomenon is seen primarily during winter and in sparsely populated areas, a fact which accentuates its peculiar character and intensifies the impression it confers. Aurorae occur most frequently at high latitudes, usually lasting from one to two hours. In the middle and southern parts of Europe and of North America, it is only rarely seen. Overall, based on a compilation over the interval 1880–1964 by W. Schröder, an average of only two observing reports were found per year. The phenomenon appeared more than 10 times each in 12 different years, but not at all in 27 different years. The typical appearance of a polar aurora in middle latitudes is different from what is shown in Figs. 22.9 and 22.10. The phenomenon is more diffuse and the dominant color is red. The polar aurora occurs in the same fashion in northern and southern polar regions, often simultaneously. The use of the archaic (but nevertheless still-used) term "northern lights" will be avoided the following discussion.

22.2.1 Explanation

Attempts to explain the aurorae are diverse and as old as the handed-down observations of the phenomenon, which date back to antiquity [22.8,9]. In this regard, the first scientific explanation was the theory proposed in 1708 by Suno Arnelius of Sweden, which held that the auroral light was sunlight reflected by ice particles high in the atmosphere. As it turned out (Sect. 22.1), this is actually the explanation of noctilucent clouds.

With current understanding, the occurrence of aurorae is in essence always linked to disturbances of the geomagnetic field, as is shown, for instance, by magnetic deflections of up to $10°$ on the compass. The currents of up to a million amperes which cause this effect flow not only through the ionosphere but also along the geomagnetic field lines between ionosphere and outer magnetosphere. Figure 22.12 illustrates how such a circuit could originate. The existence of such field-parallel currents were postulated as early as 1908 by the Norwegian researcher Kristian Birkeland, but was generally accepted only after direct satellite measurements. The flux of current itself does not suffice to cause an aurora. That requires particular circumstances leading to a strong acceleration of electrons and ions within the circuit at a height of several thousand kilometers, and energies of up to about 10 KeV. Where the current is directed outward, the electrons move downward and penetrate into the atmosphere to about 100 km above the surface, ionizing the atoms and molecules which inhabit the high atmosphere and exciting them to luminescence, which is then seen as the aurora.

The events surrounding the formation of aurorae are often illustrated by comparison with a television tube. The screen shows the image, which, however, is generated by an electron beam which has been accelerated to energies over 20 KeV and which is,

Fig. 22.9. Spiral shaped aurora, photographed on 1985 February 17/18 in Kiruna, Sweden. In the foreground are the city lights, in the background a band of clouds covers part of the aurora, and at the left edge is the Moon. Exposure time was 10 s with an aperture ratio $f/2$ and 24 DIN. By kind permission of Jacek Stegman, Stockholm.

Fig. 22.10. Polar aurora with distinct ray structure. The rays often move so quickly that they are difficult to display photographically. By kind permission of Klaus Rinnert, Lindau, Germany.

Fig. 22.11. Arc of an aurora over the moonlit Earth in the region of the tropic of Capricorn. The horizon appears doubled, the upper yellowish line originates by luminescence, called *airglow*, in the high atmosphere at an altitude of 90 to 100 km. The auroral arc, about 1000 km long, is distinctly higher, and assumes in its upper region a reddish coloring. Stars are visible close to the true Earth horizon. NASA photograph made on *Spacelab 3* by Don L. Lind.

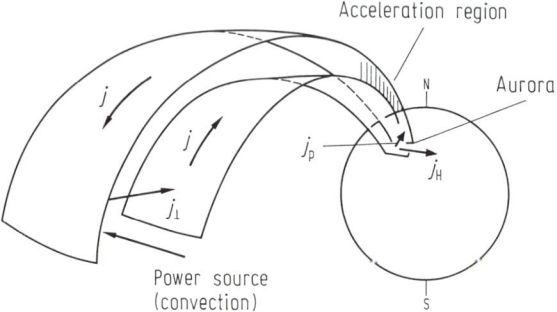

Fig. 22.12. Geometry of the electric circuit formed by magnetosphere and ionosphere. The field-parallel currents are completed to a circuit by a current directed perpendicular to the magnetic field in the magnetosphere, and by the *Pederson current* j_P in the ionosphere. There, another current perpendicular to j_P is generated. An aurora can form below the northern layer of current.

at some distance away, deflected and modulated in direction and intensity. In the case of the aurora, the formation and change of the "image" is generated in the acceleration region a few thousand km above ground. The high atmosphere with its luminescence merely reflects the complex structures and changes within the region of acceleration. Being a part of the electric circuit, however, the luminescent high atmosphere can feed back onto the upper regions, which changes the "program" and may, for instance, cause pulsations. The rays which often appear originate from electrons forced, as electrically charged particles, to follow the geomagnetic field lines. This also accounts for the distinct "drapery" structure of auroral events. The origin of an aurora will be returned to after its properties have been described.

22.2.2 Apparent Shapes

Aiming at a complete classification of the diverse auroral shapes which appear, the *International Auroral Atlas* [22.10] appeared in 1963. Visual observations have since steadily declined in importance for aurora research, and the following simplified divisions according to shapes and colors will suffice.

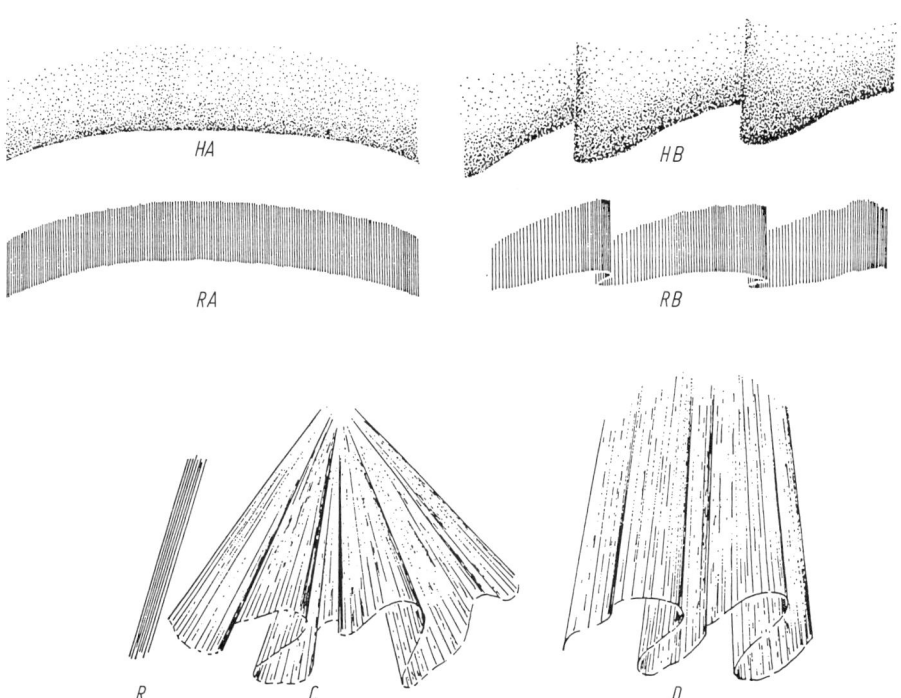

Fig. 22.13. Typical shapes of aurorae. HA: homogeneous arc; RA: rayed arc; HB: homogeneous band; RB: rayed band; R: rays; C: corona; D: drapery.

Table 22.1. Classification of aurorae according to color.

Type	Description	Base Height (km)
A	red	250
B	reddish border	80–100
C	greenish or whitish	100

The diffuse light is called a *patch* or *veil*, depending on its areal extent. The more pronounced shapes are compiled in Figure 22.13. Homogeneous arcs and bands are the most quiescent features, the others showing larger variations in brightness and more rapid motions. During one auroral phenomenon, several of the shapes mentioned may occur simultaneously or change from one into another. The most frequent colorings are listed in Table 22.1, the greenish-white coloring being the normal case. The connection between color and height of the aurora can be understood from its spectrum.

22.2.3 Spectrum, Color, and Brightness

The major constituents of the high atmosphere are atomic oxygen, molecular nitrogen, and molecular oxygen. The first two essentially determine the spectrum of the aurora (Fig. 22.14). The excitation of atoms and molecules does not take place in one step. The incident electrons yield their energy by ionizing oxygen and nitrogen, and by

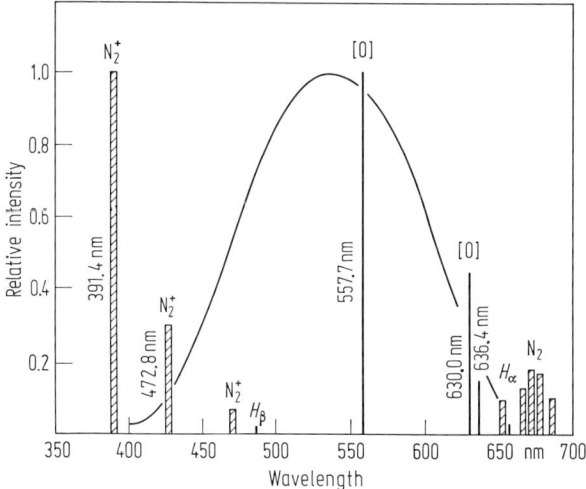

Fig. 22.14. Spectrum of an aurora in the visible range showing only the most prominent emissions. The *square brackets* indicate forbidden lines, while molecular bands are *hatched*. The *solid curve* indicates the sensitivity of the eye.

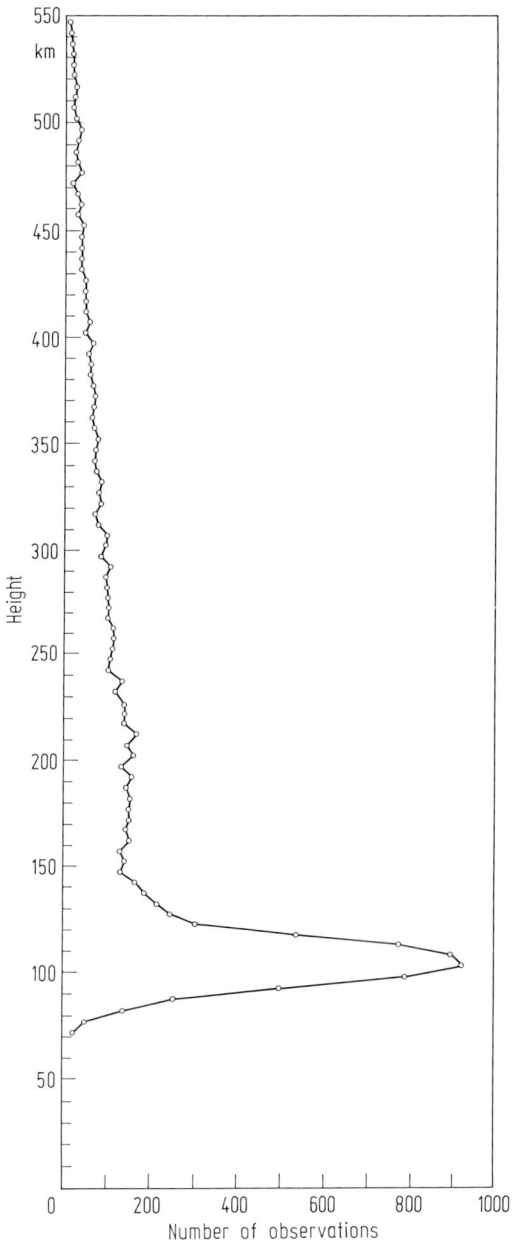

Fig. 22.15. Distribution by height of over 10 000 aurora phenomena, as measured by C.F.M. Störmer and associates.

generating secondary electrons which in turn can ionize more atoms. The ions may exchange charge with other molecules or be neutralized by electron capture. The excitation causing the auroral emissions is almost exclusively through the collision of an air molecule with one of the numerous secondary electrons. The lines of atomic oxygen dominating the spectrum are *forbidden*, which means that the excited atom requires an appreciable length of time before emitting the light and thus will do so only when not colliding during that period with other molecules, as this would cost it the excitation energy. For the red lines at 630 nm, this time interval is 110 s, but collisions will occur with this frequency at a height of 250 km, and therefore below this level the line emission is increasingly inhibited by collisions. For the green line at 557.7 nm, the time interval is 0.74 s, and the weakening occurs below only 95 km. In this range, the nitrogen band at 650 nm will dominate the color, which otherwise is outshown by the green line. This accounts for the red fringe of Type B aurorae. Similarly, the red color of Type A occurs because at high altitudes the red hydrogen dominates and is not damped by collisions.

Aurorae are observed over a large range of heights, between slightly below 100 km to about 600 km, but are mostly in the range 100 to 150 km (Fig. 22.15). The lower edge is determined by the energy of incident electrons: the lower the boundary, the higher the energy (Fig. 22.16). Thus, a simple view of the color of the aurora tells crudely how much acceleration is experienced by the exciting particles, even though they cannot be seen. A white auroral color may originate, for instance, if its intensity is too low for distinct colors to be perceived. An alternative explanation is the superposition of ordinary scattered blue light from the sky. Also, the *permitted*

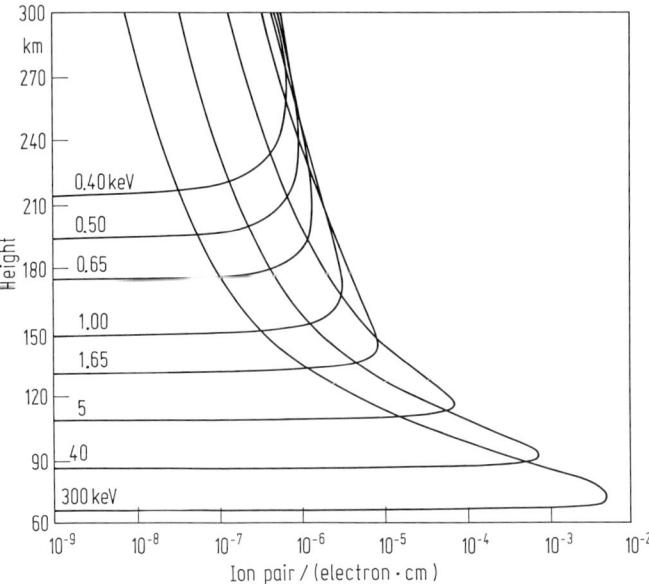

Fig. 22.16. Height profile of secondary electrons generated by incident high-energy electrons as graphed against original energy.

(i.e., not forbidden) lines of N_2^+ and N_2 are enhanced by direct excitation by sunlight. These lines produce a bluish-white coloring.

Also, ions, particularly those of hydrogen, are accelerated and contribute to the aurora by infringing on the denser atmospheric layers; in this case, the hydrogen lines in the spectrum are more distinctly pronounced. This light, however, is faint and diffuse and barely detectable by the unaided eye.

The brightness of aurorae varies over a wide range. The faintest events are below the threshold of the naked eye, while the brightest illuminate the ground with the intensity of moonlight. Nevertheless, an intensity ten times fainter than the latter is still considered quite strong. The illuminance of aurorae is comparable with the brightest noctilucent clouds observed thus far. In quantitative terms, the resulting illuminance of the ground is then 0.1 mW m^{-2} or nearly 0.1 lx; the brightness of the phenomenon itself is about 1 mW str^{-1}, 0.4 cd, or 100 kR (kilo-Rayleigh). The latter measure states that in a column with cross section 1 cm^2 drawn through the aurora, 10^{11} photons are emitted per second. These figures serve as a reminder that an aurora exacts large amounts of energy.

22.2.4 Observability

The likelihood of seeing an aurora is highest in a nearly circular region, or oval, about $20°$ around the geomagnetic pole (Fig. 22.17), where virtually every clear night provides an event. Towards the geographic pole, the frequency drops to about one-half, and the decline is much steeper toward the south, being reduced to one or two events per year at the latitudes of central Europe and central United States, and to only a few events per century in the Mediterranean and Caribbean regions. Among European countries, northern Norway has the best conditions, which illustrates the considerable contributions from that country to research on the aurora phenomenon. However, aurorae are visible only at night, which immediately excludes the summer months in that country. It is not expected that many readers will plan a trip into the polar regions specially to view aurorae, but if one has the opportunity on a flight via a northern route, Fig. 22.17 will be of some aid.

The geomagnetic pole at $78.5°$ N and $69°$ W in northwestern Greenland differs from the magnetic pole at $73.5°$ N and $100°$ W, as indicated by the needle of a compass. It is determined by the symmetry axis of the magnetic dipole of the Earth. The arrangement of auroral phenomena with repect to this axis shows their close linking with the magnetosphere. Their connection with solar activity is shown by the long-known correlation with the sunspot cycle (Fig. 22.18) and also in a periodicity with the (as seen from Earth) 27 day rotational period of the Sun. The highest incidence of aurorae occurs one to two years after sunspot maximum, and thus is expected to occur in the years around 1993 and 2004. Spectacular aurorae occur after strong solar flares. Moreover, the phenomenon can usually extends far south during such energetic events. For instance, R. Höper observed an aurora south of Munich on 1982 July 13 [22.18].

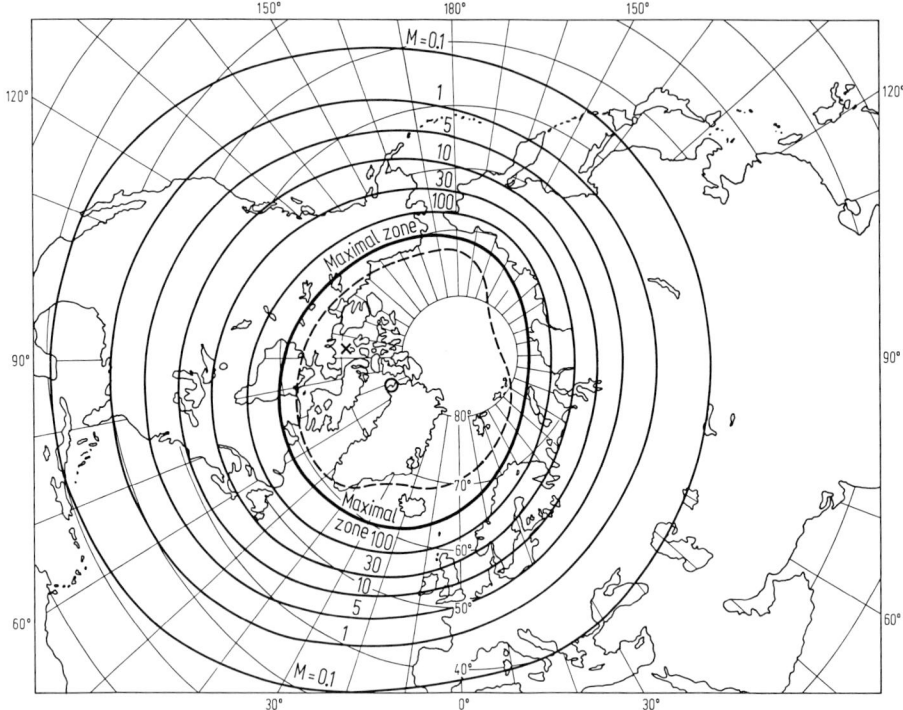

Fig. 22.17. Frequency of visibility of aurorae. The magnetic (×) and geomagnetic (○) poles are shown. After Hermann Fritz, Brockhaus, Leipzig 1881.

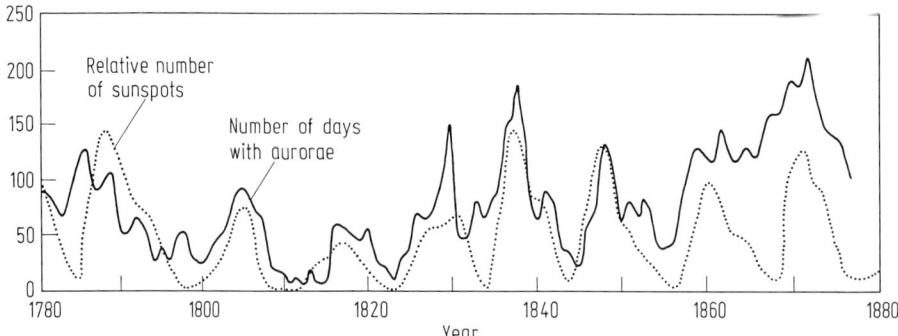

Fig. 22.18. Relation between incidence of aurorae and solar activity. After Sophus Tromholt (ca. 1880).

22.2.5 The Auroral Oval and the Geometry of the Magnetosphere

In 1982, the satellite *Dynamic Explorer I* obtained the first images of the polar region showing the *auroral oval* directly as a closed, circular ring of auroral events around the geomagnetic pole (Fig. 22.19). The auroral oval thus is more than merely a statistically characterized region; its locus is fixed by definite physical conditions. Such pictures also suggest that the cause for the sporadic occurrence of aurorae is not localized, but rather is expected to occur on a large scale. In order to understand what characterizes the auroral oval, the geometry of the Earth's magnetosphere will be considered, with Fig. 22.20 serving as an attempt at a three-dimensional presentation.

In a perfect vacuum, the geomagnetic field would display the typical dipole structure as shown by a bar magnet in in a classroom experiment. The solar wind, which

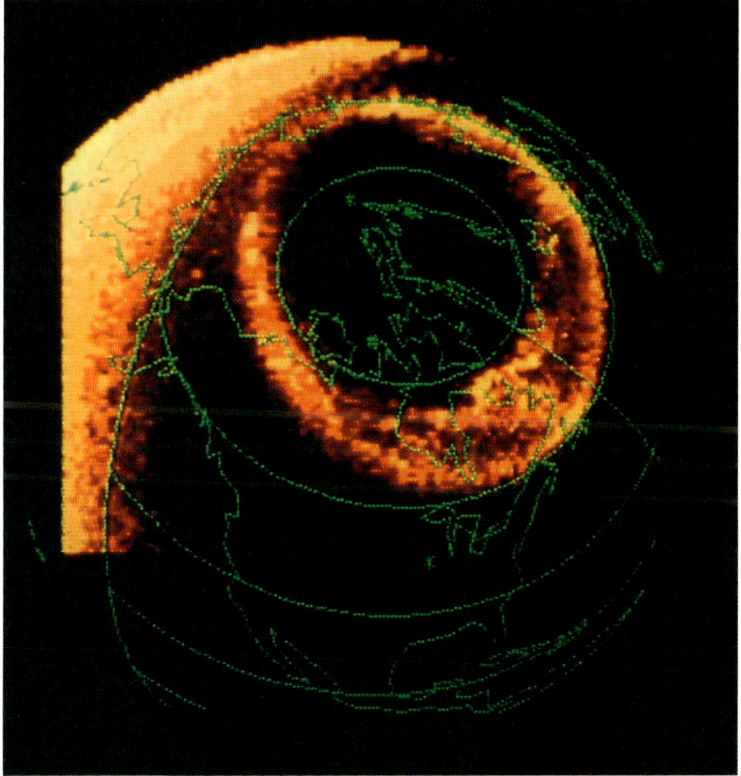

Fig. 22.19. Northern auroral oval, recorded on 1981 November 08, 03^h UT with the *Dynamic Explorer I* satellite. Geographic coast lines have been added for orientation to this false color picture. In the ultraviolet at 140 nm, where this image was made, the brightness of the auroral oval is comparable with that of the sunlit atmosphere, the primary contributions being the lines of neutral oxygen at 130.4 nm and 135.6 nm. The image was made 2 hours after the arrival of a shock wave in the solar wind. Printed by kind permission of L.A. Frank, Iowa City, USA.

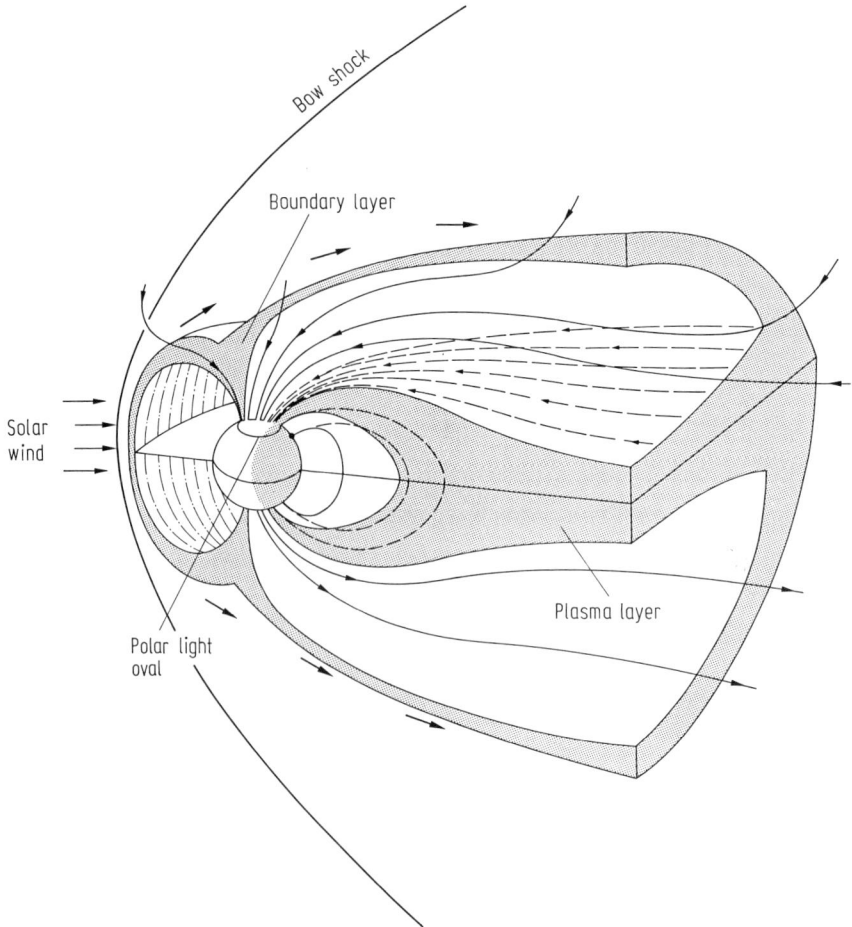

Fig. 22.20. Three-dimensional section through the magnetosphere, showing regions of origin of aurorae, the boundary layer, and the plasma layer in the tail.

arrives from the Sun at a speed of 400 km s^{-1}, is a plasma of electrons and protons which deforms the magnetosphere, compressing it on the daytime side. In addition, the field lines at high latitudes are no longer closed, but link with the magnetic field embedded in the solar wind, in particular when the latter is oriented toward south against the geomagnetic field. They are then dragged with the solar wind to the night side, forming a paraboloidal tail over 100 Earth radii long of "open" field lines. These lines extend, without linking with the Earth at their ends, into interplanetary space. Following the magnetic field lines of the boundary layer, which separates open and closed magnetic field regions, it is seen that their point of origin is in the daytime half of the auroral oval.

On the night side, a fraction of the electrons and protons moves under the influence of electric and magnetic forces toward the plane of symmetry, so that a zone of increased density is formed, the *plasma layer*. It again is a boundary layer in the sense that it separates the open magnetic field toward Earth in the north from the oppositely oriented field in the south. (The pole in the Earth's northern hemisphere is magnetically a south pole!) Following the field lines of that layer, their point of origin is on the night side of the oval. Thus, the auroral oval separates open and closed sections of the geomagnetic field, and connects the inner and outer boundary layers of the magnetosphere. A compensating current between these regions has to flow through the oval, since charged particles follow magnetic field lines. This is where the origin of aurorae is to be sought.

22.2.6 Origin

The magnetosphere is not in static equilibrium. Among other things, it is the scene of large-scale circulation of plasma and field lines. This "convection" is shown schematically in Fig. 22.21. As was mentioned in Sect. 22.2.5, the circulation starts at high latitudes with the pulling of magnetic field lines together with the surrounding plasma in the direction of the magnetic tail by the solar wind. Then, at a distance of several dozen Earth radii, opposing field lines approach each other so closely that they "reconnect." Figure 22.21 shows how field-free "neutral lines," two open field lines each with one open end in the Earth, are converted into one closed line and one open at both ends. What happens afterwards can be seen from the often useful analogy with rubber bands; two bands stretched between the thumb and index fingers of both hands, cut in the middle and reconnected by the parallel running piece in each case, make two slingshots. Correspondingly in the magnetosphere, plasma is flung from the neutral line into interplanetary space, as well as back to Earth at high speed (up to 10^3 km s^{-1}). This sends a strong current just outside the closed field lines into the auroral oval and into the daytime side of the magnetosphere. The process lasts from 1/2 to 2 hours. Part of the energy put into the magnetospheric deformation by the solar wind has thus been discharged as a "magnetic storm," and the circuit is closed. The almost weather-like variability of the energy source—which is the solar wind—is the reason that these discharges are of irregular and sporadic occurrence. Normally, the plasma of the magnetosphere is a very good conductor, but under extreme stresses of the discharge in a height of a few thousands of kilometers and for reasons not yet well understood, the conductivity collapses, and a voltage difference of many kV accumulates, which accelerates the electrons which ultimately produce the auroral light. Figure 22.21 shows that each field line along which the plasma flows back into the northern auroral oval has a symmetric counterpart in the southern oval. Auroral lights should therefore as a rule occur simultaneously in the northern and southern ovals as identical phenomena, but mirror imaged. This indeed is the case. Figure 22.22 compares the two photographs of such a pair.

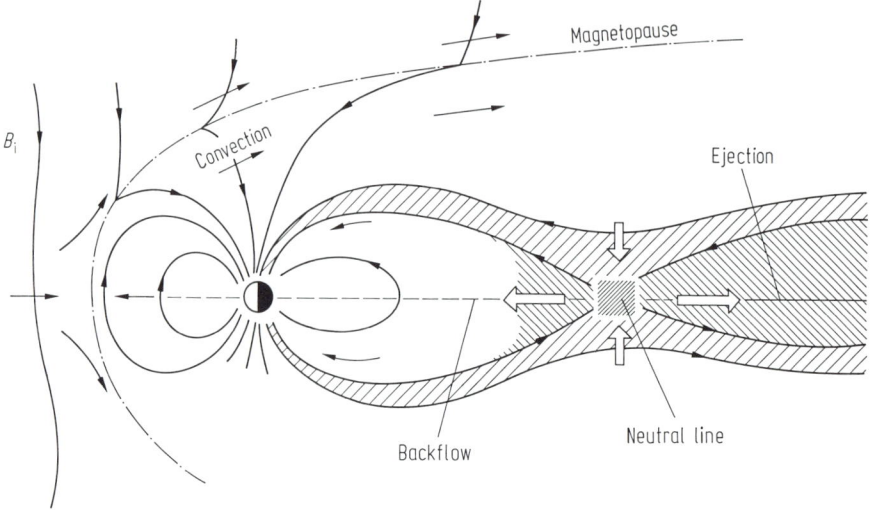

Fig. 22.21. Circulation of magnetic field and plasma in the magnetosphere (convection). B_i is the interplanetary magnetic field carried to the Earth by the solar wind. Connection with the geomagnetic field can occur easily only when the interplanetary field as shown is oriented toward the south.

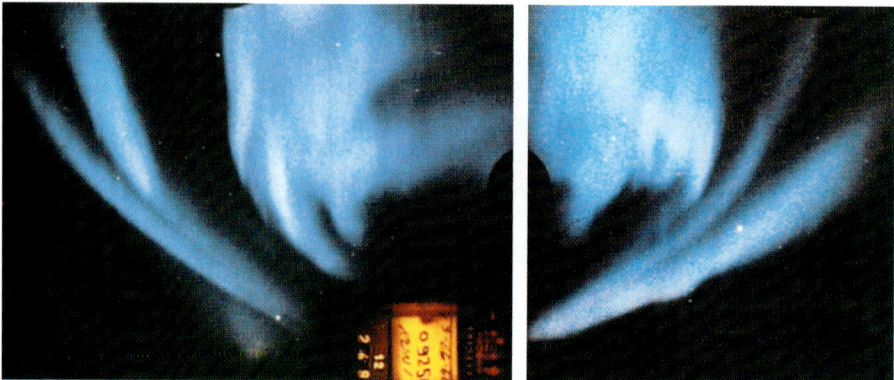

Fig. 22.22. Comparison of two photographs of aurorae simultaneously obtained at similar geomagnetic longitudes and latitudes, one from the Arctic, the other from the Antarctic. Note the mirrored, but almost identical, shape.

22.2.7 Accompanying Phenomena

Only a portion of the energy of a magnetic storm is used up in accelerating electrons, and only 3% of this energy reappears visibly in the aurora. Almost one-third goes into electromagnetic radiation of other kinds—X-rays, ultraviolet, infrared and radio—so that aurorae can also be studied in these regions. The major part of the electron energy serves to heat the conductor, which is comprised of the gases in the ionospheric layers. These currents also lead to the geomagnetic disturbances mentioned, hence the name "magnetic storm" to describe the entire phenomenon. (A magnetic storm lasts about one day—substantially longer than an aurora, and is triggered directly by high-energy particles streams from the Sun.) Substantial voltages are induced in long-distance cables which may play havoc with telegraphic transmissions and with digital data transmission. At the same time, potential differences build between various points on the Earth's surface, so that currents of up to 100 A may be fed into the electrical networks via the grounding points, causing potential blackouts. Even the durability of oil pipes can be distinctly reduced by such currents [22.19]. During a magnetic storm, the ionization of even the lowest ionospheric layer, the D-layer at a height of 80 km, may increase so much that short waves are reflected. In contrast to the tenuous E-layer, which is located 30 km higher, the denser air in the D-layer is a substantial absorber at radio wavelengths, and this may hamper or entirely disrupt radio transmissions. These events are often emphasized in order to highlight the significance of investigating solar activity and the solar wind.

22.2.8 Photographs

When photographing auroral features, it often suffices to use exposure times of a few seconds, so that even moving features can be clearly recorded; see Figs. 22.9 and 22.10. Readers who wish to contribute to statistical investigations of aurora by taking photographs may wish to contact the amateur astronomical organization in Finland, which maintains a systematic collection. The address is: Ursa Ry, Attn: Heikki Oja, Laivanvarustajankatu 9, SF-00140 Helsinki, Finland. Certainly, the observation of so marvelous an event as an aurora has its own reward apart from scientific uses.

22.2.9 Comment on the Literature

Historical surveys are given by Schröder [22.8,9], and Brekke [22.11], the latter richly illustrated work emphasizing Scandanavian events. Eather [22.12] gives a colorful introduction, while Haerendel [22.13], Hones [22.14], and Reddy [22.15] are successful in conveying the concepts of these fascinating as well as complex physically processes in an understandable manner; Vallance Jones [22.16] and Omholt [22.17] are compilatory technical books.

22.3 The Zodiacal Light

Under favorable viewing conditions, the faint, essentially colorless, pyramid-shaped *zodiacal light* may be seen (Fig. 22.23) above the western horizon in the evening about one hour after sunset, or above the eastern horizon in the morning about one hour before sunrise. As its name implies, the zodiacal light extends along the ecliptic, which is the projection of the Earth's orbital plane onto the celestial sphere. The

Fig. 22.23. The zodiacal light above the western horizon, as photographed on 1983 May 13, one hour after sunset from the peak of Mauna Kea. The setting crescent of the Moon in the evening twilight is 19° above the Sun. Venus, located at the top of the light cone, is 43° from the Sun. Exposure time was 100 s with an aperture ratio of $f/3.5$ and focal length $f = 28$ mm on Kodak CF 1000.

brightness of the zodiacal light increases sharply toward the Sun, and, at an angular distance of 30°, for example, outshines the brightest part of the Milky Way by more than a factor of two. It is seen most distinctly when its cone makes a steep angle with the horizon. Such is always the case in tropical latitudes, but at mid-latitudes this condition holds only in late February or early March for the evening sky, and in October during the morning; also, the Milky Way during these seasons is far enough away so that it does not interfere. When he discovered the zodiacal light on 1683 March 18, Giovanni Domenico Cassini [22.20] also arrived at the explanation which is still considered valid: it is created by sunlight scattered by a large number of small particles orbiting the Sun as mini-planets—the *interplanetary dust* in present terminology. However, the investigation of the zodiacal light is by no means complete. Certainly the measurements performed from space in the past 20 years have been a breakthrough in that respect; on the other hand, the force interactions to which the dust particles are subjected in interplanetary space are extraordinarily diversified.

As a result of increasing light pollution of the night sky from city and suburban lighting, coupled with the decreasing transparency near the horizon, the zodiacal light is practically no longer visible in densely populated regions at middle latitudes. On the other hand, at low latitudes ($< 40°$) and at a site located sufficiently distant from

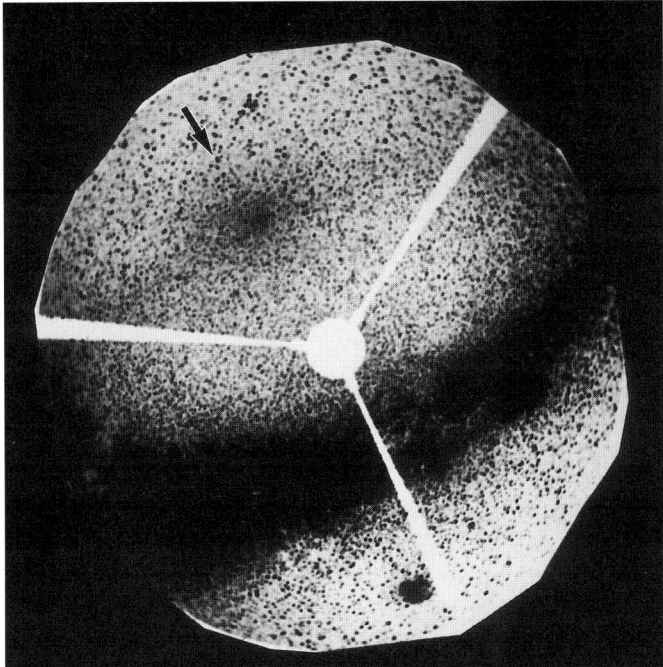

Fig. 22.24. Wide-angle photograph of the night sky with a spherical mirror of the Bochum Observatory taken from La Silla, Chile. The gegenschein, which is visible in the tenuous light bridge of the zodiacal light above the stronger Milky Way band on the bottom, is marked with an arrow. Printed by kind permission of W. Schlosser, Bochum.

any big cities and at sufficient altitude (at least 2000 m, preferably 3000 m), the zodiacal light is easy to see if one knows where to look for it. Figure 22.23 captures the zodiacal light taken under such conditions; the eye, with its superior contrast enhancement, would perceive the cone as somewhat narrower and longer. Prospective photographers may use the data from Fig. 22.23 as a guideline.

22.3.1 The Brightness Distribution

Photometric measurements have shown the zodiacal light to extend over the entire sky, though its brightness decreases with increasing distance from the Sun and from the ecliptic. At the point opposite the Sun, however, there appears a slight brightening which is barely detectable by the unaided eye. This is the *gegenschein* (Fig. 22.24), and was first discovered by Brorsen in 1876. The brightness of the zodiacal light increases toward the Sun (Fig. 22.25) and its flow—not visible from ground-based observers in twilight—continues into the solar corona, where it forms one part, the so-called *F-corona* (cf. Chap. 13). The brightness and polarization of the zodiacal light along the ecliptic are graphed in Fig. 22.26. Levasseur-Regourd and Dumond [22.21] have tabulated the brightness distribution over the sky in 5° steps (see Appendix Table B.23

Fig. 22.25. Inner zodiacal light above the lunar horizon. The ecliptic runs about vertically in the direction indicated by Mercury and Regulus, the two bright objects above the cone. The zodiacal light is seen in the distance range 5° to 20° from the Sun. The asymmetry is caused by plasma (streamers) ejected from the Sun. Photographed by Al Worden during the *Apollo 15* flight (NASA No. AS 15-98-13311).

in Vol. 3). At the ecliptic pole the brightness has only one-third the value on the light axis at the same elongation of 90°. The strong polarization of the light does not come from the Sun, but originates in the reflecting bodies; it therefore permits a conclusion regarding the surface properties of the scattering dust grains. The color of the zodiacal light corresponds nearly to that of the Sun, but with a slight reddening which the eye cannot perceive. Since very small particles would scatter the blue part of the light much more strongly, it may thus be inferred that the interplanetary dust particles are at least as large as the wavelength of the scattered light; i.e., they have diameters of at least 1 μm. Survey articles [22.22–24] give details on the measurement and interpretation of zodiacal light.

22.3.2 The Spatial Distribution of the Interplanetary Dust Cloud

The brightness distribution of the zodiacal light directly reveals how interplanetary dust is distributed in space. For instance, the concentration of the light cone toward

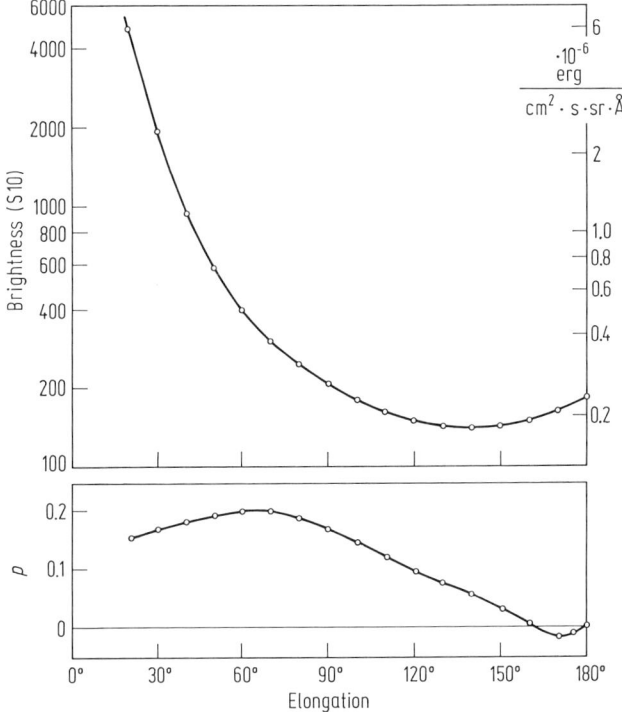

Fig. 22.26. Brightness and degree of polarization p of the zodiacal light along the ecliptic as a function of elongation from the Sun. Positive and negative polarization are taken as perpendicular and parallel, respectively, to the ecliptic. One S 10 corresponds to the brightness of a tenth-magnitude star per square degree. The numbers are for the spectral range around 500 nm.

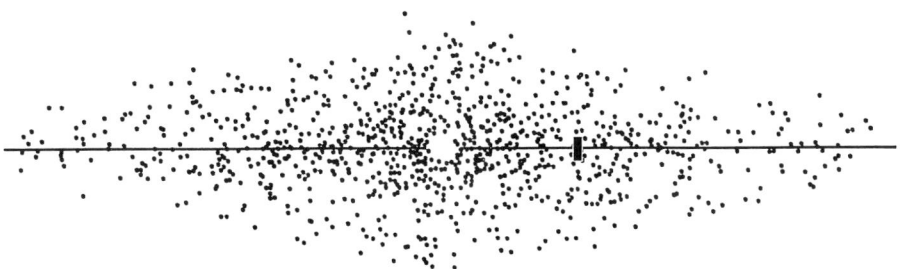

Fig. 22.27. A model for the distribution of interplanetary dust perpendicular to the ecliptic, which is shown as a *solid line*. The position of the Earth is indicated by a *bar*. The Sun should be placed in the center of the dust-free zone, the latter being displayed five times too large for the sake of clarity.

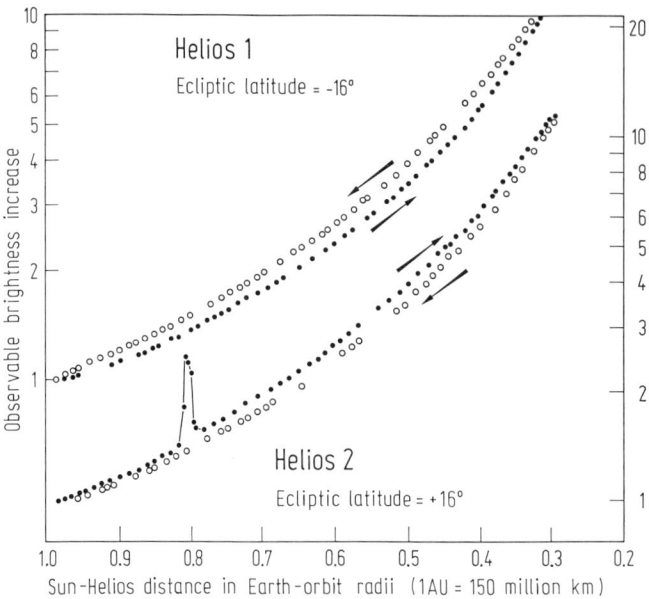

Fig. 22.28. Brightness increase in the zodiacal light on approach to the Sun. At constant density of the dust distribution, the brightness would increase by a factor of only 3.3. The excess amount is caused by the increasing space density of the interplanetary dust. Different intensities were measured on the way in (*full dots*), then on the way back (*circles*) owing to the inclination of the plane of symmetry of the dust cloud against the plane of motion of the *Helios* probes. At 0.8 AU, *Helios 2* passed near Comet West 1976.

the ecliptic leads quite directly to the flattening of the dust cloud shown in Fig. 22.27 and described approximately by an ellipse with an axis ratio of 1:7. The orbits of the interplanetary dust grains themselves are certainly not very strongly concentrated toward the ecliptic, their average inclination against that plane being no less than 32°. To express this concentration in a different fashion, 1/2 AU above the Earth's orbit the space density of the dust has dropped to 1/4 of the value in the ecliptic plane. The strong increase in brightness toward the Sun shows that the density of the interplanetary dust also increases toward the Sun. This reasoning is not entirely elementary. A part of the brightness increase results from the optical properties of the dust grains, which scatter light more efficiently forward than backward. Therefore, the radial density distribution could be deduced by Earth-based observations only after making certain assumptions regarding the properties of the dust. Only after the measurements made by the *Helios 1* and *2* spacecraft in 1972–76 (in the inner solar system) and *Pioneer 10* and *11* (in the outer system) was it learned that the density $n(r)$ of the interplanetary dust increases toward the Sun approximately inversely with the distance (see Fig. 22.28). Specifically, $n(r) \sim r^{-1.3}$, assuming that the properties of dust grains at various positions in the system are the same; otherwise, $n(r) \sim r^{-1}$ if the reflecting property of the particles is assumed to increase toward the Sun. The space density, however, is extraordinarily low—1.0×10^{-19} kg m^{-3}. This means that the interplanetary dust cloud contains, per cubic kilometer, only about one dozen particles with a typical size of 10 μm. That the reflected light is visible at all is due only to the fact that the dust distribution extends across the planetary system at least out to Jupiter, and is permanently exposed to full sunlight; along a line of sight, the contribution of a great many particles add up. For instance, the light in the cone in Fig. 22.23 is composed of roughly 10^{24} particles with individual contributions which are completely insignificant. Viewed from the distance of another star, the zodiacal light could hardly be detected next to the million-fold brighter Sun.

The inner edge of the dust distribution is set by the evaporation of the particles near the Sun, thus making a dust-free zone of about four times the solar radius. Outside 3.3 AU, the brightness of the zodiacal light has fallen so much that it cannot be identified against the general background of distant stars. However, the *Pioneer 10* and *11* space probes found evidence for dust particles far outside the orbit of Jupiter. The total mass of the interplanetary dust can thus be estimated as 10^{16} to 10^{17} kg. This is barely more than the mass of a large comet, and thus wholly insignificant with regard to the dynamics of the entire planetary system.

It would be difficult to understand how the plane of symmetry of the zodiacal light could be aligned with the orbital plane of a planet as small as the Earth. Actually, it is tilted by a few degrees with respect to that plane, 3° in the inner solar system, and somewhat less in the outer parts. Surprisingly, this deviation had already been noted by Cassini from naked-eye observations, and—whether by coincidence or not—he had already ascertained its value of 3°. This plane of symmetry coincides neither with the orbit planes of the largest planets Jupiter and Saturn, which have lesser inclinations, nor with the plane of the solar equator, the inclination of which is higher. Thus, it is not yet clear what kinds of forces have determined plane of symmetry of the dust.

It was through space travel that advances in the investigation of the interplanetary dust were originally made; the fear that the latter would present a danger to the former was shown to be unfounded. On the average, a spacecraft with a cross section of 3 m^2 would be hit by a dust grain of size 1 mm or more only once in a thousand years.

22.3.3 Properties of Interplanetary Dust Particles

Optical observations of the zodiacal light are not well-suited to accurately determine the sizes of the dust grains. Direct measurements from space are thus far neither abundant enough nor accurate enough. The solution comes from another branch of space science. A dust particle hitting the surface of the Moon, which is not shielded by an atmosphere, leaves, after impact and evaporation, a microcrater whose diameter indicates the mass of the incident particle and hence its size. The results of these studies is shown in Fig. 22.29. The distribution in size consists of two components. The larger dust particles with radii typically in the range 1 to 100 μm, but most frequently 10 μm, are responsible for the zodiacal light. (These would also qualify as "dust" in the terrestrial sense.) Particles with diameters less than 1 μm are numerous but optically insignificant; they are fragments resulting from collisions and destruction of larger dust particles. For these smallest particles, the solar radiation pressure exceeds gravity, so

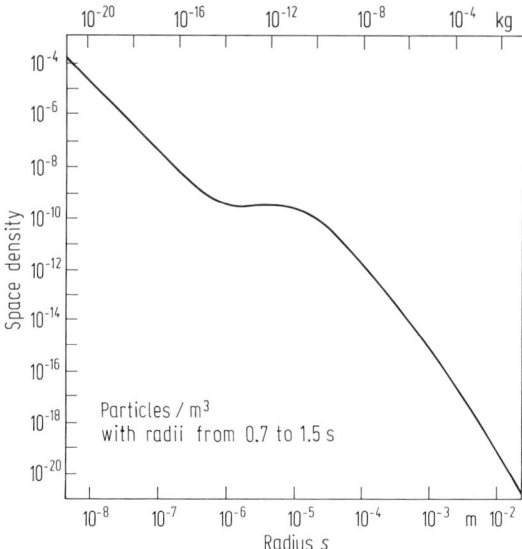

Fig. 22.29. Size distribution of interplanetary dust.

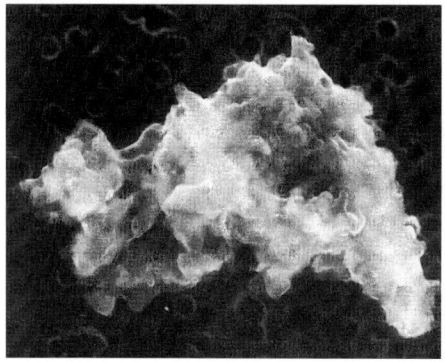

Fig. 22.30. Interplanetary dust particle of size 9 μm×6μm collected in the stratosphere. The color is brown to black. NASA photograph.

that they are, on a short time scale (about 1 year), "blown" out of the system. They are called β-meteorites (β is an often-used symbol for the force ratio of radiation pressure to gravity).

In recent times it became possible, primarily by means of the infrared-sensitive satellite *IRAS*, to reliably measure the thermal emission of the interplanetary dust. Comparison with the brightness of the reflected zodiacal light showed that the dust particles are evidently quite dark, reflecting only 7% of the incident sunlight, or about as much as a poor black varnish.

The first solid knowledge on the structure of the particles was obtained by D.E. Brownlee, who in the 1970s succeeded in collecting some interplanetary grains in the stratosphere from balloons and high-flying planes. Particles of the size found (up to 50 μm) survive the braking upon infall into the Earth's atmosphere without much damage. Their interplanetary origin was recognized from the fact that their surface layers are filled with the same gases which are present in the solar wind, especially helium. Many of the grains show a rather loose structure composed of pieces of about 0.1 μm in size (Fig. 22.30). Most of them are very dark owing to a carbon constituent of a few percent and a chemical composition similar to that of what is believed to be the most primieval class of meteorites, namely the *carbonaceous chondrites* (see Chap. 21). Their specific density is in the range 1 to 3 g cm^{-3}.

22.3.4 Lifetime

Comparison of Cassini's reports with the present-day appearance shows that the zodiacal light has not changed dramatically over the past three centuries. Reports on short-term brightness fluctuations should not be taken uncritically, as measurements by the spacecraft *Helios 1* have shown that its brightness between 1974 and 1986 has changed by at most a few percent, if at all. However, changes are to be expected on theoretical grounds. Particles revolving near the Earth's orbit with speeds of about 30 km s^{-1} receive the sunlight not radially but slightly shifted toward the front. This is the well-known *aberration of light*, which gives rise to a force component of the radiation pressure which is opposite to the direction of particle motion, thus dimin-

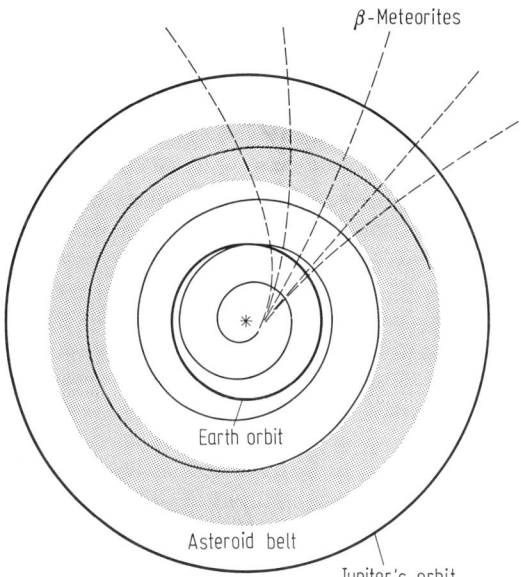

Fig. 22.31. Schematic representation of the orbital development of an interplanetary dust grain. The shrinking of the spiral during one revolution is shown greatly exaggerated. The asteroid belt is between 2 and 3.5 AU, the orbit of Jupiter at 5.2 AU.

ishing it. This process is called the *Poynting–Robertson effect* for the physicists who first demonstrated its existence. It causes the dust grains to lose energy and to gradually spiral inward toward the Sun until they are evaporated by the intense heat, or destroyed by collisions, so that they leave the system as β-meteorites (Fig. 22.31). The lifetime of a particle until evaporation is computed from its radius, its density, and the semimajor axis of the original orbit:

$$t_{PR} = 700 \; s\rho a^2 \quad \text{(years)}, \tag{22.1}$$

where s is the radius in μm, ρ the density in g cm^{-3}, and a the semimajor axis of the original orbit in AU. For a particle with size 1 to 10 μm in the asteroid belt, the lifetime thus is 10^4 to 10^5 years; this can be shortened by collisions. For a particle of the kind just mentioned, the chances of termination by collision or by evaporation are about equal. In this way, the amount of interplanetary dust diminishes by about 1 to 10 tons of solid material per second. This means that the continued presence of the zodiacal light could only have been maintained by a continuous resupply of dust particles, as it seems highly unlikely that the zodiacal light phenomenon exists only for the eyes of the present-day observer.

The Poynting–Robertson effect also permits predictions on the spatial density distribution of interplanetary dust. The result is $n(r) \sim n^{-1}$, assuming a continuous resupply of particles from outside the asteroid belt. The density distribution is steeper particularly if the supply occurs in the inner planetary system, in which case the exponent can reach the value -1.3. The good agreement of the predictions with observations— though indirect—is confirmation of the decisive importance of the Poynting–Robertson effect.

22.3.5 The Origin of the Zodiacal Light

Comets are generally held to be the source of interplanetary dust. Their dust tails are direct evidence that they can release large amounts of dust (see Chap. 20). The space probe *Giotto* analyzed cometary dust particles and found mainly chondritic composition, as is also typical for interplanetary dust. The *IRAS* satellite found enhanced dust concentrations in the orbits of some comets (Fig. 22.32). This apparently convincing picture has two flaws. First, most of the dust seen in cometary tails is rapidly driven out of the solar system by solar radiation pressure (only the largest dust grains remain on bound orbits). Thus the amount they contribute is far insufficient, at least when considering the short-period comets, which supply only 250 kg s^{-1}. Second, particles with cometary origins are thought to be very loosely structured and of low density, similiar to the residue of a compacted dust–ice mixture after the ice has been evaporated. But the direct observation of interplanetary particles as well as meteorite studies and the impact microcraters in lunar rock lead to the conclusion that only about one-fourth of interplanetary dust is of such low compactness [22.29]. A few open questions thus remain. In recent times, asteroids have again been considered as a potentially important source which, by mutual collisions, could release dust and pieces of rocks. That this is not merely a contrived idea has been shown by *IRAS* measurements (Fig. 22.33). The dust concentration was found to be noticeably higher

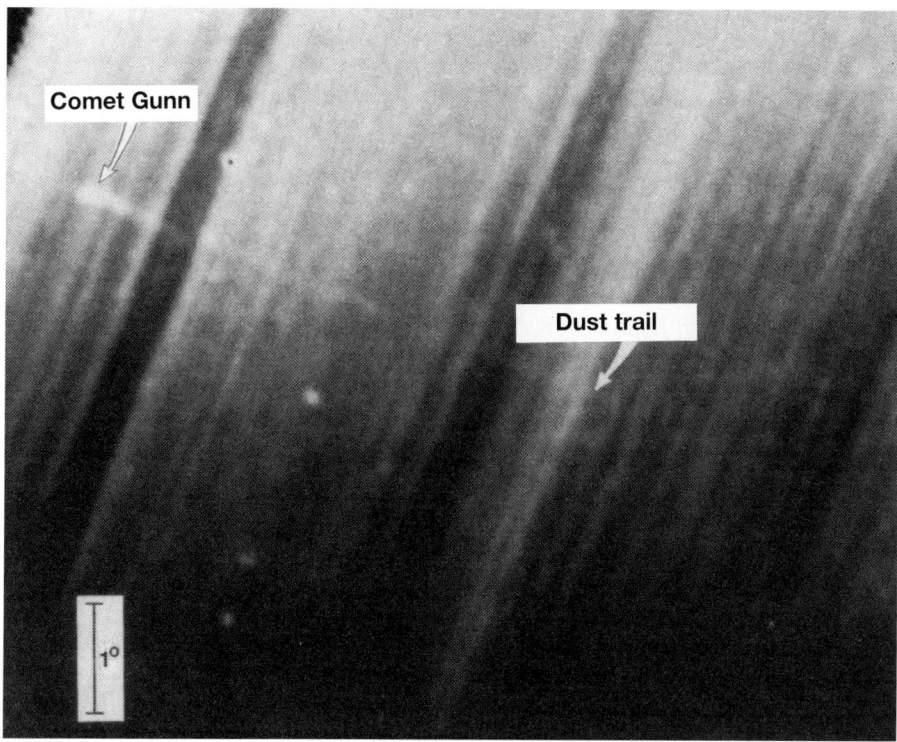

Fig. 22.32. Dust trail in the orbit of the short-period comet P/Gunn (Period = 6.80 years). It is seen that the dust distribution extends to the left in front of the comet. This must be dust that was released during earlier revolutions. Photographed at 60 μm by the IRAS satellite between 1983 May 26 and June 08.

in the ecliptic as well as in two bands located about 10° away from it on both sides. These loci can be related exactly to the three large asteroid "families" Themis, Corinis, and Eos, whose members have orbits similar to each other and consequently will suffer collisions more frequently.

The following picture of the birth and death of interplanetary dust has finally begun to emerge: after several hundred revolutions, comets decay into meteor streams, which dissolve after a comparable time span into a general scattering of meteoroids. Asteroids which have been shattered by collisions are another, but not so well determined, source of meteoroids. These meteoroids, which are roughly 1 mm to 10 cm in size, comprise the reservoir from which interplanetary dust is continuously generated by collisional processes. The continuous losses are thereby immediately replenished, and an equilibrium is reached, as is today presumably observed, in which interplanetary dust and the zodiacal light have become permanent fixtures of the planetary system. One may thus imagine a rock-grinding mill operating in interplanetary space in which small, solid bodies are repeatedly minced until they finally exit the system as vapor or, more commonly, as β-meteoroids. Zodiacal light and interplanetary dust are in

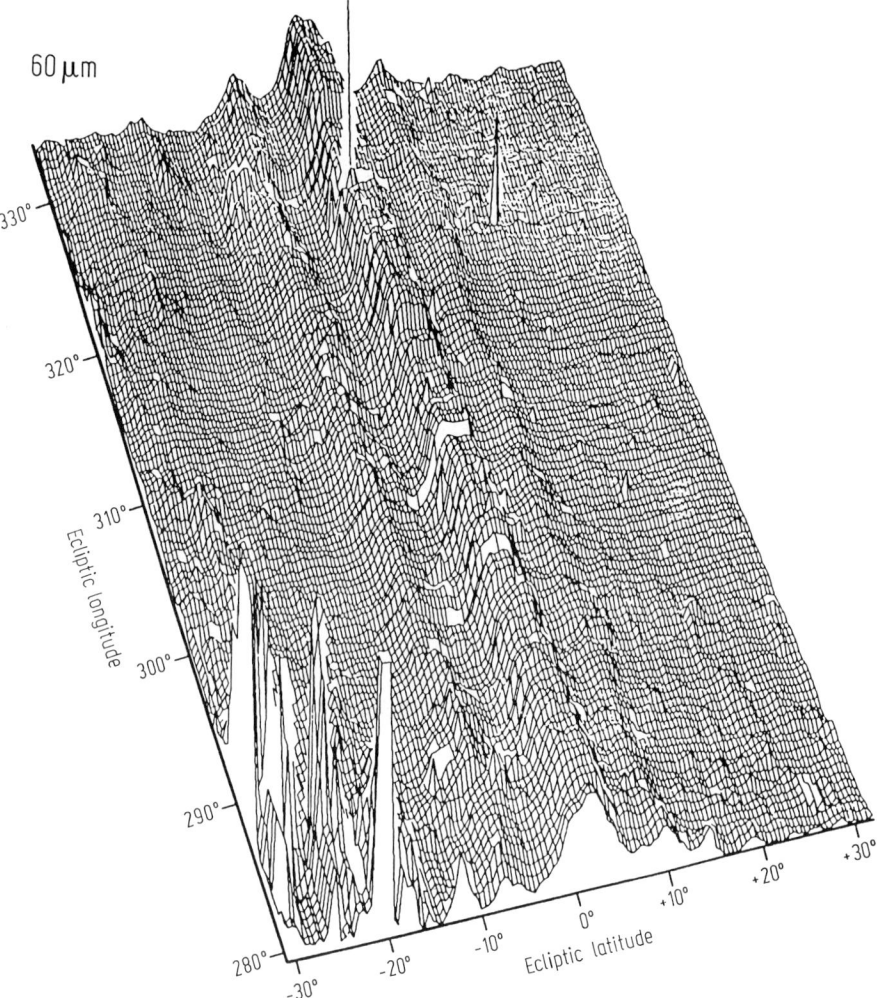

Fig. 22.33. Bands of enhanced intensity in the infrared brightness of the zodiacal light at ecliptic latitudes 0° and ±10°. The dust concentrated there is probably generated through asteroid collisions. To show the bands more clearly, the broadly distributed portion of the zodiacal light has been subtracted.

this way linked to the primordial bodies in the solar system. Owing to the intervention of multiple altering processes, those conditions which spawned the formation of solid bodies in the early planetary system can no longer be reconstructed. Finally, it must be mentioned that the Earth collects interplanetary dust as well as meteorites. The accreted mass amounts to about 40 tons per day, primarily from dust particles with sizes of 30 to 300 μm. This figure is practically negligible; even after having accumulated over the span of the Earth's age of 4.5×10^9 years, a layer only 5 cm thick would have been generated. J. Oro [22.30] pointed out the potential significance of such accretion, although he originally considered the impact of comets: the organic molecules of interplanetary or even of interstellar origin thus reaching Earth thus could have played a role in accelerating the beginning of life on Earth. Certainly this is merely a speculation, but it does emphasize the variety of connections encountered in the study of the tiny bodies in the planetary system.

22.3.6 Comment on the Literature

Leading beyond the scope of the present article but still without special knowledge are the articles by Staude and Röser [22.25] and Leinert [22.26]. More technical and detailed information on the status of research are found in McDonnell [22.27] (a highly recommended collection of survey papers), the conference report by Giese and Lamy [22.28], the paper by Grün et al. [22.29] (perhaps the best single article on interplanetary dust to have appeared in recent years), and a very recent article by Leinert and Grün [22.31].

Acknowledgements. The author acknowledges the expert advise and pictorial material contributed to this chapter by Frank Arnold, Nathan Wilhelm, Gerhard Haerendel, and Kristian Schlegel. I also thank Rita Wagner for the patient writing of all versions of this manuscript, and Karin Dorn, Martina Weckauf, Werner Neumann, and Doris Anders for the fine drafting of the often provisional sketches for figures.

References

22.1 Jesse, O.: Aufforderungen betreffend Beobachtungen der glänzenden Himmelserscheinungen, welche seit dem Sommer 1885 öfter in Mittel-Europa gesehen worden sind. *Astronomische Nachrichten* **115**, 15 (1886).
22.2 Schröder, W.: Zur Geschichte der Erforschung der Leuchtenden Nachtwolken. *Sterne* **54**, 237 (1978).
22.3 Fogle, B., Haurwitz, B.: Noctilucent Clouds. *Space Science Reviews* **6**, 278 (1966).
22.4 Gadsden, M.: Noctilucent Clouds. *Space Science Reviews* **33**, 279 (1982).
22.5 Turco, R.P., Toon, O.B., Whitten, R.C., Keesee, R.G., Hollenbach, D.: Noctilucent Clouds, Simulation Studies of their Genesis, Properties, and Global Influences. *Planet. Space Sci.* **30**, 1147 (1982).
22.6 Thomas, G.E.: Solar Mesosphere Explorer Measurements of Polar Mesospheric Clouds (Noctilucent Clouds). *J. Atmosph. Terr. Phys.* **46**, 819 (1984).
22.7 McDonnell, D.: Clouds of the Twilight. *Astronomy*, 42 (July 1987).
22.8 Schröder, W.: *Das Phänomen des Polarlichts*, Wissenschaftliche Buchgesellschaft, Darmstadt 1982.
22.9 Schröder, W.: Wandel in der Kenntnis des Polarlichts. *Sterne und Weltraum* **21**, 358 (1982).
22.10 International Union of Geodesy and Geophysics: *International Auroral Atlas*, Edinburgh University Press 1963.
22.11 Brekke, A., Egeland, A.: *The Northern Light From Mythology to Space Research*, Springer, Berlin Heidelberg New York 1983.
22.12 Eather, R.H.: *Majestic Lights*, American Geophysical Union, Washington D.C. 1980.

22.13 Haerendel, G.: Über den Ursprung des Nordlichts. *Sterne und Weltraum* **21**, 508 (1982).
22.14 Hones, E.W. Jr.: The Earth's Magnetotail. *Scientific American* **254**, 32 (March 1986).
22.15 Reddy, F.: Celestial Winds, Polar Lights. *Astronomy* **12**, No. 8, 6 (1983).
22.16 Jones, A.V.: *Aurorae*, Reidel Publishing Co., Dordrecht Boston 1974. (*Geophys. and Astrophys. Monograph No. 9.*)
22.17 Omholt, A.: *The Optical Aurorae*, Springer, Berlin Heidelberg New York 1971.
22.18 Höper, R.: Polarlicht am 13 Juli 1982. *Sterne und Weltraum* **21**, 538 (1982).
22.19 Boerner, W.-M., Cole, J.B., Goddard, W.R., Tarnawecky, M.Z., Shafai, C., Hall, D.H.: Impacts of Solar and Auroral Storms on Power Line Systems. *Space Sci. Rev.* **35**, 195 (1983).
22.20 Cassini, G.D.: Découverte de la lumière céleste qui paroist dans le zodiaque. *Mem. Acad. Roy. Sci.* Tome VIII (1666–1699), p. 119, Com. Libraires, Paris 1730.
22.21 Levasseur-Regourd, A.-Ch., Dumont, R.: Absolute Photometry of Zodiacal Light. *Astron. Astrophys.* **84**, 277 (1980).
22.22 Leinert, Ch.: Zodiacal Light—a Measure of the Interplanetary Environment. *Space Science Reviews* **18**, 281 (1975).
22.23 Weinberg, J.L., Sparrow, J.G.: Zodiacal Light as an Indicator of Interplanetary Dust. In: [22.27], p. 75.
22.24 Fechtig, H., Leinert, C., Grün, E.: Interplanetary Dust and Zodiacal Light. In: K. Schaifers and H.H. Vogt (eds.): *Landolt-Börnstein, New Series*, Vol. 2a, pp. 228–241. Springer, Berlin Heidelberg New York 1981.
22.25 Staude, H.J., Röser, S.: Zodiakallicht und Interplanetarer Staub: Neue Beobachtungen und neues Verständnis, *Sterne und Weltraum* **17**, 329 (1978).
22.26 Leinert, Ch.: Zodiakallicht-Beobachten. Neuere Ergebnisse über den interplanetaren Staub. *Naturwissenschaften* **66**, 221 (1979).
22.27 McDonnell, J.A.M. (ed.): *Cosmic Dust*, John Wiley & Sons, Chichester New York Brisbane Toronto 1978.
22.28 Giese, R.H., Lamy, P. (eds.): *Properties and Interactions of Interplanetary Dust*, D. Reidel Publ. Co., Dordrecht Boston Lancaster Tokyo 1985.
22.29 Grün, E., Zook, H.A., Fechtig, H., Giese, R.H.: Collisional Balance of the Meteoritic Complex. *Icarus* **62**, 244 (1985).
22.30 Oro, J.: Comets and the Formation of Compounds in the Primitive Earth. *Nature* **190**, 384 (1961).
22.31 Leinert, Ch., Grün, E. in: Schwenn, R. and Marsch, E. (eds.): *Physics of the Inner Heliosphere*, Vol. I, Springer, Berlin Heidelberg New York 1990, p 207.

23 The Terrestrial Atmosphere and Its Effects on Astronomical Observations

F. Schmeidler

23.1 General Remarks on the Atmosphere

The atmosphere of the Earth is a source of numerous unfavorable effects on the performance of astronomical observations for professional as well as amateur observers. These influences can be largely eliminated by, for instance, observing from a mountain top or from an airplane, rocket, or spacecraft. But such opportunities are normally not available to the average observer, and thus the impact of the atmosphere upon astronomical observations requires serious consideration. In this regard, there should be a careful distinguishing between factors which are strongly weather dependent and those which remain about the same regardless of weather conditions.

The atmospheric influence on radio observations will not be discussed here; it has been dealt with in Chap. 9 (Vol. 1). While the effects in the radio range bear some similarities to those in the optical range, they are nevertheless in many respects very dissimilar.

23.2 Weather-Dependent Phenomena

This section will concern itself primarily with the constraints imposed by various weather patterns on the possibilities of optical astronomical observations. Obviously, no optical work can be performed with an overcast sky (although radio observations are still possible), but there exist other factors which may preclude astronomical observations even under cloudless or slightly cloudy skies.

23.2.1 Assessment of Weather Patterns

Factors evident from weather maps, such as the geographic distribution of highs and lows provide some guidance as to the observing conditions on a given night, but of course not an unambiguous correlation. Weather patterns may have substantially different local effects at places less than 50 km apart, owing to the influence of, for instance, water surfaces (e.g., lakes and rivers) and upslope or downslope mountain winds. This also affects (locally) the total annual average number of clear nights. On a larger scale it is the seasonal distribution of clear nights, which is usually fairly constant with time. Large parts of western Europe as well as California, for instance, experience, as a rule, the best observing conditions in late summer and autumn, the

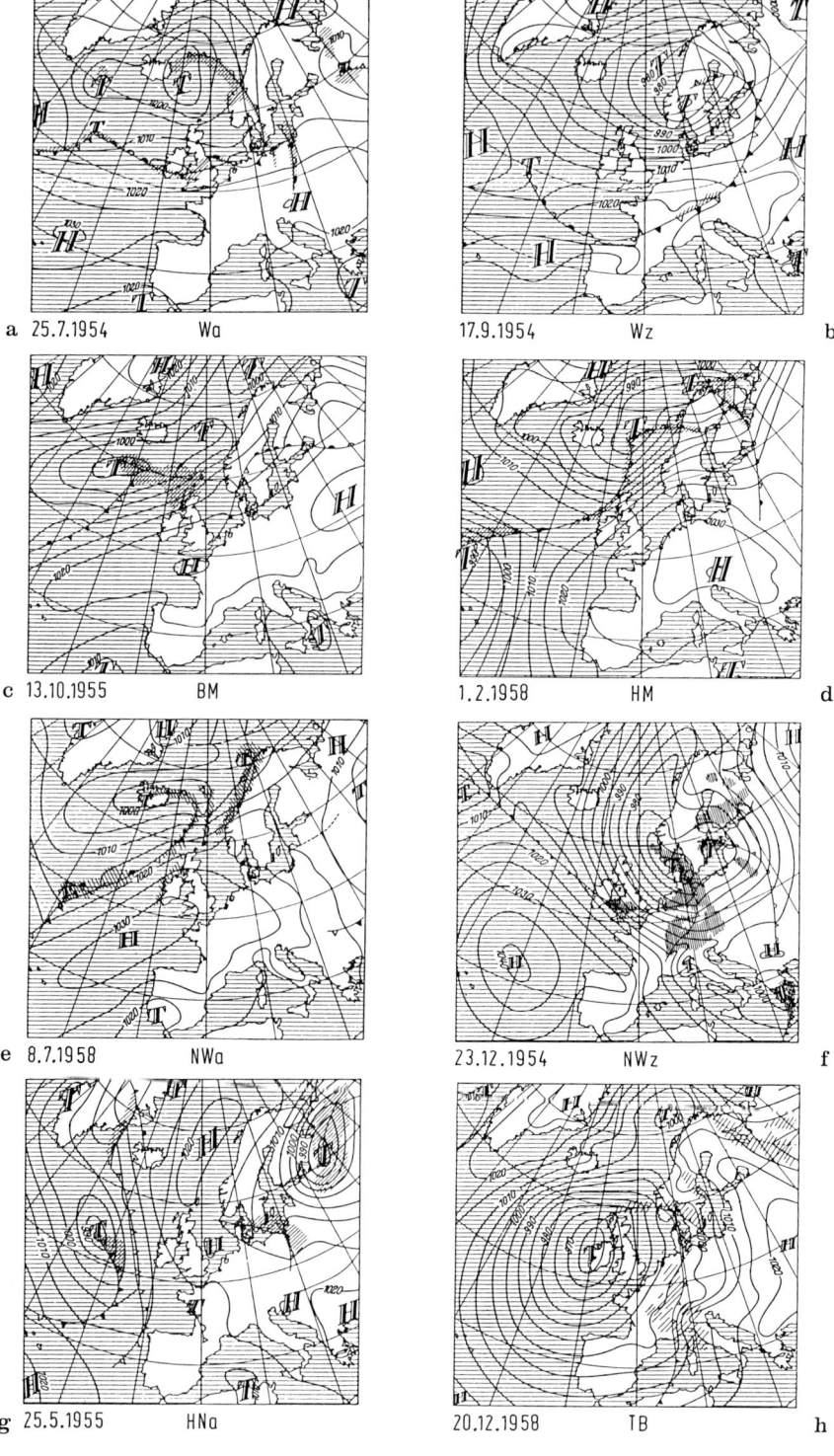

a 25.7.1954　Wa

17.9.1954　Wz　b

c 13.10.1955　BM

1.2.1958　HM　d

e 8.7.1958　NWa

23.12.1954　NWz　f

g 25.5.1955　HNa

20.12.1958　TB　h

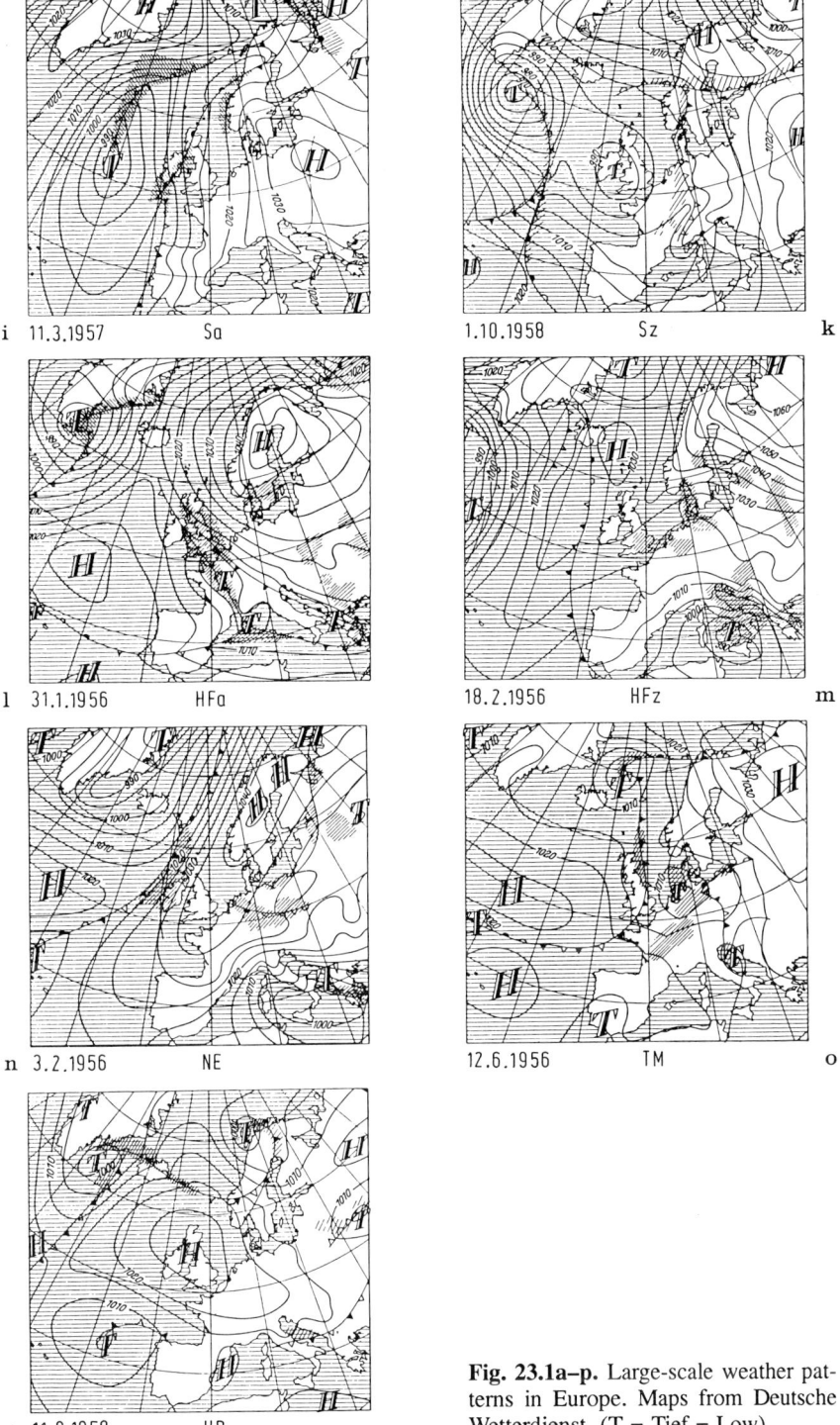

Fig. 23.1a–p. Large-scale weather patterns in Europe. Maps from Deutsche Wetterdienst. (T = Tief = Low).

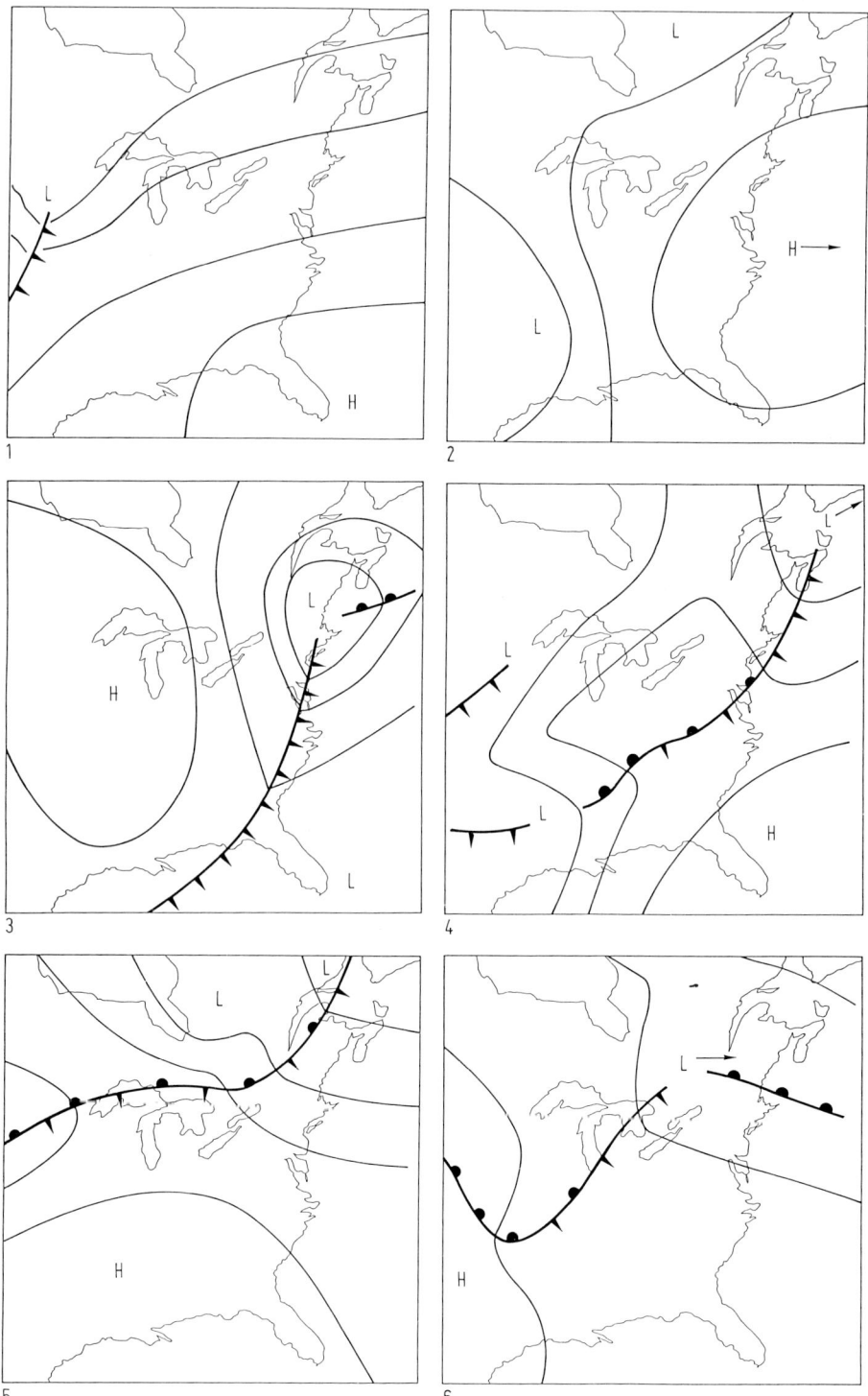

Fig. 23.2. Typical large-scale weather patterns in the eastern U.S. Refer to footnote 1 for details.

poorest in late winter, while much of the central and eastern U.S. has the worst season in the late spring and early summer.

The German Weather Service uses the following classification of large-scale weather patterns [23.1], as graphed in Fig. 23.1, characteristic for much of western Europe. (*Note*: T = Tief = Low).

W   Westerlies;
BM   High-pressure ridge over central Europe;
HM   High centered over central Europe;
SW   Southwest flow (High over southeastern Europe, Low over North Atlantic);
HN   High over Arctic Ocean;
HB   High over Great Britain;
N   North flow (High over North Sea, Low over eastern Europe);
TrM   Low pressure trough over central Europe;
TM   Closed Low over central Europe;
TB   Closed Low over Great Britain;
TrW   Trough over Western Europe;
S   South flow (Low over western Europe, High over eastern Europe);
SE   Southeast flow (High over eastern Europe, Low over Mediterranean);
HF   Closed High over Fennoscandia;
HNF   Closed High over Fennoscandia and Arctic Ocean;
NE   Northeast flow (High bridge from Azores to Fennoscandia);
Ww   Angular westerly flow.

BM, HM, and HB are always anticyclonic over central Europe (index a), while TrM, TM, TB, and TrW are always cyclonic (z). The other patterns may have either isobaric curvature.

Statistical data show that none of these patterns guarantees or prohibits cloudless skies, but there are correlations. Most favorable are HM and BM, as might be expected, whereas certain cyclonic flows like NWz and SEz are found least promising. A count of observing nights in Munich, Germany, has led to the following groups with regard to astronomical promise:

Favorable: HM, BM, Wa, HNa, SWz, TB, HB, NEa;
Average: SWa, Na, SEa, HFa, HNFa, TrW, HFz, Ww;
Unfavorable: All other patterns.

An instance of local effects is here the high incidence of clear weather in association with certain westerly flows (Wa, SWz, TB) as a consequence of downhill foehn winds north of the Alps.[1]

---

[1] *Translators' Note*: Analogous maps for the eastern United States (specifically the mid-Atlantic seaboard) are added in Fig. 23.2:

(1) *Warm Atlantic (Bermuda) High off-shore, or extending inland*: frequently good seeing but poor transparency owing to accumulating moisture.
(2) *Rear of cold High, normally traveling eastward*: usually indicates fair weather, but only for short duration.
(3) *Traveling cold front*: on its rear side, the transparency (outside the cloud range from the Great Lakes) is often excellent, the seeing invariably poor for about 36 hours.
(4) *Stationary front along shore* (particularly when held in place by a Bermuda High): expectation of some interval of inclement weather owing to probability of secondary lows forming.
(5) *Westerly weather* (particularly when supported by a High bridge over the southern states): generally good, often stable (unless a strong pressure gradient at the polar front to the north triggers cyclonic motion).

(Continued on next page)

23.2.2 Atmospheric Turbulence and Scintillation

A certain amount of turbulence is present even in cloudless skies, and causes some vibration of an image in the telescope. This unsteadiness can be so strong that it is often perceived by the naked eye as the "twinkling" phenomenon in bright stars. It is caused by small, local density and temperature differences in the atmosphere effecting changes in the refraction index. When substantial air motions are added, the variation can be quite rapid, thus causing the images to flicker wildly. Essentially the same effect can be observed on a hot summer afternoon above the ground, and in particular above hot asphalt. The book by G. Dietze [23.2] describes the optical causes in detail.

Turbulence brings about a change in both the apparent brightness and position of a star. One thus distinguishes between (1) *intensity scintillation*, or simply *scintillation*, and (2) *directional scintillation*, or *seeing*, although the term seeing is often used to include both phenomena. Directional scintillation may manifest itself as either a displacement of the entire image around a mean position, or as a diffuse apparition of a non-moving image. Some authors summarily state that the displacement dominates for small apertures, while the fuzziness or diffuseness prevails at large apertures. Not all observers concur with this assessment, however, and the experiences are too diverse for all to be placed over a common denominator.[2]

Just how much the atmospheric unsteadiness impedes astronomical observations depends on the specific purpose of the observing program. Most affected are studies which depend crucially upon resolving power, such as measurements of close double stars and visual observations of planetary surface details; many nights are completely useless for work of these kinds. Unsteadiness can be an especially noisome impediment for astronomical photography in which long focal lengths and high magnifications are needed to produce detailed pictures of the Sun, Moon, or planets. Photographic work at short to moderate focal lengths (under 2 m), on the other hand, is less sensitive to poor seeing, as are photometric observations. For observations without a telescope (e.g., of meteors and the zodiacal light), atmospheric scintillation has practically no influence.

The degree of turbulence is influenced by various factors. It is obviously larger near the horizon than near the zenith, but even this almost trivial rule is not always

---

(6) *Warm-air sector of northern traveling Low* (usually small pressure gradients in the region): frequent, but least predictable regarding cloudiness.

These samples should enable observers to recognize the most typical patterns and how they correlate with cloud cover under other weather regimes. See also E. Palmen and C.W. Newton, *Astronomical Circulation Systems*, Academic Press, New York 1969, for a more comprehensive description of weather patterns, and H.H. Lamb, in *The English Climate*, English Univ. Press, London 1964.

2 *Translators' Note*: Some observers who have worked with telescopes of different sizes have noted that images suffer greater distortion with larger apertures, because more simultaneous turbulence elements in the beam contribute to the diffraction. This is compensated by the higher image brightness, as faster eye perception and shorter recording time cut the time variation of turbulence. In a nutshell, the advantage of larger apertures is most pronounced in good seeing and in poor transparency. The most detrimental scintillation frequencies are in the range about 1 to 5 Hz, but this depends on the surface wind speed: the stronger the wind, the higher the contributing frequencies.

complied with; there are nights in which the unsteadiness increases very little or not at all from the zenith to the horizon, and there are nights when the increase is quite large. Unambiguous criteria for these cases are thus far unknown.

On a statistical average, the turbulence depends quite strongly on the time of day or night. The high degree of unsteadiness around noon is as readily understood as the observed minima of scintillation before sunrise and after sunset. What is not expected is the secondary maximum of unsteadiness that numerous observers concordantly find around midnight, but sometimes earlier or later. It is caused by a corresponding secondary maximum of atmospheric turbulence at that time, following the effects of atmospheric mass exchange. Consequently, the visual observer regularly finds that the image quality in the telescope gradually improves in the hours after sunset, but deteriorates again one or two hours before midnight.

Individual cases may differ substantially from these averages. As always in meteorological problems, there are no sure-fire rules, but rather only crude guidelines as to when to reckon with favorable or unfavorable conditions. As a rule, wind (even a weak one) is associated with a noticeable degree of unsteadiness, but this self-evident correlation also has numerous exceptions. High humidity almost always improves the image quality to the extent that observations through ground fog often make very steady images—just so long as the star is visible at all!

The seasonal variation of unsteadiness is very different, depending on locale. For instance, observers in northern Germany report the best seeing during the months of spring, while in southern Germany, experience shows the late summer and fall (August through October) to be the best months, without any readily apparent cause for this difference. The aforementioned large weather patterns show only a weak correlation with scintillation. In brief, anti-cyclonic situations (when the wind flow is curved clockwise in the northern hemisphere) are more conducive to favorable weather conditions than are cyclonic ones. An evident rule of thumb is that the stabler the large-scale state of the atmosphere, the better the astronomical image quality.

Finally, there are various local, micro-climatic influences which can modify the situation entirely, even over short distances, such as nocturnal heat from an urban climate, forests protecting a nearby town from wind, or sloping terrain causing turbulent winds.

Overall, the factors affecting image quality on a given night are so complex that even with extensive experience in a particular locale, it is not possible to make reasonably good predictions for more than 50% of the cases.

23.2.3 Halos, Rainbows, and Other Optical Phenomena

This section deals with features which are technically not of an astronomical nature, but which are naturally so often noticed by observers that a short description of their basic features is warranted. A *halo* is (usually) seen around a bright source (e.g., the Sun or Moon) as a ring-shaped apparition of light, which originates when light is refracted or reflected by ice crystals in the atmosphere. Only when these crystals are present at a certain level in the atmosphere can the halo arise, which means that the phenomenon is connected with specific meteorological patterns. The kind, distribution, and shapes of ice crystals give rise to a variety of halo types. Also, a preferential orientation must be present, because a random distribution of crystal axes would distribute the light isotropically and thus not produce any definite luminous phenomenon.

A faint ring of radius 22° around the Sun or Moon is most frequently observed as the halo, and its inner edge is distinctly reddish. Sometimes, the so-called large halo with a radius of 46° appears instead. The so-called horizontal circle lies parallel to the horizon through the position of the Sun. Its intersection with the halo proper sometimes generates particularly bright spots of light called *sun dogs*.

A technical explanation of the various phenomena is found in the previously cited book by Dietze [23.2] and will not be repeated here. Also, a paper by Leinert [23.5] gives interesting hints. Cirrus clouds are the most evident indicators of the requisite ice crystals in the atmosphere, and thus halos are most often observed in the presence of this type of cloud.

The phenomenon of the *rainbow* is much more widely known than halos. It is based on similar optical principles, but in this case the light is refracted by water droplets instead of ice crystals. Since water in liquid particles has only one symmetric shape (i.e., spherical), the rainbow does not show the confusing variety of appearances, but rather has only two shapes: the *primary rainbow* surrounds the point opposite the Sun with an angular radius of 42°, while the less frequently seen (and much fainter) *secondary rainbow* possesses a 51° radius. Both rings are best seen with the Sun at the horizon.

The distinctly visible color phenomenon in rainbows gave rise to the proverbial expression of "all the colors of the rainbow." The main bow has red on the outer and violet on the inner side, and the secondary bow reverses this order. Weak secondary bows are occasionally seen joined to one or both of these bows.

The appearance of rainbows requires, of course, the illumination of water droplets in the atmosphere by the Sun or Moon. This may happen when, for instance, during or after a rainstorm the Sun emerges from a small opening among the clouds. Even under cloudless skies, rainbows can be seen near sprinklers or water fountains. Rainbows can be caused by moonlight too, but these usually appear colorless since their low brightness does not permit the human eye to discern the colors even though they are present.

Under suitable conditions with respect to kind, distribution, and density of clouds, a few other phenomena such as the *corona* and the *glory* can appear because of the Sun or Moon. The latter are colored rings of small diameter around the point opposite the Sun. Details may again be found in the book by Dietze [23.2]; there is some fascinating information on all the phenomena described here in papers by B. Albers [23.3] and C. Leinert [23.4]. See also [23.5–7] and Chap. 22.

23.3 Permanent Atmospheric Phenomena

In contrast to the weather-dependent optical phenomena discussed thus far, there are several atmospheric influences which are always present in any weather situation

23.3.1 Refraction

Like any physical medium, atmospheric air has a refractive index, so that light rays entering the atmosphere from empty space suffer a small but measurable change in direction. Of course, this fact is significant only if the direction of the incident light (i.e., the position of the celestial body at the sphere) is of importance. The amount of refraction, which always causes a celestial light source to appear elevated above its true position relative to the horizon, depends on both the height or zenith distance of the source and on the color of the light. When two neighboring celestial objects are, owing to a small difference in zenith distance, subject to different amounts of refraction, the result is termed *differential refraction*. The fact that light from a source at a sufficiently large zenith distance is, owing to the wavelength dependence of the refractive coefficient, spread out into a spectrum is termed *atmospheric dispersion*.

The general refraction naturally vanishes in the zenith and increases toward the horizon. The rule of increase has been the subject of detailed theories. At zenith distances down to around $75°$, the amount of refraction is practically independent of the atmospheric stratification, that is, of the prevailing density and temperature decrease with altitude (this is known as the *law of Oriani and Laplace*). An approximation adequate for most observations can be derived from the simplifying assumption that the Earth's atmosphere extends with constant density to a certain upper level, and thus a light ray from space suffers a one-time refraction at that level. Consequently, the direction of the ray after refraction is observed from the ground as the apparent zenith distance z; it was $z + R$ before refraction, where R is the amount of refraction. According to the law of refraction (Snell's law; see Chaps. 4 and 7) with μ as the refractive coefficient of atmospheric air,

$$\sin(z + R) = \mu \sin z, \tag{23.1}$$

where R is a small angle so that $\cos R = 1$ and $\sin R = R$ can be substituted. Expanding the left side gives

$$\sin(z + R) = \sin z \cos R + \cos z \sin R, \tag{23.2}$$
$$R = (\mu - 1) \tan z. \tag{23.3}$$

This is the well-known rule that refraction is proportional to the tangent of the zenith distance. The constant of proportionality is very close to $1'$, which gives the rule of thumb: The amount of refraction in arcminutes equals $\tan z$.

Only at large zenith distances, or when high precision is needed, does this rule break down. More accurate theoretical work shows refraction to be expressible as a series of odd powers of $\tan z$. Detailed tables or precise calculation of refraction include terms down to the 13th power of $\tan z$.

The amount of refraction also depends on air pressure and temperature since the refractive index is a function of these quantities. If R_0 is the "mean" refraction, valid for 0°C and 760 mm Hg = 1013 mb, then

$$R = \left(\frac{b}{1013}\right)\left(\frac{1}{1+T/273}\right) R_0, \tag{23.4}$$

where b is the atmospheric pressure in millibars (mb) and T the temperature in °C.

The mentioned expansion of the refraction in powers of $\tan z$ fails entirely near the horizon where $\tan z \to \infty$. Calculating the horizontal refraction is thus a specific problem in any refraction theory, into which the vertical stratification of the atmosphere also enters. Observations reveal that horizontal refraction amounts to about 34'. A star apparently just setting is actually already one-half degree below the horizon. This fact results in an increase in the length of the day. The beginning and end of day are defined by the rising and setting of the upper limb of the solar disk. As the apparent radius of the sun is 16', then at the moment of sunrise the center of the Sun's disk is in reality $34' + 16' = 50'$ below the horizon. In mid-latitudes, this increase in the length of the day amounts to about 8–12 minutes, but in the polar regions it is substantially greater.

Another consequence of refraction (in this case, differential refraction) is the apparent elliptical shape of Sun and Moon when near the horizon. The lower limb of the disk, being closer to the horizon, suffers a stronger "lifting" by refraction, which reduces the apparent angular diameter of the Sun. Of course, this phenomenon occurs at any altitude of the Sun, but only near the horizon is the effect sufficiently pronounced to become conspicuous to the eye.

Extraordinary atmospheric conditions may cause strong distortions of the solar/lunar disks. Anomalous refraction can also lead to *mirages*, which are sometimes observed in desert regions.

23.3.2 Extinction

The attenuation or extinction of light from space by the Earth's atmosphere has two physically very different causes: (1) Light is partially absorbed by the air molecules (i.e., converted into heat), and (2) partially deflected from its direction (i.e. scattered). The total light loss is thus a combination of contributions from absorption and scattering. The amount of extinction can be expressed in terms of the *extinction coefficient*, or sometimes by its complement, the *transmission coefficient* (Appendix Tables B.1–3 in Vol. 3).

The extinction coefficient states the percentage of vertically incident radiation (star at the zenith) which is absorbed. The transmission coefficient gives the percentage of radiation reaching the ground. Both depend on the degree of *opacity* of the atmosphere, and also very much on the wavelength of the light. Strictly speaking, the definition given is valid only for monochromatic light, but only a very small error is incurred in the case of polychromatic light by assuming a simple *mean* extinction or transmission coefficient over the relevant range. The transmission coefficient of the atmosphere under average conditions and cloudless skies is about 0.8; in other words, a star near the zenith loses about 20% of its brightness. This is valid for the visual spectral range;

in the photographic (blue) range, the extinction is usually higher by a factor 1.5 or 2, and is also much more dependent on the specific atmospheric status. The extinction also depends somewhat on the color of the star.

Of course, the amount of extinction increases toward the horizon. To a very good approximation, the attenuation of starlight is proportional to the secant of the zenith distance. One does not normally express a measured stellar magnitude corrected entirely for atmospheric influences, but merely states the magnitude which the star would have at the zenith. Thus, if p is the transmission coefficient, then the amount of "reduction to zenith" which must be applied is

$$\Delta m = -2.5 \log p \; (\sec z - 1). \tag{23.5}$$

The factor $-2.5 \log p$ is about $0^{m}\!.25$ to $0^{m}\!.30$ in the visual range and $0^{m}\!.40$ to $0^{m}\!.60$ in the photographic range. This reduction is subject to factors which are rarely controllable, and therefore it is advisable to limit photometric measurements to small zenith distances (if possible, smaller than 30°). The measurements should be arranged such that stars at about equal zenith distances are compared so that only a small extinction difference needs to be determined. As expected, the atmospheric extinction at high elevations is greatly reduced. The numbers given here are for low altitudes of around a few hundred meters above sea level; for mountain-top observatories, the amount of extinction may be as much as ten times lower, but the exact amount must be determined for each location.

Measurements by Arsenijevic [23.9] a few years ago showed that substantial seasonal variations in the extinction occur on the island Hvar in the Adriatic Sea. Whether or not this is due to a one-time, local effect or a more widespread phenomenon has not yet been determined. In any event, there is good reason to regularly check the extinction coefficient when making photometric measurements.

The fact that extinction is *selective* (i.e., wavelength dependent) gives rise to a number of well-known phenomena. Short-wave light is attenuated much more than long-wave light. For this reason, the Sun, Moon, and stars appear much redder near the horizon than at the zenith. Another consequence of this is the blue color of the daytime sky, caused by scattering of sunlight by atmospheric dust. The red portion of direct sunlight is scattered very little, the blue part quite strongly. Thus, the light which dust particles at large angular distances from the Sun receive and then scatter toward the observer is predominantly of blue color. If there were no atmosphere or one without scattering, the sky would appear deep-black in the daytime, and only at the exact locations of Sun and Moon would direct light be seen at all.

The mathematical rules governing how extinction depends on wavelength differ for the various physical processes involved. The simplest is *Rayleigh's law* of scattering by air molecules, which gives the amount of the light lost by scatter as inversely proportional to the fourth power of the wavelength ($\propto \lambda^{-4}$). Since the atmosphere contains not only air molecules but also dust particles of various sizes and properties, the actual situation is more complex and also time variable. By and large, the extinction caused by dust can be taken as proportional to λ^{-1} to $\lambda^{-1.5}$. The net result is a combination of both the air and dust contributions in a manner which varies from case to case.

23.3.3 Twilight

The phenomenon of twilight is also a consequence of light scattering in the atmosphere, and can also be treated in conjunction with the extinction. Even after evening twilight ends, the sky is not perfectly dark, but retains a small, nocturnal brightness of its own. Without atmospheric scatter of light, the day would pass immediately into dark night; indeed, this transition takes place very rapidly in the desert, where the air normally contains very little dust.

The phenomenon of twilight occurs after sunset and before sunrise, when no direct sunlight reaches the observer, but does reach the upper layers of the atmosphere where it is scattered in different directions by air molecules; after one or more scatters, some fraction of this radiation reaches the eye of the observer at the ground.

There are actually several definitions of twilight. *Civil twilight* is defined as the time interval during which it is possible to read comfortably under cloudless skies. It ends or begins with the Sun 6° below the horizon. *Astronomical twilight* begins and ends with the altitude of the Sun at $-18°$, characterizing the point at which no trace of scattered sunlight is seen. The term *nautical twilight*, which is limited by a solar altitude of $-12°$ and characterized by the visibility limit of the horizon at sea, is also occasionally heard mentioned. Naturally, the actual duration of twilight is much shorter under cloudy skies. On the occasion of exceptionally clear weather, the Earth's shadow may be glimpsed during evening twilight in the eastern sky as a rather sharp boundary between dark and slightly brightened sky, and similarly during the dawn twilight in the western sky.

At mid-latitudes, civil twilight lasts about 30 to 40 minutes. It is longest in summer and winter, and shortest in spring and fall. Astronomical twilight (Appendix Table B.7) varies little in length during fall, winter, and spring. Its duration is about one and one-half hours at latitudes of 30° and two hours at 50°, but it increases at higher latitudes during summer. Latitude 49° marks the beginning of the zone of the so-called "white nights," where, during a certain period around the summer solstice, the astronomical twilight lasts all night as the Sun is never lower than $-18°$. The farther north the observer, the longer this time interval; for example, it lasts

at latitude 49° from June 11 to July 3;
at latitude 52° from May 21 to July 23;
at latitude 55° from May 9 to August 5;
at latitude 58° from April 29 to August 15.

The transition to latitudes of the white nights is so sharp that even at 48° (e.g., Munich, Vienna, Vancouver), almost nothing is seen of twilight at midnight. In some years or on certain nights, the situation may be different depending on the amount of dust in the atmosphere.

A number of interesting color phenomena, which under favorable circumstances can be seen during twilight, will not be dealt with here because of their exclusively meteorological significance. Moreover, satisfactory explanations do not yet exist for most of them. Details may be found in the repeatedly cited book by Dietze [23.2]. Twilight is, of course, of great importance to the astronomer, as it indicates when the stars will become visible (and for how long) so that he can begin (and later terminate) his observations. The exact starting and stopping times depend on the purpose of the observation, which dictates at what sky brightness this is possible. Generally speaking, photometric measurements require perfectly dark skies, while positional measurements may be made just as soon as the sought star can be seen.

This also raises the question of visibility of planets and stars during the daytime. The unaided eye can find the planet Venus at greatest brilliancy ($-4\overset{m}{.}5$) when the position is exactly known. Fainter objects are invisible during the daytime without telescopic aid, and the often-heard myth that stars can be seen during daylight hours from the bottom of a deep well is wholly false. When using a telescope, on the other hand, the background brightness of the sky will interfere less with the visibility of stars the larger the objective aperture and the smaller the focal ratio. Thus moderately bright stars can be seen during the daytime in the telescope. Under excellent conditions and full daytime brightness, a 2-inch telescope reveals stars of magnitude 1, while a 4-inch brings out nearly magnitude 3 stars. Conditions are considerably better on mountain stations. However, the observation of planets with substantial disks (Mars, Jupiter, Saturn) is pointless in the daytime because the available light, though substantial, is distributed over a relatively large area. In the case of Venus or Mercury, observability with the telescope in the daytime depends very much on its phase, and also, of course, on the magnification chosen. It is worth mentioning that even the Moon is difficult to observe in the daytime with a telescope.

23.3.4 The Brightness of the Night Sky

The faint residual brightness of the sky after the end of astronomical twilight consists of several contributions, and—even excluding the part caused by artificial illumination—its intensity is subject to time- and weather-dependent variations. The mean background brightness of the night sky corresponds roughly to that of a star of magnitude 22^m per square arcsecond. This places a practical limit on the ground-based observation of objects with faint surface brighnesses, although it is significant only for very large telescopes.

The night sky brightness can be traced to the following causes:

(a) Residue of twilight;
(b) Recombination light of air molecules;
(c) Zodiacal light;
(d) Faint stars and nebulae;
(e) Scattered light of the sources (b), (c), and (d);
(f) From time to time, aurorae or noctilucent clouds.

The causes mentioned under (c) and (d) are of an extraterrestrial nature, while the others lie in the atmosphere itself. The entire night background has a slightly reddish tinge, but this color is of course too weak for the eye to perceive. For medium- and small-size telescopes, the sky background does not interfere with observations.

23.3.5 The Polarization of Sky Light

The diffuse scattered light of the sky is polarized to a greater or lesser degree. In other words, the direction of electromagnetic vibration at right angles to the propagation of the light is not random but rather oriented in a somewhat preferential direction. Although this fact is practically insignificant for most astronomical observations, it should be included in a complete discussion of the optical phenomena connected to the Earth's atmosphere.

The cause of the polarization of the sky background light is the scatter by air molecules. The laws of optics show that, during the processes of reflection, refraction, and scattering, a certain portion of the light waves becomes polarized. The percentage of polarized compared with total light, the so-called *degree of polarization*, depends on circumstances such as the angle of incidence or the particle size. While the human eye cannot directly distinguish polarized from unpolarized light, some animals, such as bees, have been provided with this faculty by nature. The human eye, however, can see the polarization effects by viewing the incident light through a polarization filter. When rotated, the brightness of unpolarized light remains the same in any position of the filter, while for polarized light a maximum brightness occurs in that position where the vibration direction in the filter coincides with that of the incident light. The degree of polarization of diffuse sky light depends on the position of the region in the sky relative to the Sun. In a cloudless sky, there are a few (usually three) positions free from polarization; all three points lie in the vertical through the Sun, and are:

1. the *Arago Point*: 20° above the point opposite the Sun;
2. the *Babinet Point*: 10° above the Sun;
3. the *Brewster Point*: 15° below the Sun.

The Arago Point is above the horizon only when the Sun is rather low. The numbers of heights for the three points given are averages and may vary with opacity and other atmospheric parameters. The highest degree of polarization (usually some 60 to 80%) is reached at the point of the vertical through the Sun and 90° from it. Here, too, the degree of polarization and, to a lesser extent, the position of maximum polarization depend on atmospheric opacity. In general, the variation of polarization in the sky can be described by combining two rules:

1. The degree of polarization increases with the distance from the Sun up to 90° and then decreases again;
2. The degree of polarization diminishes toward the horizon.

Again, more details may be extracted from the book by Dietze [23.2].

23.3.6 The Apparent Shape of the Celestial Sphere

The fact that the celestial vault appears to the human eye not as a perfect hemisphere but rather flattened has nothing to do with the atmosphere, but it is worth briefly mentioning here. The observer has the impression that the horizon is more distant than the zenith point. Of course, this is not a genuine phenomenon, because nowhere

Table 23.1. Comparison of the true with the estimated altitude.

True Altitude	Estimated Altitude
0°	0°
15°	30°
30°	50°
45°	65°
60°	75°
75°	84°
90°	90°

does the sky as such have a well-defined "distance" from the observer; in principle, the view everywhere reaches into infinity. The phenomenon is of a physiological nature, and is connected with the structure of the human eye. This is easily recognized: an observer lying prostrate on the ground sees the zenith point of the sky much farther away than when standing upright.

An important consequence of this property of the human eye is that angles of altitude are usually substantially overestimated. Although the amount of error varies from case to case, the figures presented in Table 23.1 may be taken as averages.

Thus the height of an object not too far above the horizon may be overestimated by up to 20°. The effect is smaller at night than in the daytime.

Because of this subconscious overestimation of altitudes near the horizon, the Sun and Moon when rising and setting appear larger than when at the zenith. The same holds true for constellations. It is also a consequence of the properties mentioned of the human eye that stars seem to appear near the zenith even at true zenith distances of 10° to 20°. The observer should take this effect into account when estimating the altitudes of stars.

23.4 Site Selection for Astronomical Observations

The problem of site selection for major observatories has been the subject of discussions among astronomers for many decades. These studies were occasioned by the construction of a substantial number of large telescopes, whose high costs were justifiable only by optimum utilization in exceptionally dark, tranquil skies. As such, this problem is rarely encountered by observers who use small telescopes. Nevertheless, if an astronomer is considering several different sites at which to locate a telescope or an observatory, it might be prudent to weight the final decision according to those criteria which were gained by site testing for large telescopes. Only atmospheric criteria will be mentioned here, but other practical considerations, for instance, transportation accessibility, might also play an important role.

Of primary importance is unquestionably the frequency and kind of cloudiness at the site considered to accomodate astronomical instruments. This information may be obtained by the nearest meteorological station, but it should be kept in mind that neighboring locations often have substantial differences in cloudiness, particularly in mountainous regions. It is thus advisable to perform the crucial observations at the site itself. Also, the low cloudiness rate by itself does not make a site particularly suited for observations; the air steadiness under average conditions should also be carefully examined. Finally, the amount of dust in the atmosphere plays a role by dimming the starlight.

Although individual situations will depend on various factors difficult to separate, efforts toward comprehensive evaluation of numerous experiences have been made (see Meinel [23.10] and also an IAU report [23.11]). At best, there exist only some crude guidelines. Of course it is desirable in any event to avoid the nearness of urban areas and industry. On the other hand, to build observatories on high mountain tops, as is often recommended, does not necessarily prove advantageous for every purpose: while the atmospheric transparency on mountain tops is superior to that in low-lying regions (very much so in the infrared owing to the lesser H_2O content), high winds and large local and temporal temperature differences usually also increase the air turbulence.

In view of the above considerations, the best compromise of avoiding the atmospheric dimming as well as its inherent unsteadiness seems to be a smooth plane located at a high altitude above sea level. Indeed, the astronomical site tests made in Chile and in southwestern Africa seem to verify this statement. Another possibility which has proven very favorable is a site surrounded on several sides by water (e.g., an island). The thermal inertia of large bodies of water forestalls the buildup of strong temperature gradients in the air which would generate much turbulence.

Additional information on the influence of local atmospheric conditions on astronomical observations is provided in an article by M.F. Walker [23.12], which also gives some details on physical processes in the atmosphere that may impede the observations. The article also presents some useful hints on how the use of a few simple accessories can permit at least a qualitative testing of the situation.

References

23.1 Hess, F., Brezowksy, H.: *Berichte des deutschen Wetterdienstes*, Vol. 15, No. 113, Offenbach 1969.
23.2 Dietze, G.: *Einführung in die Optik der Atmosphäre*. Leipzig 1957.
23.3 Albers, B.: *Sterne und Weltraum* **12**, 137 (1973).
23.4 Leinert, C.: *Sterne und Weltraum* **25**, 18 (1985).
23.5 Greenler, R.: *Rainbows, Halos, and Glories*, Cambridge University Press, Cambridge 1980.
23.6 Meinel, A. and M.: *Sunspots, Twilights, and Evening Skies*, Cambridge University Press, Cambridge 1983.
23.7 Trickler, R.A.: *Introduction to Atmospheric Optics*, Elsevier, New York 1971.
23.8 Leinert, C.: *Sterne und Weltraum* **25**, 136 (1985).
23.9 Arsenijevic, J.: *Publications de l'Observatoire astronomique de Belgrade*, No. 33, 36 (1985).

23.10 Meinel: Astronomical Seeing and Observatory Site Selection. In *Telescopes: Stars and Stellar Systems, Vol. I.* (ed. G.P. Kuiper and B.M. Middlehurst), University of Chicago Press, Chicago 1960.
23.11 Reports of IAU Commission 50, in *Transactions of the IAU*.
23.12 Walker, M.F.: *Sky and Telescope* **71**, 139 (1986).

Supplemental Reading List for Vol. 2

The following is a list of suggested readings to supplement the references given at the end of each chapter in Vol. 2. The list is composed mostly of recent (up to early 1992) British and American bookprints, some of which may have already been referred to in the individual chapters. Older books are included as far as there is a fair chance of obtaining them by interlibrary loan. Few non-English books are included, since they are seldom obtainable; even university libraries carry few foreign books in the sciences because of the language barrier. Names of frequently referred-to publishers appear in abbreviated form; the coding is given in Chap. 12, Appendix A.7 in Vol. 1.

Chapter 13

- Bray, R.J., Loughhead, R.E.: *Sunspots*, Chapman and Hall, London 1964.
- Bray, R.J., Loughhead, R.E.: *The Solar Chromosphere*, Chapman and Hall, London 1974.
- Bray, R.J., Loughhead, R.E., Durrant, C.J.: *The Solar Granulation* (2nd edn.), CaUP 1984.
- Bruzek, A., Durrant, C.J.: *Illustrated Glossary for Solar and Terrestrial Physics*, ReiP 1977.
- Durrant, C.J.: *The Atmosphere of the Sun*, AdHg 1988.
- Eddy, J.A. (ed.): *A New Sun: The Solar Results from Skylab*, NASA, Washington 1979.
- Foukal, P.: *Solar Astrophysics*, JoWS 1990.
- Hufbauer, K.: *Exploring the Sun: Solar Science Since Galileo*, JHUP 1991.
- Koyama, H.: *Observations of Sunspots 1947–1984*, Kawadeshoboshinsha, Japan 1985.
- Noyes, R.: *The Sun, Our Star*, HaUP 1982.
- Pecker, J.-C.: *The Future of the Sun*, MGrH 1992.
- Priest, E.R. (ed.): *Dynamics and Structure of Quiescent Solar Prominences*, KlwA 1988.
- Shea, M.A., Smith, E.J.: *The International Heliospheric Study*, PerP 1989.
- Sonett, C.P., Giampapa, M.S., Matthews, M.S. (eds.): *The Sun in Time*, UAzP 1991.
- Stix, M.: *The Sun: An Introduction*, SpVg 1991.
- Svestka, Z.: *Solar Flares*, ReiP 1976.
- Tandberg-Hanssen, E., Emslie, A.G.: *The Physics of Solar Flares*, CaUP 1988.
- Tandberg-Hanssen, E.: *Solar Prominences*, ReiP 1974.
- Taylor, P.O.: *Observing the Sun*, CaUP 1991.
- Zirin, H.: *Astrophysics of the Sun*, CaUP 1988.

Chapter 14

- Allen, D., Allen, C.: *Eclipse*, Allen & Unwin 1987.
- Billings, D.E.: *A Guide to the Solar Corona*, New York 1966.
- Brewer, B.: *Eclipse* (2nd edn.), EarthView, Seattle 1991.
- di Cicco, D.: Photographing the Moon's Shadow. *Sky & Telescope* **53**, 323 (1977).
- Espenack, F.: Isophotes of the Sun's Corona. *Sky & Telescope* **58**, 96 (1979).

- Film the Eclipse. *Astronomy* **11**, 44 (1978).
- Leavens, P.A.: Hints on Photographing the Eclipse. *Sky & Telescope* **6**, 358 (1972).
- Littman, M., Willcox, K.: *Totality: Eclipses of the Sun*, Univerisity of Hawaii Press, 1991.
- Mahler, E.: *Die centralen Sonnenfinsternisse des XX. Jahrhunderts*, Wiener Akademie, 1885.
- Meens, J., Grosjean, C.C., Vanderleen, W.: *Canon of Solar Eclipses*, PerP 1966.
- Oppolzer, T. von: *Canon der Finsternisse*, Wiener Akademie, 1887.
- Ottowell, G.: *The Under-Standing of Eclipses*, Astronomical Workshop, Furman University, Greenville, SC 1991.
- Rao, J.: *Your Guide to the Great Total Solar Eclipse of 1991*, SkyP 1989.
- *Solar Eclipse Photography for the Amateur*, Kodak Publ. No. AM-10.
- Young, A.T.: The Problem of Shadow Band Observation. *Sky & Telescope* **43**, 291 (1972).
- Young, A.T.: Shadow Bands and the March Solar Eclipse. *Sky & Telescope* **39**, 176 (1970).
- Zirker, J.B.: Total Eclipses of the Sun. *Science* **210**, 1313 (1980).

Chapter 15

- Arthur, D.W.G., Agnieray, A.P.: *Lunar Designations and Positions*, UAzP 1964.
- Baldwin, R.B.: *The Measure of the Moon*, UChP 1963.
- Baldwin, R.B.: *The Moon — A Fundamental Survey*, MGrH 1965.
- British Astronomical Association: *Guide to Observing the Moon*, EnsP 1986.
- Cadogan, P.: *The Moon — Our Sister Planet*, CaUP 1981.
- Cain, K.: *Luna: Myth & Mystery*, Johnson Books, Boulder CO 1991.
- Chapront-Touzé, M., Chapront, J.: *Lunar Tables and Programs from 4000 B.C. to A.D. 8000*, WlmB 1991.
- Cook, A.: *The Motion of the Moon*, AdHg 1988.
- Fielder, G.: *Structure of the Moon's Surface*, PerP 1961.
- Heiken, G., Vaniman, D., French, B.M. (eds.): *Lunar Sourcebook: A User's Guide to the Moon*, CaUP 1991.
- Hill, H.: *A Portfolio of Lunar Drawings*, CaUP 1991.
- Kopal, Z., Klepešta, J., Rackham, T.W.: *Photographic Atlas of the Moon*, AcdP 1965.
- Kopal, Z., Carder, R.W.: *Mapping of the Moon Past and Present*, ReiP 1974.
- Kuiper, G.P. (ed.): *Photographic Lunar Atlas Based on Photographs taken at the Mount Wilson, Lick, Pic du Midi, McDonald, and Yerkes Observatories*, UChP 1960.
- Masursky, H., Colton, G.W., Farouk, E. (eds.): *Apollo over the Moon: A View from Orbit*, NASA, Washington 1978.
- Moore, P.: *The Moon*, Mitchell Beazley, London 1981; Rand McNally, New York 1981.
- Price, F.W.: *The Moon Observer's Handbook*, CaUP 1989.
- Rükl, A.: *Atlas of the Moon*, Paul Hamlyn Publishing, London 1990.
- Rükl, A.: *Maps of Lunar Hemispheres*, ReiP 1972.
- Wilkins, H.P., Moore, P.: *The Moon*, FaFa 1955.

Chapter 16

In addition to the literature below, a few of the books listed under *Chapter 14* contain information on both solar and lunar eclipses.

- Espenak, F.: *Fifty-Year Canon of Lunar Eclipses: 1986-2035*, available from SkyP.
- Link, F.: *Eclipse Phenomena in Astronomy*, SpVg 1969.
- Meeus, J., Mucke, H.: *Canon of Lunar Eclipses −2002 to +2525*, Astronomisches Büro, Vienna 1978.

Chapter 17

- Henden, A.A., Kaitchuck, R.H.: *Astronomical Photometry*, VNoR 1982, pp. 245–247.
- Kitchen, C.: *Astrophysical Techniques*, AdHg 1984, pp. 208–216.
- Percy, J.: *The Study of Variable Stars Using Small Telescopes*, CaUP 1986, pp. 254–255.
- Walker, G.: *Astronomical Observations: An Optical Perspective*, CaUP 1987, pp. 207–215.
- Warner, B.: *High Speed Astronomical Photometry*, CaUP 1988, pp. 35–66.

Chapter 18

- Aldrin, B., McConnell, M.: *Men from Earth*, Bantam, London 1989.
- Allen, J.A. van: *Scientific Uses of Earth Satellites*, Chapmann and Hall, London 1956.
- Baker, D.J.: *Planet Earth: The View from Space*, HaUP 1990.
- Booth, N.: *Space: The Next 100 Years*, Crown Books, 1990.
- Burrows, W.E.: *Exploring Space: Voyages in the Solar System and Beyond*, Random House, New York 1991.
- Compton, W.D.: *Where No Man Has Gone Before*, NASA, Washington, DC 1989.
- Damon, T.D.: *Introduction to Space*, Krieger Publishing Co., Melbourne FL 1989.
- Davies, J: *Satellite Astronomy: The Principles and Practice of Astronomy from Space*, JoWS 1988.
- Hartmann, W.K., Sokolov, A., Miller, R., Myagkov, V. (eds.): *In the Stream of Stars: The Soviet/American Space Art Book*, Workman Publishing Co., 1990.
- Harvey, B.: *Race Into Space: The Soviet Space Program*, JoWS 1988.
- Heyman, J.: *Spacecraft Tables 1957–1990*, Univelt, Inc., San Diego 1991.
- Hurt, H.: *For All Mankind*, Macdonald, London 1989.
- Jastrow, R.: *Journey to the Stars: Space Exploration—Tomorrow and Beyond*, Bantam, London 1990.
- Kelly, K.: *The Home Planet*, MacDonald, London 1988.
- King-Hele, D.: *Observing Earth Satellites*, McMi 1983.
- King-Hele, D.G., Pilkington, J.A., Hiller, H., Walker, D.M.C., Winterbottom, A.N.: *The Table of Earth Satellites 1957–1982* (2nd edn.), McMi 1982.
- Lewis, J.S., Lewis, R.A.: *Space Resources—Breaking the Bonds of Earth*, CoUP 1987.
- Lewis, R.S.: *Challenger: The Final Voyage*, CoUP 1988.
- Lewis, R.S.: *Space in the 21st Century*, CoUP 1990.
- Lewis, R.S.: *The Voyages of Columbia: The First True Spaceship*, CoUP 1984.
- Malin, S., Stott, C.: *Space Works—The How, Why, and Where of Artificial Satellites*, National Maritime Museum, Greenwich 1985.
- Mallove, E.F., Matloff, G.L.: *The Starflight Handbook*, JoWS 1989.
- Mansfield, R.: *Space Birds—A Computer Program for Predicting Naked-Eye Visibility of Artificial Satellites!*, SkyP.
- Marsh, P.: *The Space Business*, PngB 1985.
- McAleer, N.: *The Omni Space Almanac*, World Almanac Publishers, New York 1987.
- McDonough, T.R.: *Space: The Next Twenty-Five Years*, JoWS 1987.
- Miles, H. (ed.): *Artificial Satellite Observing and Its Applications*, FaFa 1974.
- Murray, B.: *Journey into Space: The First Thirty Years of Space Exploration*, NorC 1990.
- Needell, A.A. (ed.): *The First 25 Years in Space: A Symposium*, Smithsonian Institution Press, Blue Ridge Summit 1989.
- Ressmeyer, R.: *Space Places*, Collins, 1990.
- Shapland, D., Rycroft, M.: *Spacelab: Research in Earth Orbit*, CaUP 1984.
- Smith, A.: *Mars: The Next Step*, American Institute of Physics, 1989.
- Smith, A.: *Planetary Exploration*, Patrick Stephens, 1988.
- Spangenburg, R., Moser, D.: *Space Exploration* (4 vols.), Facts on File, 1989–90.
- Surkhov, Y.A.: *The Terrestrial Planets from Spacecraft*, JoWS 1989.

- Tatarewicz, J.N.: *Space Technology and Planetary Astronomy*, Indiana University Press, 1990.
- Veis, G.: Optical Tracking of Artificial Satellites. In: *Space Science Reviews* 2, No. 2, ReiP 1963.
- Vertregt, M.: *Principles of Astronautics* (2nd edn.), Elsevier, Amsterdam/ London/ New York 1963.

Chapter 19

- Alexander, A.F.O.'D.: *The Planet Saturn*, FaFa 1962. Reprinted by DovP 1980.
- Alexander, A.F.O'.D.: *The Planet Uranus*, FaFa 1965.
- Antoniadi, E.M.: *La Planète Mars*, Herman, Paris 1930.
- Baker, V.R.: *The Channels of Mars*, AdHg 1982.
- Batson, R.M.: *Voyager 1 and 2 Atlas of Six Saturnian Satellites*, NASA, Washington 1984.
- Batson, R.M., Bridges, P.M., Inge, J.L.: *Atlas of Mars*, NASA, Washington 1979.
- Beatty, J.K., Chaikin, A.: *The New Solar System*, CaUP 1990.
- Bergstralh, J.T., Miner, E.D.: *Uranus*, UAzP 1989.
- Bertotti, B., Farinella, P.: *Physics of the Earth and Solar System*, KlwA 1990.
- Binzel, R.P., Gehrels, T., Matthews, M.S. (eds.): *Asteroids II*, UAzP 1989.
- Blunck, J.: *Mars and Its Satellites* (2nd edn.), Exposition, New York 1982.
- Briggs, G., Taylor, F.W.: *The Cambridge Photographic Atlas of the Planets*, CaUP 1986.
- Burgess, E.: *Far Encounter: The Neptune System*, CoUP 1991.
- Burgess, E.: *Uranus and Neptune: The Distant Giants*, CoUP 1988.
- Burgess, E.: *Venus: An Errant Twin*, CoUP 1985.
- Burns, J.A., Matthews, M.S. (eds.): *Satellites*, UAzP 1986.
- Carr, M.H.: *The Surface of Mars*, YaUP 1981.
- Cattermole, P.: *Planetary Volcanism*, Ellis Horwood, Chichester 1990.
- Chamberlain, J.W.: *Theory of Planetary Atmospheres: An Introduction to Their Physics and Chemistry*, AcdP 1978.
- Davies, M.E. et al.: *Atlas of Mercury*, NASA, Washington 1978.
- Dessler, A.J. (ed.): *Physics of the Jovian Magnetosphere*, CaUP 1983.
- Doherty, P.: *Atlas of the Planets*, MGrH 1980.
- Dollfus, A.: *Moons and Planets*, North Holland, Amsterdam 1967.
- Dollfus, A.: *Surfaces and Interiors of Planets and Satellites*, AcdP 1970.
- Erickson, J.: *Target Earth: Asteroid Collisions Past and Future*, TAB Books, 1991.
- Fimmel, R.O. et al.: *Pioneer Venus*, NASA, Washington 1983.
- Francis, P.: *The Planets: A Decade of Discovery*, PngB 1981.
- Gehrels, T. (ed.): *Asteroids*, UAzP 1979.
- Gehrels, T. (ed.): *Jupiter*, UAzP 1976.
- Gehrels, T., Matthews, M.S. (eds.): *Saturn*, UAzP 1984.
- Glass, B.: *Introduction to Planetary Geology*, CaUp 1982.
- Greeley, R.: *Planetary Landscapes*, Allen and Unwin, London 1985.
- Greenberg, R., Brahic, A. (eds.): *Planetary Rings*, UAzP 1984.
- Guest, J. et al.: *Planetary Geology*, David & Charles, Newton Abbot 1979.
- Hartmann, W.K.: *Moons and Planets* (3nd edn.), WadP 1992.
- Henbest, N.: *The Planets: Portraits of New Worlds*, Viking, 1992.
- Henderson-Sellers, A.: *The Origin and Evolution of Planetary Atmospheres*, AdHg 1983.
- Hoyt, W.G.: *Planets X and Pluto*, UAzP 1980.
- Hunt, G.: *Uranus and the Outer Planets*, CaUP 1982.
- Hunt, G. (ed.): *Recent Advances in Planetary Meteorology*, CaUP 1985.
- Hunt, G., Moore, P.: *Jupiter*, Mitchell Beazley, London 1981.
- Hunt, G., Moore, P.: *Saturn*, Mitchell Beazley, London 1982.
- Hunten, D.M., Colin, L., Donahue, T.M., Moroz, V.I. (eds.): *Venus*, UAzP 1983.
- Jones, B.W.: *The Solar System*, PerP 1984.

- Kippenhahn, R.: *Bound to the Sun: The Story of Planets, Moons, and Comets*, FreC 1990.
- Kopal, Z.: *The Realm of the Terrestrial Planets*, JoWS 1979.
- Kowal, C.T.: *Asteroids: Their Nature and Utilization*, JoWS 1988.
- Kuiper, G.P., Middlehurst, B.M.: *Planets and Satellites*, UChP 1971.
- Lang, K.R., Whitney, C.A.: *Wanderers in Space*, CaUP 1991.
- Littmann, M.: *Planets Beyond: Discovering the Outer Solar System*, JoWS 1988.
- Lowell, P.: *Mars* (reprint of 1st edn. 1895), Paul W. Luther (Astronomy Books), Bernardstown 1978.
- Meeus, J.: *Planetary Phenomena 1976–2005*, Vereniging voor Sterrenkunde, Brussels 1977.
- Meeus, J.: *Astronomical Tables of the Sun, Moon, and Planets*, WlmB 1983.
- Melosh, H.J.: *Impact Cratering: A Geologic Process*, OxUP 1989.
- Miner, E.D.: *Uranus: The Planet, Rings, and Satellites*, Ellis Horwood, Chichester 1990.
- Moore, P.: *The Atlas of the Solar System*, Mitchell Beazley, London 1983.
- Moore, P.: *Guide to the Planets*, Eyre & Spottiswoode, London 1955.
- Moore, P.: *Mission to the Planets*, NorC 1990.
- Moore, P.: *The Planet Neptune*, JoWS 1989.
- Moore, P., Hunt, G.E.: *Atlas of Uranus*, CaUP 1988.
- Morrison, D. (ed.): *Satellites of Jupiter*, UAzP 1982.
- Morrison, D., Owen, T.: *The Planetary System*, AdWs 1988.
- Morrison, D., Samz, J.: *Voyage to Jupiter*, U.S. Government Printing Office, Washington 1980.
- Morrison, D.: *Voyages to Saturn*, U.S. Government Printing Office, Washington 1982.
- Murray, B.C., Malin, M., Greeley, R.: *Earthlike Planets*, FrmC 1981.
- Murray, B.C.: *The Planets: Readings from Scientific American*, FreC 1983.
- Peek, P.M.: *The Planet Jupiter* (2nd edn.), FaFa 1981.
- Pesek, R., Pesek, P., Pesek, L.: *Solar System*, Viking, New York 1978.
- Roth, G.D.: *The System of Minor Planets*, FaFa 1962.
- Roth, G.D.: *Handbook for Planet Observers*, FaFa 1970.
- Rothery, D.A.: *Satellites of the Outer Planets*, OxUP 1992.
- Runcorn, S.K. (ed.): *The Physics of the Planets*, JoWS 1988.
- Sandner, W.: *The Planet Mercury*, FaFa 1963.
- Sandner, W.: *Satellites of the Solar System*, FaFa 1965.
- Sharonov, V.V.: *The Nature of the Planets* (translated from the Russian), Israel Program for Scientific Translations, Jerusalem 1964.
- Sheehan, W.: *Worlds in the Sky: Planetary Discovery from Earliest Times Through Voyager and Magellan*, UAzP 1992.
- Slipher, C.: *The Photographic Story of Mars*, SkyP 1962.
- Slipher, C.: *A Photographic Study of the Brighter Planets*, National Geographic Society, Washington 1964.
- Spitzer, C.R. (ed.): *Viking Orbiter Views of Mars*, NASA, Washington 1980.
- Strom, R.G.: *Mercury—The Elusive Planet*, CaUP 1987.
- Tombaugh, C., Moore, P.: *Out of the Darkness—The Planet Pluto*, SkyP 1989.
- Turner, G., Pillinger, C.T.: *Diffuse Matter in the Solar System*, CaUP 1988.
- Van Dam, S.C.: *Mars Unfolds*, Van Dam, 1991.
- Vaucouleurs, G. de: *The Planet Mars*, FaFa 1954.
- Vilas, F., Chapman, C.R., Matthews, M.S.: *Mercury*, UAzP 1988.
- Whyte, A.J.: *The Planet Pluto*, PerP 1980.

Chapter 20

- Bailey, M.E., Clube, S.V.M., Napier, W.M.: *The Origin of Comets*, PerP 1990.
- Carusi, A., Perozzi, E., Valsecchi, G.B., Kresak, L.: *Long-Term Evolution of Short-Period Comets*, AdHg 1985.
- Frank, L.A., Huyghe, P.: *The Big Splash*, Birch Lane, 1990.

- Gibilisco, S.: *Comets, Meteors, and Asteroids: How They Affect Earth*, JoWS 1985.
- Hall, L.B.: *Searching for Comets: Deciphering the Secrets of Our Cosmic Past*, MGrH 1990.
- Huebner, W.F. (ed.): *Physics and Chemistry of Comets*, SpVg 1990.
- Krishna Swamy, K.S.: *Physics of Comets*, World Publishing, Singapore 1986.
- Kronk, G.W.: *Comets: A Descriptive Catalogue*, EnsP 1984.
- Lancaster-Brown, P.: *Halley and His Comet*, Blandford Press, Poole 1985.
- Marsden, G.B.: *Catalogue of Cometary Orbits*, Smithsonian Astrophysical Observatory, Cambridge MA 1982.
- Mason, J. (ed.): *Comet Halley: Worldwide Investigations, Results and Interpretations*, JoWS 1989.
- McCall, G.J.: *Meteorites and Their Origins*, David and Charles, Newton Abbott, U.K. 1973.
- Moore, P.: *Guide to Comets* (2nd edn.), Lutterworth, Guildford 1977.
- Porter, J.G.: *Comets and Meteor Streams*, Chapman and Hall, London (also JoWS) 1952.
- Reddy, F.: *Halley's Comet!*, Pan, London 1985.
- Seargeant, D.A.: *Comets: Vagabonds of Space*, Doubleday, New York 1982.
- Sears, D.W.: *The Nature and Origin of Meteorites*, OxUP 1978.
- Stephenson, F.R., Walker, C.B.F.: *Halley's Comet in History*, British Museum, London 1985.
- Tattersfield, D.: *Halley's Comet*, Blackwell, Oxford 1984.
- Whipple, F.J.: *The Mystery of Comets*, CaUP 1985.
- Wilkening, L.L. (ed.): Comets, UAzP 1982.
- Whipple, F.L.: *The Mystery of Comets*, CaUP 1985.
- Yeomans, D.K.: *Comets: A Chronological History of Observation, Science, Myth, and Folklore*, JoWS 1991.

Chapter 21

- Bagnall, P.M.: *The Meteorite & Tektite Collector's Handbook*, WlmB 1991.
- Chapman, C., Morrison, D.: *Cosmic Catastrophes*, PlnP 1989.
- Dodd, R.T.: *Meteorites*, CaUP 1981.
- Dodd, R.T.: *Thunderstones and Shooting Stars: The Meaning of Meteorites*, HaUP 1986.
- Graham, A.L, Bevan, A.W.R., Hutchison, R: *Catalogue of Meteorites* (4th edn.), UAzP 1985.
- Kerridge, J.F., Matthews, M.S. (eds.): *Meteorites and the Early Solar System*, UAzP 1988.
- Kronk, G.W.: *Meteor Showers: A Descriptive Catalog*, EnsP 1988.
- Mark, K.: *Meteorite Craters*, UAzP 1987.
- McCall, G.J.: *Meteorites and Their Origins*, David and Charles, Newton Abbott 1973.
- McSween, H.Y. Jr.: *Meteorites and Their Parent Planets*, CaUP 1987.
- Sears, D.W.: *The Nature and Origin of Meteorites*, OxUP 1978
- Spratt, C., Stephens, S.: Against All Odds — Meteorites That Have Struck Home. *Mercury* **21**, 50 (1992).
- Wassan, T.: *Meteorites — Their Record of Early Solar System History*, FrmC 1985.
- Whipple, F.L.: The Harvard Photography Meteor Program. *Sky & Telescope* **8**, 90 (1949).

Chapter 22

- Akasofu, S.I., Kamide, Y. (eds.): *The Solar Wind and the Earth*, ReiP 1987.
- Al'pert, Y.L.: *Space Plasmas*, Vol. 1: *Theory and Main Properties*; Vol. 2: *Flow, Waves, and Oscillations* CaUP 1990.
- Bone, N.: *The Aurora: Earth–Sun Interactions*, Ellis Horwood, Chichester 1991.
- Gasden, M., Schröder, W.: *Noctilucent Clouds*, SpVg 1989.
- Hargreaves, J.K.: *The Solar-Terrestrial Environment*, CaUP 1992.
- Hasegawa, A.: *Space Plasma Physics*, SpVg 1989.
- Meng, C.I., Rycroft, M.J., Frenk, L.A. (eds.): *Auroral Physics* CaUP 1990.

- Papagiannis, M.D.: *Space Physics and Space Astronomy*, GdBr 1972.
- Tohmatsu, T.: *Compendium of Aeronomy*, KlwP 1990.

Chapter 23

- Arya, S.P.: *Introduction to Micrometeorology*, AcdP 1988.
- Bohren, C.F.: *What Light Through Yonder Window Breaks?*, JoWS 1991.
- Burroughs, W.J.: *Watching the World's Weather*, CaUP 1991.
- Dickenson, T.: *Exploring the Sky by Day: The Equinox Guide to Weather and the Atmosphere*, Camden House, 1988.
- Greenler, R.: *Rainbows, Halows, and Glories*, CaUP 1990.
- Können, G.P.: *Polarized Light in Nature*, CaUP 1985.
- Lehr, P.E.: *Weather*, Golden, New York 1965.
- Lutgens, F.K., Tarbuck, E.J.: *The Atmosphere: An Introduction to Meteorology* (5th edn.), PrHl 1992.
- Meinel, A., Meinel, M.: *Sunsets, Twilights, and Evening Skies*, CaUP 1983.

Index to Volume 2

activity, solar (see solar activity)
apex (of Earth's motion) 287
area number (sunspots) 12
artificial satellites 169f
    atmospheric drag 173
    ballistic coefficient 11
    brightness of 242
    ephemerides 177
    geostationary orbits 174
    laser technique 190
    light variations 188
    Molniya orbits 175
    orbital elements 170
    payloads 169
    perturbations 171f
    phase law 184
    photography 187
    polar orbits 174
    positional observation 186
    radiation pressure 174
    simplified prediction 178
    visibility 176
artificial satellites
    Dynamic Explorer 1 313
    IRAS 262, 324
    OGO-6 303
    Solar Mesosphere Explorer 303
    Solwind 170
    Sputnik 1 169
ashen light
    of Moon 100
    of Venus 211
asteroids (see minor planets)
atmosphere (of Earth) 331f
    ionosphere 90
    mesopause 301
    meteoric dust 156
    optical phenomena 337
    ozone problem 157
    permanent phenomena 338
    temperature 301
    turbulence 336
    weather maps 332

aurora 304f
    brightness 307f
    height 308
    origin 305, 315
    photographs 317
    shapes 306
    solar-activity connection 311
    spectrum 309
    visibility 309
auroral oval 313

Bailey's beads 88
beta meteoroids 323
bolides 288

camera, solar 31, 50, 88
Carrington meridian 58, 65
celestial sphere, apparent shape 344
chondrite, carbonaceous 325
chromosphere, solar 31, 50, 88
chromospheric faculae 31f
chromospheric network 50
cinematography, solar 31, 43, 49, 87
clock rate 163
colongitude, Sun's selenographic 103, 144
color filters 74, 86, 121, 197, 271
comets 261f
    coma 262, 271
    dust contents 262, 328
    emission spectrum 279
    families 262
    formation 262
    gases 263
    halo 263
    magnitude 277
    names 267
    periodic (see Table B.22 in Vol. 3)
    tail 263, 271
comet cloud 261
comet observation 265f
    by CCD 274
    observing site 265, 269
    photography 273

photometry 275
polarimetry 277
search 265
sky scanning 265
spectroscopy 278f
structural study 271f
Committee on Space Research 170
corona, solar 32, 85
coronograph 32
coordinates
   heliographic 57, 63
   planetographic 204
   selenocentric 104
   selenographic 107
   topocentric lunar 102
craterlets 109
craters, lunar 110
   transit in eclipse 144
crossbar micrometer 268

dispersion, atmospheric 339
double star, occulted by Moon 165
drawings
   of Moon 113f
   of planets 201f

Earth's shadow
   penumbra 132f
   umbra 137
eclipse canons 135f
eclipse, of Jovian moons 244
eclipse, lunar 131f
   calculation 140
   central 134
   contact 139, 142
   Danjon classification 149f, 157
   depth of shadow 137f, 147, 154
   graphs 142
   magnitude 136f
   origin 132
   partial 133
   penumbral 133
   prediction 137f
   statistics 135
   total 132
eclipse, total solar 85f
   central 134
eclipse wind 90
etalon filter 31
extinction
   atmospheric 340 (see also Tables B.1–3 in Vol. 3)
   in lunar eclipse 138, 151
eye–ear method 164

faculae, solar

area 27
chromospheric 7f, 31f
classification 26
lifetime 26
photospheric 7f, 25f
polar 29, 57
positions 30
filaments, solar 7, 31f, 40
filter
   for Sun 12, 25
   for comets 271
   for Mars 220
fireballs 288
flares, solar 7, 44f
   classification 44f
   flux density 7
   lifetime 46
   light curve 47
   line width 46
   plage 44
   radio emission 49
   spray 44
   surge 44
   two-ribbon 44
   white-light 44
flash spectrum 88
flying shadows 89
forbidden lines
   in aurorae 307
   in comets 280
Fresnel diffraction 167

gegenschein 319
geomagnetic pole 309
granulation, solar 4, 69
Great Red Spot 237

halo, atmospheric 337
heliographic coordinates 57, 63
heliostat 86

International Planetary Patrol 193
interplanetary dust 318f
   loss rate 325
   particle size 323
   source 326
   spatial density 325
   symmetry plane 323
   thermal emission 324

Jupiter 321f
   apparent diameter 231
   atmospheric phenomena 232, 236f
   bands 234
   colors 240
   disturbances 237
   Great Red Spot 237

mapping 239
moons 244
oblateness 232
photography 240f
photometry 243
Red Spot Hollow 237
rings 233
rotation systems 231
space missions 233, 244
visibility 231
visual observation 234f
white oval spot 237
zones 234

Kepler's laws 170
Kiepenheuer scale (seeing) 4, 35

laser technique (satellites) 190
librations of Moon 103
luna incognita 97
lunar ... (see Moon)
lunar eclipses 131f
lunar nodes 132f
lunar occultations 159f
Lyot filter 33

magnetic storm 315
magnetosphere
 D–E layers 317
 dipole 313
 field lines 314
 plasma layer 315
Mars 216f
 apparent diameter 217
 axis tilt 217
 canali 219
 clouds 220
 craters 218
 dust storms 220
 Mariner missions 218, 221
 moons 223
 nomenclature 219
 oppositions 216
 photography 222
 polar caps 218
 polar cloud veil 219
 rotation 218
 seasons 217
 visual observation 218
McIntosh classification (Sun) 10
Mercury 206f
 photography 208
 transits 208
 visual observation 207
meteors 283f
meteoric dust 156

meteorites 283, 189
meteoroids 283
 orbits 289f
 velocities 287
meteor showers 286, 288
 radiant 287
micrometer 62, 115, 196
minor planets 224f
 albedo 227
 apparent brightness 224
 apparent diameters 231
 color indices 225
 families 326
 identification 230
 light variation 227
 mean opposition magnitude 226
 observation programs 227
 phase coefficient 226
 photoelectric observation 225, 230
 photographic positions 229f
Moon 95f
 absolute heights (of lunar mountains) 117
 albedo 124, 127
 brightness change 118
 colors 99, 125
 coordinate grid 106
 craters 110
 crevaces 110
 dark side 100
 domes 109
 formations 97, 109f
 geocentric coordinates 102
 ghost craters 110
 history of nomenclature 112
 infrared radiation 127
 librations 103
 luminescence 123f, 150f
 maps 96, 108
 maria 109
 meteorite impacts 100
 mountains 110f, 117
 observational programs 97
 orbit 100, 134
 parallax 102, 140, 159
 phases 102
 photometry 118, 123
 photography 119f
 polarization 126
 rays 110
 rilles 110
 ring mountains 110
 shadow measurement 115
 space missions 95
 spectrophotometry 126

terminator 103, 140f, 156
topocentric coordinates 102
topography 98, 113
   valleys 110
   visual observation 113f
   walled plains 110
moonblink 98
moons of planets (see under planets)
month
   anomalistic 134
   draconitic 134
   synodic 134
mottle, solar 50

nacreous clouds 299
Neptune 254
   Great Dark Spot 254
   ring system 254
   rotation period 254
   moons 255
   visibility 254
night sky brightness 343
noctilucent clouds 295f
   color 296
   density 296
   extent 296
   height 296
   origin 301
   photographs 297f, 303
   types 296
   visibility 298
nodes of Moon's orbit 132f
northern lights (see aurorae)

objective filter (Sun) 69
observatory, site selection 345
occultation
   by planets 229
   of planets 166
   of Jovian moons 244
occultation of stars 159f
   grazing 165
   photoelectric recording of 167
   prediction 161
   timing 162
opposition of planets 197
opposition effect 224
orbit of Moon 100, 134
orbits of artificial satellites 170f

personal equation 163
photographic material (Sun) 72
photography
   contrast enhancement 273
   of lunar eclipse 155
   of Moon 119f

   of planets 202f
   of artificial satellites 187
   solar 12, 57f, 67f, 74
   solar monocromatic 33, 75
photometry
   of lunar eclipse 147f
   of Moon 119f
photospheric faculae 7f, 25f
planetary observations 193f
   by CCD 202
   drawings 201f
   photography 202
   visual magnification 195
planets (see also under individual planets) 193f
   apparent diameters 197
   apparent illumination 199
   daytime visibility 343
   greatest elongation 197
   opposition 197
   phase angle 197f
   synoptic maps 205
Pluto 255
polarization, atmospheric 344
polarization foil 278
Poynting–Robertson effect 325
projection
   cylindrical 205
   Mercator 206
   orthographic 106, 204
projection screen, solar 14, 26, 60
prominences, solar 7, 31f
   active 34
   classification 34
   height 41
   photometry 43
   polar 39
   quiescent 34
   radial velocity 41
prominence attachment 32, 76
prominence eyepiece 32
prominence number 37
prominence telescope 32

radiant (of meteor showers) 287
radiation pressure (on satellites) 174
rainbow 338
Rayleigh's law 138, 341
rays, lunar 111
Red Planet (Mars) 219
refraction 339f
   differential 67
rille, lunar 111

Saros cycle 134

satellites (see artificial satellites, or planets, moons)
satellite radar 190
Saturn 245f
    Cassini's division 249
    Encke's division 249
    Great White Spot 246
    light variation (of Iapetus) 251
    moons 251
    oblateness 246
    photography 250
    rings 249
    rotation systems 247
    visual observation 247
    visibility 245
    Voyager missions 251
seeing scale 199
shadow depth (lunar eclipse) 137f, 147f
shadow enlargement 144f
silver-ball photometry 150f
solar activity 7, 15f, 21, 29f
    11-year cycle 13, 21
    80-year cycle 22
    Gleissberg cycle 22
    prediction 21, 25
solar eclipses, total 85f
solar filter 68f, 75
solar grid 63f
solar prism 25f, 30, 70
solar wind 313, 324
space debris 169
spectroheliograph 32
spicules (Sun) 31, 50
Spörer's law 52
sternschnuppe 283
stopwatch 162f
Sun 1f
    Bartel rotation 59
    chromosphere 31, 50, 88
    cinematography 31, 43, 49, 87
    corona 32, 85
    coronal plasma 85
    differential rotation 51, 58
    faculae 25f
    F corona 319
    filaments 7, 31f, 40
    flares 7, 44f
    granulation 4, 26, 80
    limb darkening 88
    magnetic field 7, 17
    monochromatic image 31
    mottles 26
    observing groups 80
    photosphere 7f, 14f
    pores 7
    projection grid 27
    prominences 7, 31f
    rotation period 19, 51, 58
    spicules 31, 50
    sunspots 7f, 26
    activity cycle 30
    area number 12
    averaging 22f
    bipolar 8f
    butterfly diagram 52, 55
    classification 7
    development 7f, 12
    McIntosh class 10
    motion 79
    new area/other numbers 17f
    observing network 19
    penumbra 4, 13, 16, 80
    polarity 8f
    position observations 21, 50, 61f
    surface area 10
    umbra 4, 16, 54, 80
    unipolar 9
    Waldmeier class 8f, 13, 16f
    Waldmeier rules 10, 21
    Wilson effect 14f
    Wolf number 12, 15f, 24, 79
    zonal migration 53
sunspot groups 6f, 16
    extension 9
    field structure 10
    inclination 8, 53
    lifetime 8
    light bridges 12f, 79
    secondary zone 53
Sunspot Index Data Center 17
syzygies 134

telescope
    projection finder 86
    solar 2, 17, 68
    vacuum 3, 71
terrae, lunar 109
trail method 62
transient lunar phenomena 98
transmission, atmospheric 340, 346
    volcanic influences 156
transparency scale 199
turbulence, atmospheric 336
twilight 342

uncoated reflector 68, 77
Uranus 251f
    brightness variation 253
    moons 253
    rings 252
    rotation period 252

visibility 251
Voyager 2 mission 252f

Venus 208f
  ashen light 211
  cloud structure 210
  cusps 211
  dichotomy 211
  height of atmosphere 213
  photography 214
  rotation 210
  shadings 209
  space missions 210f
  terminator 212
  transits across Sun 215
  visibility 208
  visual observation 209

Waldmeier rules (sunspots) 10, 21
weather patterns 331f
Widmannstätten patterns 289
Wilson effect 14f
Wolf number (sunspots) 12, 15f, 24, 79
  reduction factor 17

zenith attraction (meteors) 290
zenith distance (of radiant) 292
zenith prism 71
zodiacal light 85, 317f
  brightness 318
  ecliptic concentration 322